Lecture Notes in Physics

Editorial Board

H. Araki
Research Institute for Mathematical Sciences
Kyoto University, Kitashirakawa
Sakyo-ku, Kyoto 606, Japan

E. Brézin
Ecole Normale Supérieure, Département de Physique
24, rue Lhomond, F-75231 Paris Cedex 05, France

J. Ehlers
Max-Planck-Institut für Physik und Astrophysik, Institut für Astrophysik
Karl-Schwarzschild-Strasse 1, D-85748 Garching, FRG

U. Frisch
Observatoire de Nice
B. P. 229, F-06304 Nice Cedex 4, France

K. Hepp
Institut für Theoretische Physik, ETH
Hönggerberg, CH-8093 Zürich, Switzerland

R. L. Jaffe
Massachusetts Institute of Technology, Department of Physics
Center for Theoretical Physics
Cambridge, MA 02139, USA

R. Kippenhahn
Rautenbreite 2, D-37077 Göttingen, FRG

H. A. Weidenmüller
Max-Planck-Institut für Kernphysik
Saupfercheckweg 1, D-69117 Heidelberg, FRG

J. Wess
Lehrstuhl für Theoretische Physik
Theresienstrasse 37, D-80333 München, FRG

J. Zittartz
Institut für Theoretische Physik, Universität Köln
Zülpicher Strasse 77, D-50937 Köln, FRG

Managing Editor

W. Beiglböck
Assisted by Mrs. Sabine Landgraf
c/o Springer-Verlag, Physics Editorial Department II
Tiergartenstrasse 17, D-69121 Heidelberg, FRG

The Editorial Policy for Proceedings

The series Lecture Notes in Physics reports new developments in physical research and teaching – quickly, informally, and at a high level. The proceedings to be considered for publication in this series should be limited to only a few areas of research, and these should be closely related to each other. The contributions should be of a high standard and should avoid lengthy redraftings of papers already published or about to be published elsewhere. As a whole, the proceedings should aim for a balanced presentation of the theme of the conference including a description of the techniques used and enough motivation for a broad readership. It should not be assumed that the published proceedings must reflect the conference in its entirety. (A listing or abstracts of papers presented at the meeting but not included in the proceedings could be added as an appendix.)

When applying for publication in the series Lecture Notes in Physics the volume's editor(s) should submit sufficient material to enable the series editors and their referees to make a fairly accurate evaluation (e.g. a complete list of speakers and titles of papers to be presented and abstracts). If, based on this information, the proceedings are (tentatively) accepted, the volume's editor(s), whose name(s) will appear on the title pages, should select the papers suitable for publication and have them refereed (as for a journal) when appropriate. As a rule discussions will not be accepted. The series editors and Springer-Verlag will normally not interfere with the detailed editing except in fairly obvious cases or on technical matters.

Final acceptance is expressed by the series editor in charge, in consultation with Springer-Verlag only after receiving the complete manuscript. It might help to send a copy of the authors' manuscripts in advance to the editor in charge to discuss possible revisions with him. As a general rule, the series editor will confirm his tentative acceptance if the final manuscript corresponds to the original concept discussed, if the quality of the contribution meets the requirements of the series, and if the final size of the manuscript does not greatly exceed the number of pages originally agreed upon. The manuscript should be forwarded to Springer-Verlag shortly after the meeting. In cases of extreme delay (more than six months after the conference) the series editors will check once more the timeliness of the papers. Therefore, the volume's editor(s) should establish strict deadlines, or collect the articles during the conference and have them revised on the spot. If a delay is unavoidable, one should encourage the authors to update their contributions if appropriate. The editors of proceedings are strongly advised to inform contributors about these points at an early stage.

The final manuscript should contain a table of contents and an informative introduction accessible also to readers not particularly familiar with the topic of the conference. The contributions should be in English. The volume's editor(s) should check the contributions for the correct use of language. At Springer-Verlag only the prefaces will be checked by a copy-editor for language and style. Grave linguistic or technical shortcomings may lead to the rejection of contributions by the series editors. A conference report should not exceed a total of 500 pages. Keeping the size within this bound should be achieved by a stricter selection of articles and not by imposing an upper limit to the length of the individual papers. Editors receive jointly 30 complimentary copies of their book. They are entitled to purchase further copies of their book at a reduced rate. As a rule no reprints of individual contributions can be supplied. No royalty is paid on Lecture Notes in Physics volumes. Commitment to publish is made by letter of interest rather than by signing a formal contract. Springer-Verlag secures the copyright for each volume.

The Production Process

The books are hardbound, and the publisher will select quality paper appropriate to the needs of the author(s). Publication time is about ten weeks. More than twenty years of experience guarantee authors the best possible service. To reach the goal of rapid publication at a low price the technique of photographic reproduction from a camera-ready manuscript was chosen. This process shifts the main responsibility for the technical quality considerably from the publisher to the authors. We therefore urge all authors and editors of proceedings to observe very carefully the essentials for the preparation of camera-ready manuscripts, which we will supply on request. This applies especially to the quality of figures and halftones submitted for publication. In addition, it might be useful to look at some of the volumes already published. As a special service, we offer free of charge LATEX and TEX macro packages to format the text according to Springer-Verlag's quality requirements. We strongly recommend that you make use of this offer, since the result will be a book of considerably improved technical quality. To avoid mistakes and time-consuming correspondence during the production period the conference editors should request special instructions from the publisher well before the beginning of the conference. Manuscripts not meeting the technical standard of the series will have to be returned for improvement.

For further information please contact Springer-Verlag, Physics Editorial Department II, Tiergartenstrasse 17, D-69121 Heidelberg, FRG

G. Belvedere   M. Rodonò   G. M. Simnett   (Eds.)

Advances in Solar Physics

Proceedings of the Seventh European Meeting
on Solar Physics
Held in Catania, Italy, 11-15 May 1993

Springer-Verlag Berlin Heidelberg GmbH

Editors

G. Belvedere
M. Rodonò
Istituto di Astronomia dell'Università di Catania
Viale Andrea Doria 6, I-95125 Catania, Italy

G. M. Simnett
Department of Physics and Space Research
of the University of Birmingham
Edgbaston, Birmingham B15 2TT, United Kingdom

ISBN 978-3-662-13984-4   ISBN 978-3-540-48420-2 (eBook)
DOI 10.1007/978-3-540-48420-2

CIP data applied for

This work is subject to copyright. All rights are reserved, whether the whole or part of the material is concerned, specifically the rights of translation, reprinting, re-use of illustrations, recitation, broadcasting, reproduction on microfilms or in any other way, and storage in data banks. Duplication of this publication or parts thereof is permitted only under the provisions of the German Copyright Law of September 9, 1965, in its current version, and permission for use must always be obtained from Springer Fachmedien Wiesbaden GmbH.

Violations are liable for prosecution under the German Copyright Law.

© Springer-Verlag Berlin Heidelberg 1994
Originally published by Springer-Verlag Berlin Heidelberg New York in 1994
Softcover reprint of the hardcover 1st edition 1994

SPIN: 10080353   55/3140-543210 - Printed on acid-free paper

Foreword

A series of triennial meetings are currently organised by the Board of the Solar Physics Section of the Astronomy and Astrophysics Division of the European Physical Society. As is customary, the Board met in the year following the sixth meeting to decide on the topic and location of the seventh. On this occasion, we enjoyed the hospitality of Arcetri Observatory in Florence, and decided to concentrate on the whole Sun, from the core to the corona, with the emphasis on the way magnetic fields influence the phenomena we observe. It was a conscious decision not to discuss solar flares, although it proved difficult to maintain this rigorously in some of the coronal sessions. To convey the scientific breadth of the meeting, we decided on the simple title "Advances in Solar Physics". The choice of a location was a difficult one and we were reminded that the previous meetings in the series were held in Florence, Toulouse, Oxford, Noordwijkerhout, Titisee and Debrecen. Possibly influenced by the ambience of Arcetri, but definitely by the quality of the bid from the Astrophysical Observatory and the University of Catania, the Board unanimously decided to hold the Seventh European Meeting on Solar Physics in Catania, from 11 to 15 May 1993.

The Board organised the meeting into six main science sessions, followed by a seventh occasion on the new instrumentation that we hope will be available in the near future. An opening session, chaired by one of us (MR), was devoted to a lecture by J.C. Pecker on "The Sun Today". There was a balance between theory and observations, and throughout the week the focus shifted from the solar core, which we study via neutrino emissions and helioseismology, through the interface regions where it is believed the large-scale magnetic fields are generated, to the corona, where most of the high temperature phenomena characteristic of this region may be studied directly. As energetic particles play such a vigorous role in this part of the Sun, a separate session was devoted to their transport and storage in the corona.

The bulk of the work now fell to the Scientific and Local Organising Committees. The Scientific Organising Committee consisted of W. Mattig (co-chair) and B. Schmieder (co-chair), G. Belvedere, G. Godoli, P. Heinzel, P. Hoyng, A. Righini, M. Rodonò, G. Simnett, J. Staude, G. Trottet, J.C. Vial, L. Vlahos and Yu.D. Zhugzhda. The session chairmen, P. Hoyng, G. Belvedere, B. Schmieder, G. Trottet, L. Vlahos, G. Simnett and W. Mattig were responsible for the organisation and balance of the seven main sessions. Their work was admirably co-ordinated by Brigitte Schmieder.

The Local Organising Committee included S. Catalano, L. Paternò, D. Spadaro, R. Ventura and R.A. Zappalà. Their collaboration with the LOC chairmen (GB and MR) in the months before the meeting enabled all the participants to relax in the convivial atmosphere of the conference facilities of La Perla Ionica Hotel and concentrate on Solar Physics. Special thanks go to the Mayor of Acireale, who hosted us one evening in his City Hall. This was followed by a visit to the world famous "Opera dei Pupi".

One day was devoted to excursions. On the morning of Thursday, May 13 we visited the ancient city of Taormina, while in the afternoon we were taken on a spectacular drive up the local volcano, Mount Etna. The local spirit must have felt threatened by this invasion of so many eminent scientists studying phenomena far more energetic than anything this earth could provide, and we were greeted by ferocious winds at the highest point we reached. We duly retreated a little to the relative comfort of the Mountain Station of Catania Astrophysical Observatory, where we were given a tour of the facilities. This was followed by the conference dinner at the Rifugio Sapienza of the Club Alpino Italiano.

The meeting was deliberately organised with each session consisting of 90 minutes of invited review talks (generally 2 topics) plus 60 minutes of invited contributions (generally 4 topics), the latter selected from the papers submitted for the meeting. This format worked well, and there was ample time for discussion, and even for some speakers to overrun their allocated time slightly!

The remaining contributions were presented as posters, which were on permanent display near the area where the coffee break was taken. A prize, offered by Solar Physics, was awarded to the best poster, which was judged by a special committee who met between the two excursions on Thursday - there was no respite for some! The winner was Geil Halberstadt who, together with his co-authors J.P. Goedbloed and S. Boedts, gave a "reader-friendly" account of "Resonant Heating of Coronal Loops".

The meeting would not have been possible without the sponsorship of the following: European Physical Society, European Astronomical Society, European Space Agency, Catania Astrophysical Observatory, University of Catania, Italian National Research Council, Italian Astronomical Society, Sicilian Region, Catania Province and City, City of Acireale, Azienda Autonoma della Stazione di Cura e Turismo di Acireale, Sip, Credito Italiano, MetroKart. To these organisations we express our deepest gratitude.

The daily running of the meeting was admirably handled by the Catania Observatory staff: C. Anastasi, A. Calì, P. Massimino, D. Randazzo, M.L. Rapisarda, D. Recupero, L. Santagati, S. Sardone, C. Sorge, V. Stancanelli and A. Wanausek. Their generous dedication and sterling work made the conference possible. We would like to thank also the La Perla Ionica hotel staff, in the name of the directors P. Piergiovanni and F. Tropea for their extremely efficient collaboration.

It has been the task of the editors to bring the scientific contributions to a wider audience. We are gratefully indebted to Wolfgang Mattig for making the publication arrangements with Springer-Verlag with great enthusiasm, and to Daniela Recupero for invaluable help in editing these Proceedings. We have tried to allow the authors their freedom to present their ideas in their individual styles; we have intervened only for the sake of consistency and accuracy. We have allotted about 12 pages for the invited reviews and about 6 for the invited contributions. All the papers have been submitted to referees, whose competent advice is gratefully acknowledged.

Catania, February 1994   G. Belvedere, M. Rodonò, G.M. Simnett

Contents

LIST OF PARTICIPANTS

ABOUDARHAM Jean	Institute of Space Astrophysics Orsay	France
ABRAMENKO Valentina	Crimean Astrophysical Observatory Nauchny	Ukraine
ALISSANDRAKIS Constantine	Department of Physics University of Athens	Greece
ANASTASIADIS Anastasios	Department of Physics University of Thessaloniki	Greece
BALLESTER Jose Luis	Department of Physics University of Palma de Mallorca	Spain
BAUDIN Frederic	Institut of Space Astrophysics Orsay	France
BELLOTTI Enrico	Department of Physics University of Milan	Italy
BELY-DUBAU Françoise	Nice Observatory Nice	France
BELVEDERE Gaetano	Institute of Astronomy University of Catania	Italy
BENTLEY Robert	Mullard Space Science Laboratory Dorking	U.K.
BIESECKER Douglas	Space Science Center University of New Hampshire, Durham	U.S.A.
BOCCHIALINI Karine	Institute of Space Astrophysics Orsay	France
BONMARTIN Jean	Meudon Observatory Meudon	France
BOROVIK Valery	Main Astronomical Observatory St. Petersburg	Russia
BRAJSA Roman	Kiepenheuer Institute for Solar Physics Freiburg	Germany
BRANDENBURG Axel	HAO/NCAR Boulder	U.S.A.
BREKKE Paal	Institute of Theoretical Astrophysics Oslo	Norway
BRIAND Carine	Institute of Space Astrophysics Orsay	France
CARBONE Vincenzo	Department of Physics University of Calabria, Cosenza	Italy
CATALANO Santo	Astrophysical Observatory Catania	Italy

CHARBONNEL Corinne	Geneva Observatory Sauverny	Switzerland
CHERTOK Ilia	Solar Radio Laboratory, IZMIRAN Troitsk	Russia
CHIAN Abraham	DAMTP University of Cambridge	U.K.
CHIUDERI Claudio	Dept. of Astronomy and Space Science University of Florence	Italy
CLETTE Frederic	Belgium Royal Observatory Bruxelles	Belgium
CORNILLE Marguerite	Meudon Observatory Meudon	France
CUTISPOTO Giuseppe	Astrophysical Observatory Catania	Italy
DAILLET Sylviane	CNES Toulouse	France
DARA Helen	Research Center for Astronomy Athens	Greece
DEMOULIN Pascal	Meudon Observatory Meudon	France
DESAI Mihir	School of Physics and Space Research University of Birmingham	U.K.
DEUBNER Franz	Institute for Astronomy and Astrophysics University of Würzburg	Germany
DIALETIS Dimitris	National Observatory of Athens Athens	Greece
DOMINGO Vicente	ESA - Space Science Department Noordwijk	The Netherlands
DRAGO Franca	Department of Astronomy and Space Science University of Florence	Italy
DUBAU Jacques	Meudon Observatory Meudon	France
DUMITRACHE Cristiana	Astronomical Institute Bucharest	Rumania
DZIEMBOWSKI Wojchek	Copernicus Astronomical Center Warsaw	Poland
EGIDI Alberto	2nd University of Rome Rome	Italy
EINAUDI Giorgio	Dept. of Astronomy and Space Science University of Florence	Italy
FAUROBERT Marianne	Nice Observatory Nice	France
FIORENTINI Giovanni	Department of Physics University of Ferrara	Italy

FOSSAT Eric	Department of Astrophysics University of Nice	France
FOSTER Valerie	Dept. of Physics, Queen's University Belfast	U.K.
FRANK Anna	Institute of General Physics Moscow	Russia
FREIRE FERRERO Rubens	Astronomical Observatory Strasbourg	France
FRISCH Helene	Nice Observatory Nice	France
GAIGÉ Yves	Midi-Pyrénées Observatory Toulouse	France
GARCIA Adriana	Astronomical Observatory University of Coimbra	Portugal
GAVRYUSEVA Elena	Institute for Nuclear Research Moscow	Russia
GODOLI Giovanni	Dept. of Astronomy and Space Science University of Florence	Italy
GOOSSENS Marcel	Center for Plasma Astrophysics Heverlee	Belgium
HALBERSTADT Giel	Plasma Physics Institute Nieuwegein	The Netherlands
HARRA Louise	Dept. of Physics, Queen's University Belfast	U.K.
HASLER Karl-Heinz	Astrophysical Institute Potsdam	Germany
HEINZEL Petr	Astronomical Institute Ondrejov	Czech. Republic
HENOUX Jean-Claude	Meudon Observatory Meudon	France
HERISTCHI Djamshid	Meudon Observatory Meudon	France
HERMANS Dirk	School of Mathematics and Statistics University of Birmingham	U.K.
HILDERBRANDT Joachim	Astrophysical Institute Potsdam	Germany
HOFMANN Axel	Astrophysical Institute Potsdam	Germany
HORN Thomas	Astrophysical Institute Potsdam	Germany
HOYNG Peter	Laboratory for Space Research Utrecht	The Netherlands
JOARDER Parthasarathi	School of Mathemathics & Statistics University of Birmingham	U.K.
JORDAN Carole	Department of Physics University of Oxford	U.K.

KIM Iraida	Sternberg Astrophysical Observatory Moscow	Russia
KLEIN Karl-Ludwig	Meudon Observatory Meudon	France
KOCHAROV Leon	St. Petersburg Technical University St. Petersburg	Russia
KOSOVICHEV Alexander	Crimean Astrophysical Observatory Naunchny	Ukraine
KOUTCHMY Serge	Institute of Astrophysics Paris	France
KUČERA Ales	Kiepenheuer Institute for Solar Physics Freiburg i.B.	Germany
LAING Gordon	Department of Mathematical Sciences University of St. Andrews	U.K.
LANDI DEGL'INNOCENTI Egidio	Dept. of Astronomy and Space Science Firenze	Italy
LANG James	Space Science Department Didcot	U.K.
LANZA Antonino	Astrophysical Observatory Catania	Italy
LANZAFAME Alessandro	Armagh Observatory Armagh	U.K.
LANZAFAME Giuseppe	Astrophysical Observatory Catania	Italy
LEMAIRE Philippe	Institute of Space Astrophysics Orsay	France
LEPELTIER Thomas	Cea SAp Gif-sur-Yvette	France
LORENC Marian	Astronomical Observatory Hurbanovo	Slovakia
LORENTE Rosario	Astrophysical Department University of Madrid	Spain
LUKAČ Bohuslav	Astronomical Observatory Hurbanovo	Slovakia
MARKOVA Eva	Úpice Observatory Úpice	Czech Republic
MATTIG Wolfgang	Kiepenheuer Institute for Solar Physics Freiburg i.B.	Germany
MESSEROTTI Mauro	Astronomical Observatory Trieste	Italy
MONTESINOS Benjamin	ESA-IUE Satellite Tracking Station Madrid	Spain
MULLER Richard	Pic du Midi Observatory Bagnères de Bigorre	France
OLIVER Ramon	Department of Physics University of Palma de Mallorca	Spain

ORAEVSKY Victor	IZMIRAN Troitsk	Russia
PAGANO Isabella	Astrophysical Observatory Catania	Italy
PATERNÒ Lucio	Institute of Astronomy University of Catania	Italy
PECKER Jean-Claude	College of France Paris	France
PERES Giovanni	Astrophysical Observatory Catania	Italy
PETROVAY Kristof	Department of Astronomy Eötvös University, Budapest	Hungary
PICK Monique	Meudon Observatory Meudon	France
PINTER Teodor	Astronomical Observatory Hurbanovo	Slovakia
POLETTO Giannina	Arcetri Observatory Florence	Italy
PRIMAVERA Leonardo	Dept. of Physics, Univ. of Calabria Cosenza	Italy
RIBES Elisabeth	Meudon Observatory Meudon	France
RICCA Renzo	Dept. of Mathematics University College, London	U.K.
ROBERTS Bernard	University of St. Andrews St. Andrews	U.K.
ROCA CORTES Teodoro	Institute of Astrophysics of the Canary Islands La Laguna	Spain
RODONÒ Marcello	Institute of Astronomy University of Catania	Italy
ROMPOLT Bogdan	Astronomical Institute University of Wroclaw	Poland
RONAN Robert	Stanford University Stanford, California	U.S.A.
ROXBURGH Ian	Queen Mary and Westfield College London	U.K.
RUDAWY Pawel	Astronomical Observatory University of Wroclaw	Poland
RÜDIGER Günther	Astrophysical Institute Potsdam	Germany
RUSIN Vojtek	Astronomical Observatory Tatranska Lomnica	Slovakia
RUŽDJAK Vladimir	Hvar Observatory Zagreb	Croatia
RYAN James	Space Science Center, Univ. of New Hampshire Durham	U.S.A.

SAHAL-BRECHOT Sylvie	Meudon Observatory Meudon	France
SAMAIN Denys	Institute of Space Astrophysics Orsay	France
SAYLE Ken	School of Physics and Space Research Univ. of Birmingham, Edgbaston	U.K.
SCHMIEDER Brigitte	Meudon Observatory Meudon	France
SCHMITT Dieter	Göttingen Observatory University of Göttingen	Germany
SCHRIJVER Karel	Astronomical Institute University of Utrecht	The Netherlands
SCHUTGENS Nick	Astronomical Institute University of Utrecht	The Netherlands
SCHWENN Rainer	Max-Planck Institute for Aeronomy Lindau	Germany
SEEHAFER Norbert	Astrophysical Istitute Potsdam	Germany
SERIO Salvatore	Astronomical Observatory Palermo	Italy
SIMNETT George	School of Physics and Space Research University of Birmingham	U.K.
SOFIA Sabatino	Center for Solar and Space Research Yale University, New Haven, Connecticut	U.S.A.
SOMOV Boris	Sternberg Astronomical Institute Moscow	Russia
SPADARO Daniele	Astrophysical Observatory Catania	Italy
STAATH Eric	Institute of Space Astrophysics Orsay	France
STAUDE Jürgen	Astrophysical Institute Potsdam	Germany
SYKORA Ilius	Astronomical Observatory Tatranska Lomnica	Slovakia
SYLWESTER Janusz	Space Research Centre Wroklaw	Poland
TERNULLO Maurizio	Astrophysical Observatory Catania	Italy
THOMPSON Michael	Queen Mary & Westfield College London	U.K.
TIKHOMOLOV Evgeniy	Institute of Solar-Terrestrial Physics Irkutsk	Russia
TORELLI Maria	Astronomical Observatory Rome	Italy
TROTTET Gerard	Meudon Observatory Meudon	France

VAN DER LINDEN Ronald	Dept. of Mathematics & Computer Sciences University of St. Andrews	U.K.
VAUCLAIR Sylvie	Midi-Pyrénées Observatory Toulouse	France
VAZQUEZ Manuel	Institute of Astrophysics of the Canary Islands La Laguna	Spain
VELTRI Pierluigi	Department of Physics University of Calabria, Cosenza	Italy
VENTURA Rita	Astrophysical Observatory Catania	Italy
VIAL Jean Claude	Institute of Space Astrophysics Orsay	France
VILMER Nicole	Meudon Observatory Meudon	France
VLAHOS Loukas	Department of Physics University of Thessaloniki	Greece
WENZEL Klaus	ESA Space Science Department Noordwijk	The Netherlands
WÖHL Hubertus	Kiepenheuer Institute for Solar Physics Freiburg i.B.	Germany
YERLE Raymond	Midi-Pyrénées Observatory Toulouse	France
ZAITSEV Valery	Institute of Applied Physics Nizhny Novgorod	Russia
ZAPPALÀ Rosario Aldo	Astrophysical Observatory Catania	Italy
ZHARKOVA Valentina	Department of Physics and Astronomy University of Glasgow	U.K.
ZHUGZHDA Yuzef	IZMIRAN Troitsk	Russia
ZLOBEC Paolo	Astronomical Observatory Trieste	Italy
ZUCCARELLO Francesca	Institute of Astronomy University of Catania	Italy

Opening Session

The Sun Today

Jean-Claude Pecker

Collège de France, Annexe, 3 rue d'Ulm, 75231 Paris Cedex 05, France

Abstract: A few ideas emerge from modern developments: (a) the distinction between the quiet and active Sun is misleading and obsolete; (b) the theoretical treatment of the Sun as an isotropic sphere, without magnetism, is obviously not valid; (c) multilayer active phenomena are important as proxies for deeper phenomena more than *per se*; (d) the solar cycle cannot be described simply by the level of activity, but it needs a consideration of the latitudinal migrations of activity. These points emerge following a systematic look at each of the four parts of this paper, which deal with: 1) The Sun as a variable star (irradiance, radius, neutrino flux). 2) The coupling between rotation, magnetism, convection. 3) The active phenomena in outer layers. 4) The solar cycle and migrations. The conclusion is naturally oriented towards the development of more complete observations, both on the ground, coordinated over the Earth, and in space, using multiple satellites.

Introduction

The membership of the three solar IAU Commissions is 774 members. The specialization of solar physicists is increasing every year; and one has to rejoice when we see that, sometimes, like now in Catania, they try to understand each other. To introduce a meeting dealing with the whole Sun is therefore not easy. I want to apologize for my ignorance, the literature is enormous; the main question is how to follow its progress with a sufficiently critical eye? This led me to either misquote, or not quote, some of the more important references. Each field of solar physics often ignores what is done in other fields, and in other countries. I shall be no better than others; so pardon me for being perhaps global-Sun biased and French-parochial; but please consider that we should all try to change our attitudes in this respect.

In what respect does the Sun differ from what it was a few years ago? Several important ideas, which were in the minds already, but never strongly formalized, are the following:

(a) The distinction between the quiet and active Sun is misleading and obsolete. The Sun is never quiet; this pure abstraction had heuristic qualities, but was a brake to research on solar magnetism. Even the quiet corona has its origin in processes not included in the theory of the quiet Sun. (b) The theoretical treatment of the Sun as an isotropic sphere is obviously not valid, either in the upper atmosphere or in the convective layers. It is perhaps no more valid even in the deep interior. (c) Any layer, any wave-length domain, is more or less coupled with some other.

3

Therefore, some well observed phenomena are proxies for other phenomena, either difficult or impossible to observe, or are tests for some modeling. This is the best motivation for the study of individual phenomena. (d) The classical approach to the solar cycle has to be modified. Solar activity needs several indices, to be described and correlated with planetary phenomena. In particular, the latitudinal migrations of activity have to be considered.

1 The Sun as a Variable Star

Very slowly, the solar mass decreases. Its structure changes. But much quicker changes occur, affecting the essential parameters of the Sun.

1.1 The Luminosity Variation of the Sun

Variations in the solar irradiance have been observed for the last decades (Fröhlich & al., 1991). One has often translated them into changes of the luminosity. This is not implied, -but not excluded either. The main reason for the variation of irradiance is the passage of spots across the disk. Admitting this to be the only cause for the irradiance variability, we are confronted with a problem. If the darkening of the spots is due to the magnetic blocking, then the energy output must be redistributed; it is likely that it will appear in the facules. But when a spot crosses the disk, the facules do it too, and, even if there is some spectral redistribution, the irradiance should not change. Therefore, one is forced to admit either some brightness redistribution in areas affected by foreshortening (polar regions), or some global change of the luminosity. This last alternative seems rather unlikely, as energy does take some 100,000 years to travel from the core to the surface. Therefore, we are led to think that the latitude redistribution of brightness is real (Kuhn, 1988, 1989). The same result was inferred several years ago (Pecker, 1986) by the author from the comparison between the infrared results (Kneubühl & al. 1981, Müller & al., 1975) and visible data. Without giving too much weight to the analysis, which may be not unique, we can only state that the radiation flux is function of the latitude, and of the phase of the solar cycle.

1.2 The Variation of the Solar Radius

The origin of the luminosity variations has been discussed by Spruit (1991), in connexion with the variations of the solar radius. Our argument above seems to suggest that variations of radius could be also latitude-dependent. What about the observations? We tend to be convinced by the quality of the data obtained by Laclare and others (Ribes & al., 1991). They indicate a significant variation of the radius, during the last two decades. A 22-year period seems almost obvious; butthere are also significant fluctuations of shorter period. Data concerning the XVII-th century have been discussed by Ribes & al. (1988, 1989): during the Maunder minimum, the Sun's radius is smaller. The variation in radius is accompanied by a latitudinal variation of the radius. Does it go in the direction of the Sun having a constant volume? Or are the radii at the pole and the equator changing in phase, and not in antiphase? In both cases, is there a redistribution of energy which affects only the solar radiative envelope and not its core? Or are the core and the radiative envelope affected as well? But a true question is: How deep have we to go to find some strictly spherical behaviour? The usual reply is below the convective zone, where the Sun behaves like a rigid body. But there is no reason why the radiative flux should be anisotropic only from the convective zone upwards. The shape of the energy producing central core could hardly be affected by rotation and magnetism, but the transfer of radiation outside the core may be strongly affected. Difficulties are obvious: how quick is the response of outer layers to changes in the deep-rooted distribution of the generation of energy? Where are the couplings between, say, evolution of magnetism and changes in structure taking place? How to build such non-spherical models of the solar interior is an interesting and young domain of solar research.

1.3 The Neutrino Solar Astronomy and the Solar Models

The spectral investigation of solar neutrinos (Wolfsberg, Kocharov, 1991) has brought new data, some of them surprising. One may note as essential: (1) The fact that the calculated neutrino fluxes are 2 to 3 times that of observed neutrinos. (2) The (controversial) correlation between the activity and the neutrino flux (Davis, Cox, 1991). (3) The (controversial) correlation between the radius and the neutrino flux (Delache & al., 1993).

Discrepancies are assigned either to the inadequacy of models, or to the inadequacy of neutrino physics. Improving the modeling is not excluded; but the tricks to be performed (such as WIMPS) need confirmation. While waiting for it, one is oriented towards the solutions implying the physics of neutrino, instead of towards the solar non-standard models. On the side of physics, Rosen (1991) reviews the solutions associated with the neutrino oscillations along the so-called MSW theory (Mikheyev & Smirnov, 1986, Wolfenstein, 1978). They seem at first to be coherent with the observations; however the results of the various experiments are contradicting this first impression.

In a recent paper, Schatzman and Berthomieu (1992), taking a better account of the Gallex data, tend therefore to mellow the rejection of new modelling. The facts that the models are static, that they do not take into account the production of internal waves, or of dynamical instabilities, leave the door open for better solar models. The fact that the neutrino flux seems to be correlated with activity and with radius variations, orients us in a similar direction. One is not yet able to decide. Neutrinos are produced isotropically; this isotropy is not affected either by the geometric distribution of the productive processes, or by any transfer process. But, in this case, we have still to understand the modulation of the neutrino flux. Luminosity and radius variations can be accounted for (Schatzman, 1990) by the consideration of the varying structure of convective zone, the compressibility γ being increased by a strong entangled magnetic field; this phenomenon is a kind of breathing. The difficulty of obtaining a quantitative theory for this effect has been underlined by Schatzman, but Durney (1990) has managed to predict the required amplitude of the breathing by a proper treatment of the interaction of differential rotation with magnetic tubes. How would this affect the neutrinos and explain their behaviour across the magnetic cycle? A change in luminosity seems to be necessary there; can the suggested changes in the structure of Sun affect its luminosity?

But as long as we are feeling safer about the suggested variations of luminosity, radius, and neutrino-flux, the neutrino non-adiabatic oscillations between the three flavors may be the better solution to the neutrino problem.

2 Rotation, Magnetism and Convection

2.1 Helioseismology

Helioseismology has brought to us, in a few decades, splendid data (Gough & Toomre, 1991). During the coming days, we shall hear about this, and about the efficiency of helioseismological networks, such as IRIS; hence I shall be quite short.

The problem of gravity waves seems of course still wide open. But the progress in the field of acoustic waves has been indeed so important that it seems that we know almost everything about them. We know in particular that their modal structure is linked with the rotation of the deep layers of the Sun. From inverting the problem equations, one can derive the variation with depth of the rotational velocity. Although there is a large uncertainty as to the uniqueness of the inversion process, the result shows strikingly that up to about $r/R = 0.55$, the rotation seems to be rigid, whenever differential rotation affects layers above that level. Needless to say, this conclusion invites confirmation. A discontinuity in rotation rate appears at the transition between the core and the radiative envelope. The data from IRIS (Fossat, 1991) suggest an increase of a factor of about 1.5 to 2 times.

Apart from the rotation rate, the seismological data may give valuable information on the model in the core, as shown by Gough & Kosovichev (see Christensen-Dalsgaard, 1992).

2.2 Rotation, Differential Rotation and Activity

Observations over the solar cycle refine the picture derived from helioseismology. Spot measurements (Nesme-Ribes & al., 1993a), magnetic field data (Snodgrass, 1991, Snodgrass, Ulrich, 1990, Komm & al., 1993), show the rotation as more rigid in periods of maximum activity, more differential in periods of minimum activity, the difference being of the order of 1 to 2%. The case of the Maunder minimum (Ribes & al., 1989) gives an extreme example of this behaviour. A similar difference exists between the two solar hemispheres, the more active hemisphere being perhaps the more rigid (Nesme-Ribes & al., 1993a).

The variation of the rotation rate with depth (Nesme-Ribes & al., 1993b) is linked with a shear dv/dr which departs from zero in a transitional layer ($0.5 < r/R < 0.7$). A strong positive shear allows convective motions to drag out magnetic tubes of force. Thus by invoking a low level of magnetic activity one links the differential rotation and the small (average) rate of rotation. The acoustic waves are time-dependent during the activity cycle (Gelly & al., 1988, Roudier, 1986): this seems in good agreement with the known decrease of granular size with decreasing activity, and with the variation of radius during the same period; it might allow us to study variations of the rotation rate.

On what depth do the tracers of rotation give us information? Does the rigid rotation start as high as $r/R = 0.7$ or instead near $r/R = 0.5$? Quite clearly, the coronal holes are indicators of rigid- rotation; this was shown from their geomagnetic effects by Pecker and Roberts (1954). Actually, all tracers (spots, supergranulation, magnetic fields) display a more rigid rotation than does the photospheric plasma, as already demonstrated years ago. Young spots are emerging such that they avoid, at all latitudes, privileged longitudes (Maunder & Maunder, 1905): they are deep tracers, not affected yet by the differential rotation. The evolution of prominences displays also the deeply rooted pivot-points (Mouradian & al., 1987).

2.3 The Structure of the Convective Zone

One could refer the reader to the extensive review by Spruit, Nordlund, Title (1990).

The usual theory of the thermal structure of the convective zone assumes it to be only slightly superadiabatic. However, the very appearance of granulation shows that, at least in its outer layers, thermal homogeneity is not achieved, whatever the cause of it. The granulation is a rich source of information (Topka, Title, 1991); the theory (Nordlund, 1985) of granule dynamics is satisfactory. New studies at Pic-du-Midi (Muller et al., 1992, Muller & Roudier, 1992) have shown the existence of bright filamentary regions in some of the dark areas; these filigrees seem to coincide with the feet of spicule bushes. The network bright points (NBP) are associated with kilogauss magnetic fields; the NBP have a lifetime longer than the usual lifetime of small granules. They occur very near the boundary of the supergranules. It seems that the NBP are formed following some compression of matter in the intergranular medium: this illustrates the formation and emergence of magnetic fields such as would result from magnetic enhancement in compressed volumes. The granule size is smaller at phases of small activity, as shown years ago by R.C. Macris, or R. Muller and associates; in magnetic areas, the granular structure is strongly perturbed as a consequence of interactions between magnetism and convection.

Granulation is one of the tracers of the convective zone. Other tracers are the young spots. The latter have shown, at different periods of the cycle, the rolling structure of convective motions, in rolls parallel to the equatorial plane. These motions, suggested by Ribes et al. (1985), have been confirmed and studied across the solar cycle (Ribes, 1992). They describe the meridian circulation. The roll structure is of course a function of the phase in the solar cycle. One can describe it in details. The meridian circulation has as a consequence the differential rotation at the surface due to the transport of momentum, and the magnetic activity, due to the pumping of the deep magnetic tubes of force.

How, dragged by the convection, do the magnetic fields emerge and migrate? What is the mechanism of the transport of momentum? How do we account for the five scales of convective cells (McIntosh, 1992)? The theory is far behind the empirical deductions. One can find a very thorough discussion on some of these questions in Dumey (1992), as well as a good bibliography.

To quote V. Trimble (1993): "Notoriously, modern astrophysics still has no proper theory of convection. But in the solar case we can perhaps say, at least, that all the pieces of the puzzle have been laid on the table right side up".

3 Active Phenomena in All Layers

The convective motions, combined with static buoyancy, drag out magnetic fields as seen above. They explode as active phenomena, and they control the physics of the chromospheric and coronal plasmas. Their observation has progressed substantially in recent years (Faurobert-Scholl et al., 1992), and is likely to progress more with THEMIS, or LEST or others, in sensitivity and in accuracy, as well as in angular resolution on the solar surface. The measurement of the 4 Stokes components will become possible; but already, the use of only the line profiles allows an unambiguous 2- D fine description of the fields, such as in the MSDP (multichannel substractive double pass spectrograph) (Mein, 1991). The 3-D description comes next from a theoretical extension of the surface data; but in the corona, the 3-D structure of the fields can sometimes be determined directly (Koutchmy, Molodensky, 1991). It is impossible to review these instrumental developments and to cover all the literature concerning the active phenomena. We shall concentrate on those which display the coupled nature of the layers of the Sun, being proxies for deeper phenomena and linked with the emergence of magnetism.

The coupling of magnetic fields with granulation and convection, and with rotation, luminosity, and radius has been shown. These relations are clear when analysing the solar behaviour during the cycle. Another recent discovery is the close correlation between the frequencies, amplitudes, and even the damping factors of the acoustic waves; and the general level of magnetic activity (*e.g.* Gough & Toomre, 1991).

3.1 The Multilayer Flare

Let us start from the flares. Recent reviews are those by Haish & al. (1991), Kahler (1991), Zirin & al. (1991), Ramaty & Simnett (1991).

A flare is a combination of a release of energy, a strong acceleration of particles, mass ejections and significant epiphenomena.

Remarkable images have been obtained up to energies of 100 keV (Kosugi & al., 1992). The mechanism of global restructuring is very clear (Tsuneta & al., 1992): An helmet-shaped arch appears in the corona long before the flare (several hours), topped by a vertical structure (Y-shaped feature). It is associated with a continuous expansion and restructuring of the active-region magnetic structure. It has been well known for some time that a reorganization of the magnetic field accompanies the release of energy. How does it proceed? Hénoux and Somov (1987) suggest as essential the effects of the systems of currents generated from photospheric motions in a photospheric multipolar magnetic region. The topological complexity of the fields is instrumental in the process. This is actually supported by observations (Hénoux & al., 1992, Mandrini & al., 1993). Note that it is in the corona that the phenomenon takes place, at the expense of the coronal magnetic field shear. Wang (1992) finds an increase of the photospheric magnetic shear after a flare. The study of the evolution of surges or flares within active regions confirms that idea (Schmieder & al., 1993).

The classical view of particle acceleration is based upon the optical and radio features of flares and bursts and upon the high energy data. An important group of recent observations concern the flare extension up to the corona. The Yohkoh data are particularly relevant in the domain of hard and soft X-rays. The satellites SMM, Compton, GRANAT, GRAMMA1, GRO, as well as Yohkoh, have provided important spectrometric data for energies up to 100 keV, often much above. But some doubts exist as to the accelerating mechanisms.

Coronal flaring events measured in soft X-rays by Yohkoh are striking, and linked to structures in the field. Coronal transients are well studied, and their evolution is also intimately linked (Somov, 1991) with the evolution of Y-shaped magnetic field structures. The coronal mass ejections are associated with radiobursts (metric) which allow their study (Gopalswamy, Kundu, 1993). As the CME is often close to a magnetic field streamer, interactions do often occur, giving place to thermal and non-thermal emission; an interesting point is that the CME is not caused by the flare, but that a particle acceleration high up in the corona is acting (Simnett & Harrison, 1985).

In any case, the flare is a very complex multilayer phenomenon. Associated with it are the beams of accelerated particles, which radiate a variety of radiations (Pick, Klein, Trottet, 1990). The energetic electrons are injected into magnetic structures of different scales, for which the proxies are the developments of the radiated burst. The growth of a flare illustrates the interaction between magnetic structures from the low to the middle corona. How, in this condition, do we understand the sequence of the primary events? Does the reconnection itself follow some accelerating mechanism, then to be linked with the progressive evolution of the magnetic topology? Or not?

3.2 Prominences in the Corona

Just as in the case of flares, eruptive prominences are associated with coronal mass ejections of plasmides, and produce radioemission. A causal relation between the dense coronal matter and the non- thermal electrons emitting the radiowaves is suspected (Klein & Mouradian, 1990).

The filaments are signatures of a slowly evolving magnetic situation. Their formation requires magnetic loops and a preexisting dip in the magnetic structure (Démoulin & Priest, 1993, Démoulin, 1993). The dip may come from a twist of the feet of the loops, perhaps linked with photospheric motions or to the influence of a higher magnetic loop. The origin of a filament may be (Schmieder et al., 1991b) photospheric vortices in the magnetic regions. In arch filaments, matter seems to drain out from the filament, towards the footpoints; in the meanwhile, the whole structure of the filament is slowly rising (Schmieder & al., 1991a, Tsiropoula & al., 1992).

The disappearance of a filament (Mc Allister & al., 1992) is typically associated with an X-ray source. The study of the event leads to question the eruption-reconnection standard model of the double-ribbon flare. Much of the bright structure comprises heated pre-existing loops. The field stays contained by arcades. It is likely that most of the filament mass will stay in the heated, unwound axial magnetic field.

The fine structure of filaments is made of small threads (Engvold & al., 1989, Zirker & Koutchmy, 1990, 1991, Mein & Mein, 1991). This is the basis for the quantitative study of how matter is supplied to prominences, and how their behaviour is controlled by the magnetic field.

3.3 The Corona

The corona is closely coupled to the solar interior, more than to the photosphere, as if the dragging out of deep magnetic fields, which bloom in the corona, was the major phenomenon responsible for its origin. Actually, a single global picture seems to describe the supply of mass to the corona and the coronal heating. The spicules seem to be the main provider of matter to the corona, their inclination links them to the phospheric magnetism (Heritschi, Mouradian, 1992). Another essential point is that the NBP footpoints of the coronal magnetic field, seem to move fast (Choudhuri & al., 1993); the energy implied by this motion seems sufficient to heat the less active (quiet) parts of the corona. The heating of the corona, like the obvious heating of chromosphere, seems due to dissipation of magnetic energy generated much lower down in the Sun.

Among the most interesting discoveries concerning the corona is the abundance anomaly of elements with a first ionization potential (FIP) smaller than 10 eV or so (Feldman, 1992, Meyer, 1993), which are more abundant by a factor 3 to 4 in the corona than in the photosphere. Can we explain this? Following the previous attempts of Delache (1967), ordinary diffusion of neutrals in the gravitational field (Vauclair and Meyer, 1985), or across an upward jet (von Steiger & Geiss, 1989) have been invoked. But a more efficiently-driven diffusion mechanism, involving dynamic plasma effects, seems indeed necessary (see Hénoux & Somov, 1992, for a bibliography). These authors suggest the separation as due to transverse electric currents flowing in magnetic flux tubes. Hénoux & Somov convincingly argue that the phenomenon can very well be considered as the very origin of the existence of the chromosphere.

Many papers describe the structural features of the corona from their evolution during the cycle. One should mention the great progress accomplished in mapping the magnetic fields in the corona, both from indirect inferences (Hoeksema, 1989), and also from direct determinations, or even measurements (Klein, 1992) using Faraday rotation. And let us praise the beautiful movies obtained by Yohkoh displaying a solid rotation of the coronal structures, and many fine 3-D pictures of the corona.

The X-ray study of the corona provides us with very important new avenues. The high temperatures in the loops are well determined (Hara & al.). Similar temperatures exist in much less dense regions. The active regions of the corona expand continuously, at velocities of a few to a few tens of km/s; this result was unexpected (Uchida & al., 1992). Apart from flares (see 3.1), some very interesting features of coronal activity are the bright points observable in X-rays (Strong & al., 1992), and also the X-ray jets (Shibata & al., 1992).

An important feature of the large scale corona is the structure of the neutral line separating inward and outward polarity regions, such as displayed in the studies by Hoeksema (1989), or now studied regularly through the radioheliographic data (Lantos & al., 1992).

10

4 The Solar migrations; the Cycle

Every phenomenon depends upon the general level of activity: luminosity and radius, granulation and convection, rotation and differential rotation, and of course all magnetic and active features, the corona, the solar wind. New considerations pertaining to the solar migrations of activity have recently emerged, not only related to the observation of solar features, but also to terrestrial observations (which we shall not discuss here).

Should we see the migrations to be always equatorward, up to the poles (the butterfly diagram!), as suggested by many authors (Pecker, 1988)? The convective rolls are moving poleward. So are the polar filaments, and the general magnetic field. One sees on the large-scale magnetic field charts, for example, the poleward flows of magnetic waves, and more generally, an obvious poleward behaviour (Mouradian & Soru-Escaut, 1993). The convective description of convection by toroidal rolls ascending in latitude leads the quoted authors to propose an asymmetrical dynamo, animated by the behaviour of the bipolar magnetic islands. However, the fact that the neutral line separating positive and negative polarities (Harvey, 1991) moves poleward is simply associated with the reversal of the dipole, and may be the cause of some poleward appearances. But we feel that most of these poleward migrations may be caused by deeper phenomena associated with the dynamo model (*e.g.* Belvedere & al., 1991).

A way to approach this question is to introduce, as Howard and Labonte (1984) did, the concept of torsional oscillations. These oscillations are located on the Sun at the edge of the convective rolls.

Quite striking is the evidence for an extended solar cycle of 16 to 22 years (from a tracer to another) (Simon, 1979, Harvey, 1991, Mouradian, Soru-Escaut, 1991, 1993). The migrations appear asymmetrical, associated with the north-south asymmetry of the rotation.

The connexion of successive cycles, the idea of extended cycles, the complementarity of the notion of migration with that of the cycle became clear. Combining the cyclic behaviour of the neutral sheet with the related solar wind and geomagnetic data, Legrand & Simon (1991), Simon & Legrand (1992) suggest a two-component solar cycle. In essence, a dipolar field, typical of minimum activity, controls the coronal field, and associated phenomena (geomagnetism). Upon this dipolar field is superimposed, with about 5-6 years delay, a toroidal field, spot-producing; hence the maximum is delayed by 5-6 years compared to the dipolar field. It explains why the sunspot maximum number correlates closely with the geomagnetic activity which actually occurred 5-6 years later (Ohl, 1966, Dodson-Prince, 1968). A similar conclusion is perhaps reached by McIntosh (1991), from the study of the time-behaviour of large-scale surface patterns.

Semi-empirical models of the cycle, derived from surface phenomena by continuity equations, within the reference frame of an $\alpha - \omega$ dynamo model, display separately features of the dipolar field and of the toroidal field (*e.g.* Belvedere & al. 1991, Brandenburg & al., 1991). Many alternatives have been suggested. Obviously, the question is wide open.

To conclude, for the time being, no model is still completely satisfactory. But the observations are so numerous, so accurate, that the right questions start to be asked. All parts of the Sun are intimately coupled, and coupled also with its history from its very formation. An understanding of the solar machinery has to start with the idea that the magnetism and the rotation are coming from pre-stellar times, as is the mass itself. If one does not put the solar study in an evolutionary reference frame, I am afraid that we are giving ourselves an impossible problem.

Samuel Butler, quoted by memory, said: "A magnificent theory, beautiful in all respect, is widely murdered by a very minute fact"... Gathering the ugly facts is therefore a basic duty of solar physicists: - on the ground where the operation of world- scale networks is a warrant of continuity in data; - in space where the harvest is enormous; we Europeans expect much from SOHO and ULYSSES; and the latter should only be the first of a series of out-of-the-ecliptic missions... Simultaneously, much effort has to be devoted to the study of solar-type stars (at different stages of their evolution), and magnetic stars... This will help us to understand the evolution of the rotation and magnetism to the evolutionary stage of the present-day Sun.

Acknowledgements

I want to thank all those, who have been so kind as to provide me with many published or unpublished papers, and to comment for me the main avenues of the solar research, namely: R. M. Bonnet, V. Domingo, E. Fossat, J.-C. Hénoux, R. Kandel, K.L. Klein, S. Koutchmy, F. Laclare, P. Lantos, S. Laloë, J.-P. Legrand, P. Mein, Z. Mouradian, J.-P. Meyer, E.A. Müller, R. Muller, E. Ribes-Nesme, E. Schatzman, B. Schmieder, M. Semel, P. Simon, G. Trottet, Y. Uchida, M. Vilmer. Most of this information has been the basis of a longer review, already prepared for the Catania meeting, of which the present paper is itself a shorter abstract. This longer paper is available on request to the author.

References

Several colloquia, or symposium volumes have been devoted to the Sun in the recent years, as well as some review papers (including the authors). They have been frequently used, and some contain a very extensive list of references. In a chronological order, they are as follows (but the list is far from complete) and properly referenced in the list: Pecker, 1988; Priest, Kristian, 1990; Berthomieu, Cribier, 1990; Cox, Livingston, Matthews, 1991; Sonett, Giampapa, Matthews, 1991; Harvey, 1991; Domingo, 1992; Yohkoh, 1992 (papers collected in Publ. Astron. Soc. Japan, **44** n.5); SCLERA, 1992; IAU Coll. **141**, 1993.

Belvedere, G., Proctor, M.R.E., Lanzafame, G., 1991, Nature **350**, 481-483

Berthomieu, G., Cribier, M., eds., 1991, Inside the Sun, IAU Coll. n.121, Kluwer Acad. Publ., Dordrecht

Brandenburg, A., Moss, D., Tuominen, I., 1991, in: Harvey, ed. 1991, 536-542

Choudhuri, A.R;, Auffret, H., Priest, E.R., 1993, Solar Phys. **143**, 49-68

Christensen-Dalsgaard, 1992, in Priest, Kristian, eds., 23-44

Cox, A.N., Livingston, W.C., Matthews, M.S., eds., 1991, Solar Interior and Atmosphere; The Univ. of Arizona Press, Tucson

Davis, R., Jr., Cox, A.N., 1991, in Cox et al., eds., 51-85

Delache, Ph., 1967, Ann. Astrophys. **30**, 827-860

Delache, Ph., Gavryusev, V., Gavryuseva, E., Laclare, F., Régulo, C., Roca Cortés, T., 1993, Astrophys. J. **407**, 801-805

Démoulin, P., Priest, E.R., 1993, Solar Phys. **144**, 283-305

Démoulin, P., 1993, in Proceedings COSPAR meeting, year 1992

Démoulin, P., Hénoux, J.-C., 1992, Solar Phys. **139**, 105-123

Dodson-Prince, H., 1968, private communication quoted in Legrand & Simon, 1991

Domingo, V., ed., 1992, Coronal Streamers, Coronal Loops and Coronal and Solar Wind Composition, First Soho Workshop, Annapolis, ESA-SP **348**

Durney, B., 1990, private communication quoted in Schatzman, 1990

Durney, B., 1992, SCLERA, in press

Engvold, O., Jensen, E., Zhang, Y., Brynilsen, N., 1989, Hvar Observ.Bull., **213**, n.1, 205

Faurobert-Scholl, M., Frisch, H., Mein, N., eds., 1992, Méthodes de détermination des Champs Magnétiques, Atelier Observatoire de Paris

Feldman, U., 1992, Physica Scripta, **46**, 202-220

Fossat, E., 1991, Solar Phys. **133**, 1-12

Fröhlich, C., Foukal, P.V., Hickey, J.R., Hudson, H.S., Willson, R.C., 1991, in Sonett et al., 1991, 1129

Gelly, B., Fossat, E., Grec, G., 1988, Astron. Astrophys. **200**, L29-31

Gough, D.O., Toomre, J., 1991, Ann. Rev. Astron. Astrophys. **29**, 627- 686

Gopalswamy, N., Kundu, M.R., 1993, Solar Phys. **143**, 327-343

Haisch, B.M., Strong, K.T., Rodonò, M., 1991, Ann. Rev. Astron. Astrophys. **29**, 274-324

Hara, S., Tsuneta, S., Acton, L.W., Lemen, J.R., Uchida, Y., 1992, Yohkoh, L135-140

Harvey, K., ed., 1991, The Solar Cycle, Astr. Soc. Pacific. Ser. **27**, 562

Harvey, K., 1991, in Harvey 1991, p. 337-365

Hénoux, J.-C., Somov, B.V., 1987, Astron. Astrophys. **185**, 306-314

Hénoux, J.-C., Somov, B.V., 1992, in Domingo, 1992, p.325

Hénoux, J.-C., Démoulin, P., Mandrini, C.H., Rovira, M.G., 1992, IAU Colloquium **141**, in press

Heritschi, D., Mouradian, Z., 1992, Solar Phys. **142**, 21-34

Hoeksema, J.T., 1991, Adv.Space Res. **11**, 15-24

Howard, R., Labonte, B.J., 1992, Astrophys J. **239**, L33-36

IAU Colloquium 141, 1993, Bejing, China, Sept. 1992, in press

Kahler, S.W., 1992, Ann.Rev.Astron.Astrophys. **30**, 113-142

Klein, K.L., Mouradian, Z., 1990, in: Dynamics of Solar Flares Workshop, Chantilly, Oct. 1990, Schmieder, B. and Priest, E., eds., p. 185-186

Klein, K.L., 1992, in, Faurobert-Scholl et al., p. 113-130

Kneubühl, F.K., Rast, J., Stettler, P., Müller, E.A., Huguenin, D., 1981, in: Fourth Intern. Conf. on IR and mm Waves and their Applications, Miami Beach, Dec. 1979, p. 177-178

Komm, R., Howard, R., Harvey, J.W., 1993, Solar Phys. **145**, 1-10

Kosugi, T., Sakao, T., Masuda, S., Makishima, K., Ina, M., Murakami, T., Ogawara, Y., Yaji, K., Matsushita, K., 1992, Yohkoh

Koutchmy, S., Molodensky, M.M., 1992, Nature, **360**, 717-719

Kuhn, J.R., 1988, Astrophys J., **331**, L131-134

Kuhn, J.R., 1989, Astrophys J., **339**, L45-47

Laclare, F., 1987, C.R. Acad. Sc. Paris, **305**, II, 451-454

Lantos, P., Alissandrakis, C.E., Rigaud, D., 1992, Solar Phys. **137** 225-256

Legrand, J.P., Simon, P., 1991, Solar Phys. **131**, 187-209

Mc Allister, A., Uchida, Y., Tsuneta, S., Strong, K.T., Acton, L.W., Hiei, E., Bruner, M.E., Watanabe, Ta., Shibata, K., 1992, Yohkoh

Mc Intosh, P.S., 1991, in: Harvey, 1991, p. 14-33

Mandrini, C.H., Démoulin, P., Hénoux, J.-C. Machado, M.E., 1991, Astron.Astrophys. **250**, 541-547

Mandrini, C.H., Rovira, M.G., Démoulin, P., Hénoux, J.-C., Machado, M.E., Wilkinson, L.K., 1993, Astron.Astrophys. **272**, 609-620

Maunder, M., Maunder, A.S.D., 1905, Monthly. Not. Roy. astr. Soc. **65**, 813

Mein, P., 1991, Astron.Astrophys. **248**, 669-676

Mein, P., Mein, N., 1991, Solar Phys. **136**, 317-333

Meyer, J.P., 1985, Astrophys. J. Suppl. **57**, 173

Meyer, J.P., 1993, in: Origin and Evolution of the Elements, Paris, June 1992, Prantzos, Vangioni-Flam, Casse, eds., Cambridge Univ. Press, p. 26-62

Mikheyev, S., Smirnov, A. Yu., 1986, Nuovo Cimento, **9C**, 17-26

Mouradian, Z., Martres, M.J., Soru-Escaut, I., Gezstelyi, L., 1987, Astron. Astrophys. **183**, 129-134

Mouradian, Z., Soru-Escaut, I., 1991, Astron. Astrophys. **251**, 649- 654

Mouradian, Z., Soru-Escaut, I., 1993, Astron. Astrophys., in press

Müller, E.A., Stettler, P., Rast J., Kneubühl, F.K., Huguenin, D., 1976, in: First European Meeting on Solar Physics, Florence, Feb. 1975, Chiuderi, C., Landini M., Righini A., eds., Osservazioni e Memorie dell'Osservatorio Astrofisico di Arcetri, n.105

Muller, R., Auffret, H., Roudier, Th., Vogneau, J., Simon, G.W., Frank, Z., Shine, R.A., Title, A.M., 1992, Nature **356**, 322-325

Muller, R., Roudier, Th., 1992, Solar Phys. **141**, 27-33

Nesme-Ribes, E., Ferreira, E.N., Mein, P., 1993, Astron. Astrophys., in press

Nesme-Ribes, E., Ferreira, E.N., Sadourny, R., Le Treut, H., Li, Z.X., 1993, Astron. Astrophys., in press

Nordlund, A., 1985, Solar Phys. **100**, 209-235

Ohl, A.I., 1966, Soln. Dann., n. 12, 84-85

Pecker, J.-C., 1988, Irish Astron.J. **18**, n.3, 135-146

Pecker, J.-C., 1986, Annuaire Collège de France, 171-184, p. 175

Pecker, J.-C., Roberts, W.O., 1954, J. Geophys. Res. **60**, 33

Pick, M., Klein, K.-L., Trottet, G., 1990, Astrophys. J. Suppl. Series **73**, 165-175

Priest, E., Kristian, J., eds., 1990, Basic Plasma Processes on the Sun, IAU Symposium n.142, Kluwer Acad. Press, Dordrecht

Ramaty, R., Simnett, G.M., 1991, in: Sonett et al., 1992, p. 232-259

Raulin, J.P., Willson, R.F., Kerdraon, A., Klein, K.-L., Lang, K.R., Trottet, G., 1991, Astron. Astrophys. **251**, 298-306

Ribes, E., Mein, P., Mangeney, A., 1985, Nature **318**, 170-171

Ribes, E., Beardsley, B., Brown, T.M., Delache, Ph. Laclare, F., Kuhn, J.R., Leister, N.V., 1991, in Sonnett et al., 1991

Ribes, E., Laclare, F., 1988, Geophys. Astrophys. Fluid Dyn. **41**, 171-180

Ribes, E., Merlin, Ph., Ribes, J.C., Barthalot, R., 1989, Annales Geophysicae, **7** (4), 321-329

Ribes, E., 1992, J. Astron. Fr., n.43, 2-10

Rosen, S.P., 1991, in: Cox et al., 1991, p. 86-111

Roudier, T., 1986, Thèse de doctorat, Univ. Paul Sabatier, Toulouse

Schatzman, E., 1990, in: Berthomieu, Cribier, 1990, p. 5-20

Schatzman, E., Berthomieu, G., 1992, J.Astron.Fr., n.43, 11-14

Schmieder, B., Raadu, M.A., Wüh, J.E., 1991, Astron. Astrophys. **253**, 353

Schmieder, B., Van Driel-Gesteleyi, L., Hénoux, J.-C., Simnett, G.M., 1991, Astron. Astrophys. **244**, 533

Schmieder, B., Van Driel-Gestelyi, L., Gerlei, O., Simnett, G.M., 1993, Solar Phys. **146**, 163-176

SCLERA (Santa Catalina Laboratory for Experimental Relativity by Astrometry) 1992 Symposium Proceedings, Singapore, in press

Shibata, K., Ishido, Y., Acton, L.W., Strong, K.T., Hirayama, T., Uchida, Y., Mc Allister, A.H., Matsumoto, R., Tsuneta, S., Shimizu, T., Hara, H., Sakurai, T., Ichimoto, K., Nishino, Y., Ogawara, Y., 1992, Yohkoh, L173-180

Simnett, G.M., Harrison, R.A., 1985, Solar Phys. **99**, 291-311

Simon, P.A., 1979, Solar Phys. **63**, 399-410

Simon, P.A., Legrand, J.-P., 1992, Solar Phys. **141**, 391-410

Snodgrass, H.B., 1991, Astrophys. J. **383**, L85-87

Snodgrass, H.B., Ulrich, R.K., 1990, Astrophys.J. **351**, 309-316

Somov, B.V., 1991, Adv.Space Res. **11**, 179-185

Sonett, C.P., Giampapa, M.S., Matthews, M.S., eds., 1991, The Sun in Time, The University of Arizona Press, Tucson

Spruit, H.C., 1991, in: Sonett et al., 1991, p. 118-158

Spruit, H.C., Nordlund, A., Title, A., 1990, Ann.Rev.Astron.Astrophys. **28**, 263-301

Strong, T., Harvey, K.L., Hirayama, T., Nitta, N., Shimizu, T., Tsuneta, S., 1992, Yohkoh, L161-166

Thompson, W.T., Schmieder, B., 1991, Astron. Astrophys. 243, 501- 511

Topka, K.P., Title, A.M., 1991, in: Cox et al., 1991, p. 727-747

Trimble, V., 1993, Publ.Astr.Soc.Pacific, **105**, 1-21

Tsiropoula, G., Georgakilas, Z.A., Alissandrakis, C.E., Mein, P., 1992, Astron.Astrophys. **262**, 587-596

Tsuneta, S., Takahashi, T., Acton, L.W., Bruner, M.E., Harvey, K.L., Ogawara, Y., 1992, Yohkoh, L211-214

Uchida, Y., Mc Allister, A., Strong, K.T., Ogawara, Y., Shimizu, T., Matsumoto, R., Hudson, H.S., 1992, Yohkoh, L155-160

Vauclair, S., Meyer, J.P., 1985, in: Proc.19th Intern. Cosmic Rays Conf., La Jolla, 233-236

Von Steiger, R., Geiss, J., 1989, Astron. Astrophys. **225**, 222

Wang, H., 1993, in: IAU Coll. 141, 1993, in press

Wang, H., Zirin, H., 1992, Nature, in press

Wolfenstein, L., 1978, Phys.Rev., **17D**, 2369-2374

Wolfenstein, L., 1978, Phys.Rev., **18D**, 958-960

Wolfsberg, K., Kocharov, G.E., 1991, in: Sonett et al., p. 288-310

Zirin, H., Mac Kinnon, A., McKenna-Lawlor, S.M.P., 1991, in: Cox et al., p. 964-995

Zirker, J.B., Koutchmy, S., 1990, Solar Phys. **127**, 109-118

Zirker, J.B., Koutchmy, S., 1991, Solar Phys, **131**, 107-118

I

Solar Internal Structure

Recent Results from Helioseismology

M. J. Thompson

Astronomy Unit, Queen Mary & Westfield College, London E1 4NS

Abstract: The very accurately measured frequencies of the Sun's global oscillation are providing us with a detailed picture of the solar interior. The adiabatic sound speed is of foremost importance in determining the observed p-mode frequencies and hence is readily accessible to seismic analysis of the frequency data. It has thus been known for some time that the sound speed in the bulk of the radiative interior is greater than in theoretical models. Much of this discrepancy is removed by the latest Livermore opacities, but a significant discrepancy remains. The cause of this discrepancy may be settling and diffusion of helium in the solar interior. This would also explain the discrepancy between seismic measurements of the envelope helium abundance and the abundance required to produce correctly calibrated evolution models. The frequencies have also enabled us to pinpoint the base of the convection zone and to investigate the extent of overshoot into the underlying convectively stable region. The variation of the p-mode frequencies with time is discussed, as are the current helioseismic determinations of the Sun's internal rotation. Finally future prospects are considered.

1 The Nature of Global Solar Oscillations

This short review will discuss a selection of the recent results from helioseismology, giving just enough theoretical background to indicate how the results are obtained. The interested newcomer to the field is referred to some other recent reviews to make up for the deficiencies of the present brief treatment: in particular those by Christensen-Dalsgaard (1986); Vorontsov & Zharkov (1989); Gough & Toomre (1991). For more detail on theoretical aspects, see Christensen-Dalsgaard & Berthomieu (1991) and Gough (1993); for more detail on helioseismic inversion see Gough (1985) and Thompson (1991). Briefly, and slightly simplistically, what we observe on the solar surface is the superposition of many normal modes of oscillation, each of which depends on the colatitude θ, longitude ϕ and time t as $Y_l^m(\theta, \phi) \exp(i\omega t)$, where Y_l^m is a spherical harmonic of degree l and (azimuthal) order m. Here ω is the frequency of the mode, though observations are normally described in terms of the cyclic frequency $\nu \equiv \omega/2\pi$. The modes are probably excited by turbulence in the upper convection zone. They permit us to do seismology of the solar interior because different modes are sensitive to different parts of the interior: in particular they penetrate to different depths and are trapped in different latitudes.

Most of the observed global modes have the character of acoustic waves, and the adiabatic sound speed c plays a crucial role in determining their frequencies. The square of the sound speed is given by $c^2 = \Gamma_1 p/\rho$, where p and ρ are pressure and density respectively, and Γ_1 is $\partial \log p/\partial \log \rho$ at constant specific entropy. For a perfect gas, $c^2 = \Gamma_1 \mathcal{R} T/\mu$, where T is temperature, μ is mean molecular weight and $\mathcal{R}$ is the gas constant. The sound speed increases with depth inside the Sun because of the increase in temperature. However, in the core the gradual conversion of hydrogen to helium changes μ; and in the ionization zones of the most abundant elements Γ_1 may vary significantly. Thus among other things we may hope to learn about energy transport, chemical evolution, abundances and equation of state from mode frequencies. Waves propagating into the interior are refracted back towards the surface at the lower turning point, at which the phase speed ω/k_h equals the local sound speed ($k_h \equiv [l + 1/2]/r \equiv L/r$ being the local horizontal wavenumber and r the radial coordinate)[1]. Hence the lower turning point is at radius $r = r_t$, where

$$\frac{\omega}{L} = \frac{c(r_t)}{r_t}. \tag{1}$$

Note that the penetration depth of the mode thus depends on the mode frequency and degree only in the combination ω/L; and that in particular, at fixed frequency, modes of low degree penetrate most deeply into the interior.

The modes have another turning point near the surface, where the decreasing density scale height reflects upward-propagating waves. This happens where the vertical wavenumber is of the order of the reciprocal of the density scale height H, viz $\omega/c \sim 1/H$. A more precise calculation for a polytropically stratified atmosphere gives that the upper turning point is where

$$\omega = \omega_c(r) \equiv \frac{c(r)}{2H(r)} \left(1 - 2\frac{dH}{dr}\right)^{1/2} \tag{2}$$

(Deubner & Gough 1984). Modes are principally sensitive to conditions within the acoustic cavity bounded by their upper and lower turning points. This sensitivity is reflected in their frequencies. So, for example, because the acoustic cut-off frequency $\omega_c(r)$ increases towards the surface (cf. Christensen-Dalsgaard 1986) the lower-frequency p-modes have upper turning points somewhat beneath the photosphere and are therefore less sensitive to surface conditions than are modes of higher frequency.

If one takes two models that differ in their internal structure, how are their mode frequencies related? The answer from simple asymptotics is that the difference $\delta\omega$ between the frequencies of corresponding modes is given by

$$S(w)\frac{\delta\omega}{\omega} \simeq H_1(w) + H_2(\omega), \tag{3}$$

---

[1] In the identification of k_h with L/r, L has been defined here as $l + 1/2$. Some authors use the more obvious $L = \sqrt{l(l+1)}$. However, $l + 1/2$ gives more satisfactory agreement between asymptotic and exact solutions for $l = 0$, as has been known for many years in quantum mechanics (cf. Kemble 1937); and the numerical difference between the two is negligible in practice, except for very low l.

(Christensen-Dalsgaard *et al.* 1988, 1989) where $w \equiv \omega/L$, δc is the difference in sound speed at fixed r between the models,

$$S(w) = \int_{r_t}^{R} \left(1 - \frac{L^2 c^2}{\omega^2 r^2}\right)^{-1/2} \frac{dr}{c}, \quad \text{and} \quad H_1(w) = \int_{r_t}^{R} \left(1 - \frac{L^2 c^2}{\omega^2 r^2}\right)^{-1/2} \frac{\delta c}{c} \frac{dr}{c}.$$

Here R is the photospheric radius. The H_1 term arises from the sound travel time in the interior changing from one model to the other, while the H_2 term arises from changes to surface structure. Note that when the frequency differences are appropriately scaled by S, the contributions H_1 and H_2 from interior and surface can be separated by their different dependences on ω/L and ω.

2 Inversions for Solar Structure

2.1 Radially symmetric hydrostatic stratification

Eq. (3) can be used to infer the difference in internal sound speed between the Sun and a model: first H_1 can be isolated from the difference between the observed frequencies and those computed for the model, using the different functional dependencies of H_1 and H_2; and then the function H_1 can be inverted analytically to infer $\delta c/c$ (Christensen-Dalsgaard *et al.* 1989). Using this or related asymptotic techniques, it was found as early as the mid-1980s that the Sun is hotter (more precisely, has a higher sound speed) in the outer part of the radiative interior than was the case for the solar models of the time (Christensen-Dalsgaard *et al.* 1985). It was suggested at the time that this discrepancy could be resolved by increasing the opacities in the model (cf. Korzennik & Ulrich 1989). The recently published OPAL opacities (Iglesias & Rogers 1991; Rogers & Iglesias 1992; Iglesias *et al.* 1992) are indeed somewhat higher than the older opacities and using them in solar models substantially reduces the discrepancy: however there is a residual difference which indicates some other error in the models (see Section 2.3).

Not all helioseismic inversions for structure are based on asymptotics. A lot of recent effort has gone into inversions which start instead from a variational principle for the mode frequencies (e.g., Gough & Kosovichev 1988; Dziembowski *et al.* 1990, 1992; Däppen *et al.* 1991). The variational principle can be written schematically as

$$\omega^2 = \mathcal{K}/\mathcal{I}; \tag{4}$$

here $\mathcal{K}$ and $\mathcal{I}$ are integrals over the interior of the Sun which depend on the eigenfunction of the mode and on the hydrostatic structure of the solar interior, which may be described by three variables, say p, ρ and Γ_1. By formally perturbing eq. (4) and linearizing, one obtains the frequency difference between the Sun and a reference model as a weighted integral of the differences δp, $\delta \rho$ and $\delta \Gamma_1$, say. But hydrostatic support provides a relation between δp and $\delta \rho$, so the number of unknown functions can be reduced to two, so for example

$$\frac{\delta \omega}{\omega} = \int_{0}^{R} \left(K_{u,\Gamma_1} \frac{\delta u}{u} + K_{\Gamma_1,u} \frac{\delta \Gamma_1}{\Gamma_1}\right) dr, \tag{5}$$

where $u = p/\rho$, and K_{u,Γ_1}, $K_{\Gamma_1,u}$ are known functions of radius. Various techniques can then be used to estimate δu and $\delta \Gamma_1$ from the frequency differences (e.g. Gough &

Thompson 1991). Results from such inversions are presented in the above-mentioned references.

2.2 Helium abundance in convective envelope

If one supposes further that the equation of state is known then $\delta\Gamma_1$ can also be eliminated before performing the inversion. Supposing also that the mix of heavy elements is known, but allowing for an uncertainty in the helium abundance Y, eq. (5) yields

$$\frac{\delta\omega}{\omega} = \int_0^R \left(K_{u,Y}\frac{\delta u}{u} + K_{Y,u}\delta Y \right) dr .$$ (6)

where again $K_{u,Y}$, $K_{Y,u}$ are known functions. Eq. (6) may now be inverted to obtain an estimate of δY. Recalling that the term in δY originated in a term in $\delta\Gamma_1$ in eq. (5), it is not surprising that we only measure δY in regions where a change of composition would cause a change in the adiabatic exponent, which is principally in the partial ionization zones of the abundant elements. Thus specifically we can use eq. (6) to measure the helium abundance at $r \simeq 0.98R$. Assuming that the convection zone is homogenized, this gives us an estimate of the helium abundance Y_{env} of the convective envelope. Best estimates from such an approach, assuming the so-called MHD equation of state, give $Y = 0.23 - 0.24$ (Dziembowski *et al.* 1991; Kosovichev *et al.* 1992). Alternatively, methods based on asymptotic analyses using something like the term H_2 in eq. (3), have yielded $Y \simeq 0.25$ (Christensen-Dalsgaard & Pérez Hernández 1991; Vorontsov *et al.* 1991). The difference between the two kinds of estimate are probably not significant, but it *is* significant that both are substantially lower than the initial helium abundance ($Y \simeq 0.28$) required to produce a standard solar model with the newest opacities. A resolution of this is suggested in Section 2.3.

The sensitivity of the frequencies to $\delta\Gamma_1$ (cf. eq. [5]) provides the opportunity to do more than just measure Y_{env}. In principle we may measure aspects of the equation of state itself: work on this is underway (see Christensen-Dalsgaard & Däppen 1992; Vorontsov *et al.* 1992).

2.3 Gravitational settling of helium

A physical effect not generally taken into account in solar models in the past is the gravitational settling of helium and the heavier elements. The effect of gravitational settling on solar oscillation frequencies has recently been investigated by Christensen-Dalsgaard *et al.* (1993) and Guzik & Cox (1993). In particular, the former authors conclude that the remaining discrepancy between the solar and model sound speeds in the outer part of the radiative interior can mostly be accounted for by the effects of helium diffusion. Diffusion also depletes the helium in the convective envelope: over the lifetime of the Sun this is sufficient to explain why the measured envelope helium abundance is lower than the value of Y required to construct non-diffusive solar models by about 0.03. Thus the determinations of helium abundance and sound speed both support the notion of gravitational settling of helium in the solar interior.

2.4 Convective overshoot

Another effect not generally included in solar models is the penetration of convective motions from the convection zone into the stably stratified radiative region beneath. If there is such overshoot in the Sun, several theoretical investigations suggest that the effect would be to extend the nearly adiabatically stratified region beyond that predicted by the Schwarzschild criterion, with a rather abrupt transition to subadiabatic stratification beneath (e.g. Zahn 1991). The effect of a sharp change in the stratification on the mode frequency depends on the spatial phase of the mode at the location of the abrupt transition (e.g. Gough 1990a). Suitably scaled, the effect will be the same on all modes that have the same spatial phase at that location. This gives rise to a characteristic signature in the frequencies. Various investigations along these lines have found no strong evidence for overshoot: indeed Basu *et al.* (1993) and Monteiro *et al.* (1993) estimate the extent of overshoot to be no more than about 0.1 pressure scale heights. However, all such investigations have assumed spherical symmetry: it is quite possible that the boundary of any overshoot region is not spherically symmetric, in which case its spherically symmetric average may be rather less abrupt than supposed and this would cause the extent of overshoot to be underestimated.

3 Changes in Mode Frequencies over Time

It is now well established that the frequencies of solar p-modes change with time (Elsworth *et al.* 1990; Libbrecht & Woodard 1990). What causes these changes, in particular is the change occurring deep in the solar interior or near the surface? Libbrecht & Woodard found that for l up to 60 the frequency shifts are only weakly dependent on degree, so at least some of the change must occur in the outer 15% or so of the Sun, which is common to the acoustic cavities of all the observed modes with $l < 60$. In fact the relative frequency shifts are found to be well described by $H_2(\omega)/S(w)$, which by eq. (3) implies that the *dominant* change (as far as the acoustic waves are concerned) is confined to the outer layers. Additional strong evidence comes from the frequency dependence, i.e. the form of H_2 itself. This is small for frequencies below about 1.6mHz but then increases rapidly with frequency. The explanation of this behaviour is that the dominant acoustic changes are occurring in the photosphere or even higher in the atmosphere. Modes with low frequency have upper turning points somewhat below the photosphere and are therefore only weakly sensitive to changes in the very outer layers: thus H_2 is small as low frequencies.

It is not yet certain what aspect of the near-surface structure is responsible for the frequency shifts, but it is likely that magnetic fields play a role. Indeed the shifts are well correlated with changes in the observed magnetic field, on the timescale of about one month (Woodard & Libbrecht 1993; Bachmann & Brown 1993). This is not surprising, since an acoustic wave will tend to feel an extra restoring force due to the presence of a magnetic field. The frequency changes are also tied in with solar activity by the apparent latitudinal variation of the temporal changes (Libbrecht & Woodard 1990). One possibility is that the frequencies are shifted by a photospheric fibril field (Gough & Thompson 1988; Goldreich *et al.* 1991). Alternatively, changes in the chromospheric field could be responsible (Evans &

Roberts 1992): indeed a combination of changes in both the magnetic field and the chromospheric temperature may explain not only the principal increase in H_2 with frequency but also its observed drop-off above 4mHz (Jain & Roberts 1993).

4 Inversions for Solar Rotation

Differences in the frequencies of modes of different azimuthal orders m (called frequency splittings) have been used to obtain helioseismic determinations of the rotation rate as a function of latitude at fractional radii $0.5 \lesssim r/R \lesssim 0.85$, where r is the radial coordinate and R is the photospheric radius (Brown et al. 1989; Christensen-Dalsgaard & Schou 1988; Dziembowski, Goode & Libbrecht 1989; Goode et al. 1991; Korzennik et al. 1988; Kosovichev 1988; Schou 1991; Thompson 1990). The principal conclusions that have been drawn thus far from such inversions are (i) that the rotation rate deep in the convection zone exhibits similar latitudinal variation to what is observed at the surface, i.e., there is little radial gradient in the rotation rate through the convective envelope, and (ii) that there is a transition near the base of the convection zone to latitudinally independent rotation in at least the outer part of the radiative interior. These findings are contrary to the expectation from numerical simulations that the rotation rate in the convection zone would be roughly constant on cylindrical surfaces aligned with the rotation axis.

Intriguing hints as to the origin of surface magnetic fields come from the inversions for rotation in the equatorial plane, using modes with l up to 500 (Korzennik et al. 1990). These suggest that the rotation at $r \simeq 0.8R$ is about 462nHz, which is consistent with the surface magnetic feature rate (Snodgrass 1983). This leads one to speculate that the magnetic features are anchored in the lower convection zone. The rotation increases towards the surface and reaches a local maximum of about 475nHz at $r \simeq 0.92R$. It then decreases even closer to the surface and extrapolation to the photosphere would give a value close to the spectroscopic determination of 452nHz (Snodgrass 1984).

The rotation rate in the core is still very uncertain. The problem is that only the most deeply penetrating low degree p-modes are sensitive to conditions in the core; and even for these modes the frequency is mostly determined by conditions in the bulk of the Sun, the core having a relatively small effect. Thus even a modest increase, say, in the rotational splitting in low-degree modes, relative to that in the higher degree modes, would indicate that the core is rotating substantially faster. The IPHIR data (Toutain & Fröhlich 1992) apparently indicate that the splitting of low modes is some 20% larger than the splitting of other modes, which by a naive estimate would lead one to conclude that the core could be rotating some $2\frac{1}{2}$ times faster than the rest of the interior (Thompson 1993): this is confirmed by more precise inversions of the same data (Goode et al. 1992). However, because of the great sensitivity of this conclusion to small changes in the low-degree splittings, the question of the core's rotation must still be considered an open question (cf. Gough 1990b).

5 Prospects

To answer definitively questions such as how the core is rotating we need more and even better data, particularly at low degree. A very exciting prospect is that we shall soon have a large quantity of excellent data from several imminent international collaborations. Much of this will come from the IRIS and GONG networks, each using of the order of six observing stations distributed in longitude. The idea of such a network is to minimize the amount of time that the Sun is not observed, and in particular to avoid regular night-time gaps in the observations which cause spurious peaks in the derived power spectrum due to aliasing. Another way to avoid night-time gaps is to observe from space, which has the added advantage of overcoming atmospheric seeing problems. Three helioseismology experiments are to fly on board the SOHO satellite. Just to pick a couple of aspects of the forthcoming datasets, GONG will provide individual frequency splittings, whereas most current observations only provide information about the frequencies' m-dependence in terms of a parametrization: this will greatly improve the latitudinal resolution of our inversions for rotation (Schou *et al.* 1992), as well as permitting investigations of the latitudinal variation of the internal structure. SOI-MDI on board SOHO is a high-resolution instrument which will enable us to observe waves on small spatial scales: these are confined to the near-surface layers (cf. eq. [1]) and will provide valuable information about this region.

I shall close by mentioning three new areas of observation which offer new challenges and prospects. The first is the discovery that p-mode power extends to much higher frequencies than was previously realized (Duvall *et al.* 1991). Such observations may not be modes in the usual sense (Kumar *et al.* 1990), but could nonetheless provide a probe of the nature and location of the mode excitation mechanism (Kumar & Lu 1991). The second is so-called time-distance helioseismology (Duvall *et al.* 1993). This is similar to much geoseismology in that what is used is the travel time between events. It offers another diagnostic of the solar interior, especially the outer layers. The third is the scattering and absorption of p-modes by sunspots and active regions (e.g. Braun *et al.* 1990, 1992; Bogdan *et al.* 1993) which offers the opportunity to probe the internal structure and depth dependence of sunspots. These new areas will be pursued alongside the more developed activities such as helioseismic inversion, applied to the forthcoming data, and will continue to make helioseismology an exciting endeavour.

References

Bachmann, K.T., Brown, T.M. (1993): "p-mode frequency variation in relation to global solar activity", Astrophys. J. **411** L45

Basu, S., Antia, H.M., Narasimha, D. (1993): "Helioseismic measurement of the extent of overshoot below the solar convection zone", Mon. Not. R. Astr. Soc., submitted.

Bogdan, T.J., Brown, T.M., Lites, B.W., Thomas, J.H. (1993): "The absorption of p-modes by sunspots: variations with degree and order", Astrophys. J **406** 723

Braun, D.C., LaBonte, B.J., Duvall, T.L. (1990): "The spatial distribution of p-mode absorption in active regions", Astrophys. J. **354** 372

Braun, D.C., Duvall, T.L., LaBonte, B.J., Jefferies, S.M., Harvey, J.W., Pomerantz, M.A. (1992): "Scattering of p-modes by a sunspot", Astrophys. J. **391** L113

Brown, T.M., Christensen-Dalsgaard, J., Dziembowski, W.A., Goode, P., Gough, D.O., Morrow, C.A. (1989): "Inferring the Sun's internal angular velocity from observed p-mode frequency splittings", Astrophys. J. **343** 526

Christensen-Dalsgaard, J. (1986): "Theoretical aspects of helio- and asteroseismology", in Seismology of the Sun and the distant Stars, ed. by D.O. Gough (Reidel, Dordrecht) pp. 23–53

Christensen-Dalsgaard, J., Berthomieu, G. (1991): "Theory of solar oscillations", in Solar interior and atmosphere, ed. by A.N. Cox, W.C. Livingston & M. Matthews (Space Science Series, University of Arizona Press) pp. 401–478

Christensen-Dalsgaard, J., Duvall, T.L., Gough, D.O., Harvey, J.W., Rhodes, E.J. (1985): "Speed of sound in the solar interior", Nature **315** 378

Christensen-Dalsgaard, J., Däppen, W. (1992): "Solar oscillations and the equation of state", Astron. Astrophys. Rev. **4** 267

Christensen-Dalsgaard, J., Gough, D.O., Pérez Hernández, F. (1988): "Stellar disharmony", Mon. Not. R. Astr. Soc. **235** 875

Christensen-Dalsgaard, J., Gough, D.O., Thompson, M.J. (1989): "Differential asymptotic sound-speed inversions", Mon. Not. R. Astr. Soc. **238** 481

Christensen-Dalsgaard, J., Pérez Hernández, F. (1991): "Influence of the upper layers of the Sun on the p-mode frequencies", in Challenges to theories of the structure of moderate-mass stars, Lecture Notes in Physics Vol. 388, ed. by D.O. Gough & J. Toomre (Springer, Berlin) pp. 43–50

Christensen-Dalsgaard, J., Proffitt, C.R., Thompson, M.J. (1993): "Effects of diffusion on solar models and their oscillation frequencies", Astrophys. J. **403** L75

Christensen-Dalsgaard, J., Schou, J. (1988): "Differential rotation in the solar interior", in Seismology of the Sun & Sun-like Stars, ed. by E.J.Rolfe (ESA SP-286) pp. 149–153

Deubner, F.-L., Gough, D.O. (1984): "Helioseismology: Oscillations as a diagnostic of the solar interior", Ann. Rev. Astron. Astrophys. **22** 593

Duvall, T.L., Harvey, J.W., Jefferies, S.M., Pomerantz, M.A. (1991): "Measurements of high-frequency solar oscillation modes", Astrophys. J. **373** 308

Duvall, T.L., Jefferies, S.M., Harvey, J.W., Pomerantz, M.A. (1993): "Time-distance helioseismology", Nature **362** 430

Dziembowski, W.A., Goode, P.R., Libbrecht, K.G. (1989): "The radial gradient in the Sun's rotation", Astrophys. J. **337** L53

Dziembowski, W.A., Pamyatnykh, A.A., Sienkiewicz, R. (1990): "Solar model from the helioseismology and the neutrino flux problem", Mon. Not. R. Astr. Soc. **244** 542

Dziembowski, W.A., Pamyatnykh, A.A., Sienkiewicz, R. (1991): "Helium content in the solar convective envelope from helioseismology", Mon. Not. R. Astr. Soc. **249** 602

Dziembowski, W.A., Pamyatnykh, A.A., Sienkiewicz, R. (1992): "Seismological tests of standard solar models calculated with new opacities", Acta Astron. **42** 5

Däppen, W., Gough, D.O., Kosovichev, A.G., Thompson, M.J. (1991): "A new inversion for the hydrostatic stratification of the Sun", in Challenges to theories of the structure of moderate-mass stars, Lecture Notes in Physics Vol. 388, ed. by D.O. Gough & J. Toomre (Springer, Berlin) pp. 111–120

Elsworth, Y., Howe, R., Isaak, G.R., McLeod, C.P., New, R. (1990): "Variation of low-order solar oscillations over the solar cycle", Nature **345** 322

Evans, D.J., Roberts, B. (1992): "Interpretation of solar-cycle variability in high-degree p-mode frequencies", Nature **355** 230

Goldreich, P., Murray, N., Willette, G., Kumar, P. (1991): "Implications of solar p-mode frequency shifts", Astrophys. J. **370** 752

Goode, P.R., Dziembowski, W.A., Korzennik, S.G., Rhodes, E.J. (1991): "What we know about the Sun's internal rotation from solar oscillations", Astrophys. J. **367** 649

Goode, P.R., Fröhlich, C., Toutain, T. (1992): "Rotation of the Sun's core", in The Solar Cycle, ASP Conf. Series Vol. 27, ed. by K.L. Harvey, pp. 282–285

Gough, D. O. (1985): "Inverting helioseismic data", Solar Phys. **100** 65

Gough, D.O. (1990a): "Comments on helioseismic inference", in Progress of seismology of the sun and stars, Lecture Notes in Physics Vol. 367, ed. by Y. Osaki & H. Shibahashi (Springer, Berlin) pp. 283–318

Gough, D.O. (1990b): "Open questions", in Proc. IAU Colloquium No 121, Inside the Sun, ed. by G. Berthomieu & M. Cribier, (Kluwer, Dordrecht) pp. 451–475

Gough, D. O. (1993): "Linear adiabatic stellar pulsation", in Astrophysical Fluid Dynamics, ed. by J.-P. Zahn & J. Zinn-Justin (North-Holland, Amsterdam), in press.

Gough, D.O., Kosovichev, A.G., (1988): "An attempt to understand the Stanford p-mode data", in Seismology of the Sun & Sun-like Stars, ed. by E.J. Rolfe (ESA SP-286) pp. 195–201

Gough, D.O., Thompson, M.J. (1988): "On the implications of the symmetric component of the frequency splitting reported by Duvall, Harvey and Pomerantz", in Proc. IAU Symposium No 123, Advances in helio- and asteroseismology, ed. by J. Christensen-Dalsgaard & S. Frandsen (Reidel, Dordrecht) pp. 175–180

Gough, D. O., Thompson, M. J. (1991): "The inversion problem", in Solar interior and atmosphere, ed. by A.N. Cox, W.C. Livingston & M. Matthews (Space Science Series, University of Arizona Press) pp. 519–561

Gough, D.O., Toomre, J. (1991): "Seismic observations of the solar interior", Ann. Rev. Astron. Astrophys. **29** 627

Guzik, J.A., Cox, A.N. (1993): "Using solar p-modes to determine the convection zone depth and constrain diffusion-produced composition gradients", Astrophys. J. **411** 394

Iglesias, C.A., Rogers, F.J. (1991): "Opacities for the solar radiative interior", Astrophys. J. **371** 408

Iglesias, C.A., Rogers, F.J., Wilson, B.G., (1992): "Spin-orbit interaction effects on the Rosseland mean opacity", Astrophys. J. **397** 717

Jain, R., Roberts, B. (1993): "Do p-mode frequency shifts suggest a hotter chromosphere at solar maximum?", Astrophys. J., in press

Kemble, E.C. (1937): "The fundamental principles of quantum mechanics" (McGraw Hill, London)

Korzennik, S.G., Cacciani, A., Rhodes, E.J., Tomczyk, S., Ulrich, R.K. (1988): "Inversion of the solar rotation rate versus depth and latitude", in Seismology of the Sun & Sun-like Stars, ed. by E.J. Rolfe (ESA SP-286) pp. 117–124

Korzennik, S.G., Cacciani, A., Rhodes, E.J., Ulrich, R.K. (1990): "Contribution of high-degree frequency splittings to the inversion of the solar rotation rate", in Progress of seismology of the sun and stars, Lecture Notes in Physics Vol. 367, ed. by Y. Osaki & H. Shibahashi (Springer, Berlin) pp. 341–347

Korzennik, S.G., Ulrich, R.K. (1989): "Seismic analysis of the solar interior. I. Can opacity changes improve the theoretical frequencies?", Astrophys. J. **330** 1144

Kosovichev, A.G. (1988): "The internal rotation of the Sun from helioseismological data", Sov. Astron. Lett. **14** 145

Kosovichev, A.G., Christensen-Dalsgaard, J., Däppen, W., Dziembowski, W.A., Gough, D.O., Thompson, M.J. (1992): "Sources of uncertainty in direct seismological measurements of the solar helium abundance", Mon. Not. R. Astr. Soc. **259** 536

Kumar, P., Duvall, T.L., Harvey, J.W., Jefferies, S.M., Pomerantz, M.A., Thompson, M.J. (1990): "What are the observed high-frequency solar acoustic modes?", in Progress of seismology of the sun and stars, Lecture Notes in Physics Vol. 367, ed. by Y. Osaki & H. Shibahashi (Springer, Berlin) pp. 87–92

Kumar, P., Lu, E. (1991): "The location of the source of high-frequency solar acoustic oscillations", Astrophys. J. **375** L35

Libbrecht, K.G., Woodard, M.F. (1990): "Solar-cycle effects on solar oscillation frequencies", Nature **345** 779

Monteiro, M.J.P.F.G., Christensen-Dalsgaard, J., Thompson, M.J. (1993): "Seismic study of overshoot at the base of the solar convective envelope", Astron. Astrophys, in press.

Rogers, F.J., Iglesias, C.A. (1992): "Rosseland mean opacities for variable compositions", Astrophys. J. **401** 361

Schou, J. (1991): "An inversion for the rotation rate in the solar interior", in Challenges to theories of the structure of moderate-mass stars, Lecture Notes in Physics Vol. 388, ed. by D.O. Gough & J. Toomre (Springer, Berlin) pp. 81–86

Schou, J., Christensen-Dalsgaard, J., Thompson, M.J. (1992): "The resolving power of current helioseismic inversions for the Sun's internal rotation", Astrophys. J. **385** L59

Snodgrass, H.B. (1983): "Magnetic rotation of the solar photosphere", Astrophys. J. **270** 288

Snodgrass, H.B. (1984): "Separation of large-scale photospheric Doppler patterns", Solar Phys. **94** 13

Thompson, M.J. (1990): "A new inversion of solar rotational splitting data", Solar Phys. **125** 1

Thompson, M.J. (1991), "Helioseismic inversion", in Challenges to theories of the structure of moderate-mass stars, Lecture Notes in Physics Vol. 388, ed. by D.O. Gough & J. Toomre (Springer, Berlin) pp. 61–80

Thompson, M.J. (1993): "Seismic investigation of the Sun's internal structure and rotation", in GONG 1992: Seismic Investigation of the Sun and Stars, ASP Conf. Series Vol 42, ed. by T.M. Brown, pp. 141–154

Toutain, T., Fröhlich, C. (1992): "Characteristics of solar p-modes: Results from the IPHIR experiment", Astron. Astrophys. **257** 287

Vorontsov, S.V., Baturin, V.A., Pamyatnykh, A.A. (1991): "Seismological measurement of solar helium abundance", Nature **349** 49

Vorontsov, S.V., Baturin, V.A., Pamyatnykh, A.A. (1992): "Seismology of the solar envelope: towards the calibration of the equation of state", Mon. Not. R. Astr. Soc. **257** 32

Vorontsov, S.V., Zharkov, V.N. (1989): "Helioseismology: theory and interpretation of experimental data", Sov. Sci. Rev. E. Astrophys. Space Phys. **7** 1

Woodard, M.F., Libbrecht, K.G. (1993): "Solar activity and oscillation frequency splittings", Astrophys. J. **402** L77

Zahn, J.-P. (1991): "Convective penetration in stellar interiors", Astron. Astrophys. **252** 179

The Structure of the Solar Core

W.A. Dziembowski

Astronomy Unit, Queen Mary and Westfield College, University of London
and
Copernicus Astronomical Center, ul.Bartycka 18, 00-716 Warsaw, Poland

Abstract: The efforts to resolve the solar neutrino problem resulted in a considerable progress in reliability and accuracy in modelling the Sun's interior. Implications from helioseismic sounding supports the standard picture of the solar evolution. A comparison of neutrino fluxes measured in various experiments seems to rule out a non-standard solar core as the solution of the neutrino problem.

1 Introduction

When at the beginning of the nineteen seventies the existence of a large difference between the observed and predicted neutrino flux was first realized, the theory of stellar internal structure was commonly blamed for the discrepancy. Indeed, the construction of stellar interior models involved many approximations and assumptions that could be questioned. During the subsequent years, various modifications of the standard treatment of stellar evolution leading to solar models producing lower neutrino fluxes have been proposed. Since the detectable neutrino flux is much more temperature sensitive than the photon flux, the natural direction of these efforts was to look for solar models with a lower central temperature and a higher neutrino flux. This goal has been achieved either by allowing element mixing in the core or by enhancing energy transport there. I do not believe that any of these proposals is a viable solution of the solar neutrino problem, but I do not think either that this has been all wasted effort. These works resulted in a better justification and understanding of the basic assumptions of stellar evolution theory.

The solar neutrino problem also gave a stimulus for improving the macroscopic physics needed for solar model construction within the framework of the standard theory. Significant progress has been made in the accuracy of the equation of state and opacity. The effect of gravitational settling, once thought to be unimportant in solar evolution, is now being included as part of the standard models. There is also continuous improvement in the calculations of nuclear reaction cross-sections.

In recent years the observational tools for probing the Sun's interior have been *greatly* expanded. In addition to the chlorine detector continuously operated since 1972 (Davis *et. al*, 1990), we have three new neutrino observatories: the water

Cerenkov detector at Kamiokande (Hirata *et al.* 1990) and two gallium detectors GALLEX (Anselmann *et al.* 1992) and SAGE (Abazov, 1991). Even more important were developments in the field of helioseismology (e.g. M. Thompson, in this volume). Measurements of p-mode frequencies yield much more direct information about the solar interior structure, including the core, than the neutrino fluxes.

It is indeed fair to say that the solar core has become an object of direct observations. However, modelling remains the main method of learning about its structure and observations must be used for testing the various hypotheses.

2 Standard Solar Models

The concept of the *Standard Solar Model* (SSM) is used to denote any solar model built according to the following prescription.

Use

    (i) best measurements of mass, radius, luminosity, age
       and abundance of elements heavier than helium relative to hydrogen
    (ii) best available microscopic physics data on nuclear reaction rates,
       equation of state, opacity and diffusion coefficients
    (iii) Mixing Length Theory (or another theory containing one free
       parameter) to describe the convective transport.

Assume

    (i) that at zero age the sun had uniform chemical composition
    (ii) it has evolved maintaining mechanical equilibrium under the action
       of only gravity and pressure forces
    (iii) mixing of chemical elements has occurred only within convective
       regions.

Adjust

    (i) the convection theory parameter and
    (ii) helium abundance

to fit present-day solar radius and luminosity.

In Table 1 I list the central temperatures, helium abundance - present, Y_c and initial, Y_0 - as well as the neutrino flux, $\Phi_{\nu,Cl}$ for solar models published in last two years. None of these models was built following exactly the above prescription and therefore, strictly speaking, none of them may be regarded as the standard solar model. The closest is the BPD model. It uses the most up-to-date opacities and recent data on the nuclear reaction cross-sections. It takes into account gravitational settling of helium, but ignores that of heavier elements. The equation of state is not the most accurate available.

The only other model taking helium settling into account is PM, but it uses less reliable opacities. Model BPN is calculated in exactly the same way as BPD except that it ignores effect of helium settling. There are various reasons for the differences between the BPN model and other models ignoring this effect. In the case of the TCD model the likely main cause is the different choice for for nuclear reaction parameters. In the case of S0 model this is the choice of the iron abundance

Table no 1. Solar Models

Model	$T_c[10^6\mathrm{K}]$	Y_c	Y_0	$\Phi_{\nu,\mathrm{Cl}}[\mathrm{SNU}]$	reference
GC	15.40	0.631	0.270	8.5	Guzik and Cox (1991)
PM	15.79	0.646	0.280	...	Proffitt and Michaud(1991)
CD	15.68	...	0.281	8.9	Christensen- Dalsgaard (1992)
BPN	15.59	0.627	0.272	7.2	Bahcall and Pinsonneault (1992)
BPD	15.69	0.638	0.273	8.0	Bahcall and Pinsonneault (1992)
CDF	15.76	0.649	0.295	9.1	Castellani et al. (1992)
BPML	15.55	0.656	0.273	7.4	Berthomieu et al.(1993)
TCD	15.43	...	0.271	6.4	Turck-Chieze and Lopes (1993)
S0	15.71	0.643	0.283	8.2	Dziembowski et al. (1993)

from spectroscopic rather than from meteoritic determinations, which are regarded as more reliable.

3 The Core Structure in a Standard Solar Model

We now focus on the BPD model. Bahcall and Pinsonneault (1992) provide a detailed numerical description of this model. I have used their TABLE XVI data to make the plots shown in Fig. 1.

The part of the solar model shown in Fig. 1, covering the inner 25% of the solar radius and about 50% of its mass, produces 99% of the total luminosity. Thus, somewhat arbitrarily, it may be regarded as the solar core. As a result of the gravitational settling of helium, which is included in this model, the terminal rise of X to the surface value of 0.733 occurs in the outer part of the radiative interior. The bottom of the convective zone in the BPD model is at $r = 0.707R_\odot$.

Among other parameters Bahcall and Pinsonneault (1992) provide the differential neutrino fluxes. In Fig. 2 the plots of the three fluxes most important for the evaluation of the count rates for existing detectors are shown. The three fluxes shown in Fig. 2 are produced in the proton-proton cycle. Differences between the three curves reflect different temperature sensitivity of the relevant reactions and reveal the potential role of the neutrino flux measurements in differential probing of the solar core. The evaluated count rates from the BPD model are as follows.

For the Cl experiment: (8.0 ± 3.0) SNU. The measured rate (Homestake) is $(0.26\pm0.04)\times$ BPD value. The contributions from individual reactions are : pp–0.0, pep–0.2, ^{7}Be–1.2, ^{8}B–6.2, ^{13}N– 0.1, ^{15}O–0.3.

For the Ga experiment 131.5^{+21}_{-17} SNU. The mesured rate (GALLEX+SAGE)is $(0.54 \pm 0.12)\times$ the BPD value. The contributions from individual reactions are: pp–70.8, pep–3.1, ^{7}Be–35.8, ^{8}B–13.8, ^{13}N- -3.0, ^{15}O–4.9.

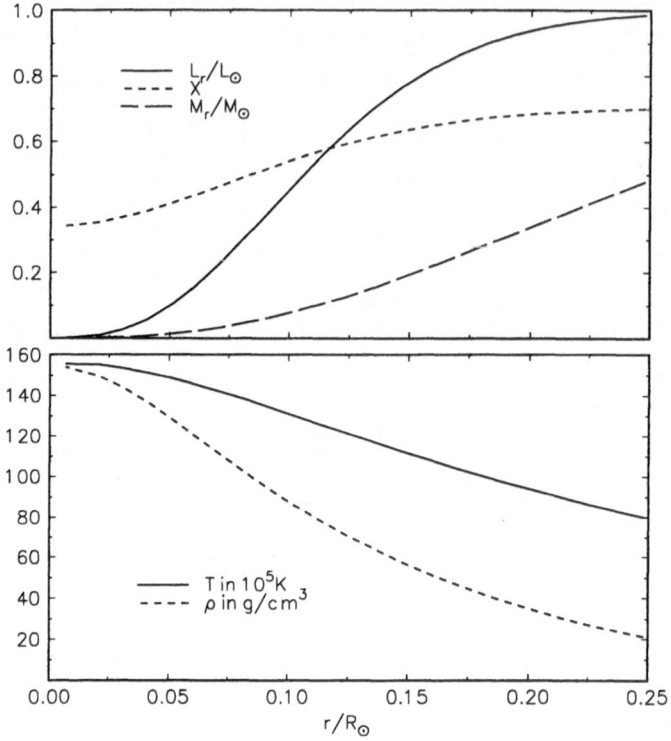

Fig. 1. (*top*) Fractional mass, $M_r/M_\odot$, radiative flux, $L_r/L_\odot$, and the hydrogen abundance parameter, X, (*bottom*) Temperature, T and density, ρ, plotted against fractional radius, $r/R_\odot$

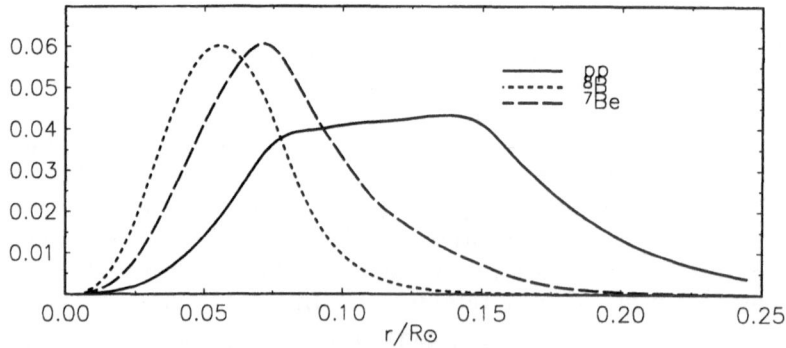

Fig. 2. The differential neutrino fluxes

The Cerenkov water experiment detects only the ^{8}B neutrinos. The predicted flux is $\Phi(^8\text{B}) = 5.1(1 \pm 0.43) \times 10^6 \text{cm}^{-2}\text{s}^{-1}$. The measured flux (Kamiokande II + III) is $(0.50 \pm 0.07) \times$ BPD value.

4 Recent Progress in the Macroscopic Physics for Solar Models

4.1 Nuclear Reaction Rates

The nuclear cross-sections adopted in solar model calculations are not directly measured parameters. They are obtained by means of extrapolation of experimental data to much lower energies and therefore there is room for uncertainty. Bahcall and Pinsonneault (1992) regard the uncertainty in the rate of ^{7}Be $+^1$ H as the most severe for calculated neutrino fluxes (1.7 SNU for the Cl and 3.9 for the Ga experiments). They point out that the difference in the adopted cross-section was primarily the source of the difference in predicted neutrino flux between Bahcall and Ulrich (1988) and Turck-Chieze et al. (1988). The BPD model uses an intermediate value recently determined by Johnson et al. (1992).

For the model structure calculation, by far the most important reaction cross-section is that of ^{1}H $+$ ^{1}H Turck-Chieze and Lopes (1993) contemplate possibility of up to a 10% modification of currently used rates. The resulting change in the sound speed would certainly be detectable by means of helioseismology.

A large (factor ~ 9) upward revision of the ^{3}He $+^3$ He reaction rate leading to a *nuclear physics solution of the solar neutrino problem* was proposed by Castellani et al. (1992) and by Paternó and Scalia (in this volume). Such a change would cause a decrease in ^{7}Be and ^{8}B by the same factor ~ 3. It is interesting that this proposal may be nearly ruled with current seismic data. (Dziembowski et al., 1993).

4.2 Equation of State and Opacity

The MHD (Mihalas, Hammer, and Däppen, 1988) equation of state and the OPAL opacities (Iglesias and Rogers 1991, Iglesias, Rogers, and Wilson 1992) are now widely available state-of-art data for use in stellar model construction. The MHD data brought about a significant improvement in modeling the solar envelope, which was particularly important for interpretation of the helioseismic data. In the core the gas is very nearly perfect and thus the changes are much less significant. The BPD model uses a simpler treatment of the equation of state.

The OPAL opacities are typically larger than the Los Alamos opacities used earlier. The difference is largest (factor 3) at $T \sim 200000$ K and it is caused by effect of previously neglected Fe lines. This large change is unimportant for the Sun because it occurs within the adiabatically stratified convective envelope. In the solar interior the difference is much smaller, but it is significant. It leads to an increase in central temperature and consequently to an increase in the calculated neutrino flux $\Phi_{\nu,\text{OPAL}} - \Phi_{\nu,\text{LAOL}} = 0.8$ and 3 SNU for the Cl and Ga experiments, respectively.

The OPAL opacities and, consequently, the calculated Φ_ν values are very sensitive to the Fe abundance. According to Bahcall and Pinsonneault (1992), we have $\Phi_{\nu,\text{OPAL,ph}} - \Phi_{\nu,\text{OPAL,m}} = 1.3$ and 6 SNU for the Cl and Ga experiments, respectively. The subscripts ph and m refer to the photospheric and meteoric mixture

of heavy elements, respectively (Anders and Grevesse 1989). In the BP model the
meteoric value was adopted.

4.3 Gravitational Settling

Gravitational settling of the elements is an effect introduced only recently to standard solar models (Cox, Guzik, and Kidman 1989, Proffit and Michaud 1991). The
effect of helium settling is important. Its inclusion causes a 0.6% increase in T_c,
and consequently increases for the Φ_ν calculated for the various detectors: 11.7% for
Kamiokande, 10.5% for Homestake (chlorine), and 3.1% for the gallium detectors
(Bahcall and Pinsonneault 1992).

There is some uncertainty in the diffusion coefficients. The difference in the
amount of helium settling between the BPD value of 0.0261 and 0.0284 found by
Proffit and Michaud (1989) is likely due to the difference in the coefficients. The
consequences for the core structure are most likely negligible.

The effect of helium settling is similar to that of an opacity increase. It implies a
higher central temperature, helium abundance and predicted neutrino flux. Taking
into account the heavy elements acts in the same direction, because it implies that
the heavy element abundance in the interior is higher than in the photosphere and
consequently that the interior opacity is higher.

5 Instabilities in Solar Core

There are no well-established instabilities in the solar core. In the standard treatment
the convective instability (both according to the Ledoux and Shwarzschild criteria)
is not found in the core of the evolving Sun. A possibility of the occurrence of a small
convective core in early solar evolution has been considered by Shaviv and Salpeter
(1971). They noted that such a core could be sustained only for a very short time.

The thermal stability, *i.e.* stability against perturbations that do not affect mechanical equilibrium, of the solar core has been verified by Roseblut and Bahcall
(1973).

A potential instability that has attracted the most attention was a vibrational
instability of some low-order g-modes of $l = 1$ spherical harmonic degree, proposed
by Dilke and Gough (1972). These authors suggested that the unstable modes may
develop into large amplitude circulatory motion which could cause mixing of the
nuclear burning products and therefore could explain the neutrino flux deficit. Roxburgh (1985) suggested that even at lower amplitudes, a nonlinear effect of the
excited modes could lead to a reduction of the neutrino flux. According to his picture the effect of the oscillatory motion would drive the ^{3}He out of equilibrium
leading to an enhancement of the reaction rate and consequently to a reduction of
the central temperature and the neutrino production rates. This possibility was discussed further by Gough (1991), De Rújula and Glashow (1992) and by Bahcall and
Kumar (1993). The authors of the last paper showed that the nonlinear feedback
would be important only at the amplitudes implying the surface radial velocity amplitudes by factor 10^4 larger than the observational upper limit. This result clearly
rules out any significance of the nonlinear effect. Furthermore, Bahcall and Kumar

questioned even the existence of the instability. They pointed out that ignoring the effects of convection in the linear stability calculations cannot be justified for the relevant modes.

6 Effects of Rotation and Magnetism

The dynamical effect of rotation is measured by the ratio of the averaged centrifugal to gravity force

$$\lambda = \frac{\Omega(r)^2 r^3}{2GM_r},$$

where $\Omega(r)$ is the local angular velocity of rotation. To get $\lambda = 10^{-2}$ in the core, $\Omega(r)$ would have to be over $10^4 \times \Omega(R_\odot)$.

Mixing by the rotationally-induced meridional circulation may be regarded as important if

$$\tau_\odot v_{mc} \geq 0.1r$$

where the meridional circulation velocity, v_{mc} (see e.g. Shatzman 1991) is approximately given by

$$v_{mc} = \frac{\lambda L_r r^2}{GM_r^2(1 - \nabla/\nabla_{ad})} P_2(\cos\theta),$$

where ∇ and ∇_{ad} denote, respectively, adiabatic and radiative temperature gradients. This requires the rotation rate to be over $60\times$ the surface rate.

Such a rapid rotation rate may be ruled out by means of helioseismology. Recently, Loudagh et al. (1993) used their measurements of the rotational frequency splitting of some $l = 1$ p-modes, to set a limit on the mean rotation rate in the core. They found that it is at most $3\times$ the surface equatorial value. This excludes any noticeable effect of rotation in the mean solar structure now through most of its past evolution.

The dynamical importance of the magnetic field is best measured by the ratio of magnetic to gas pressure, $\beta = \frac{B^2}{4\pi p}$. The upper limit on β in the interior obtained from helioseismology is about 30 MG (Goode and Thompson, 1993). This is good enough to eliminate a significant dynamical effect ($\beta < 0.01$), but not to evaluate the importance of magnetically-induced circulation.

There is a good evidence that the Sun was a rapidly rotating and magnetically active star during its early evolution. Rotation and magnetism could also affect significantly its interior evolution, but it seems unlikely that we will find any traces of this phase in the present core structure.

7 What Helioseismology Tells us About the Core

The helioseismic inversions yield $p(r)$ and $\rho(r)$ in the Sun's interior but not directly $T(r)$ and $X(r)$. However, modification in the SSM structure leading to changes in T_c and $\Phi_{\nu,\text{Cl}}$ leave a clear signature in the behavior of the quantity

$$u(r) = \frac{p}{\rho} \propto T(1.25X + 0.745),$$

which is the square of the isothermal sound-speed.

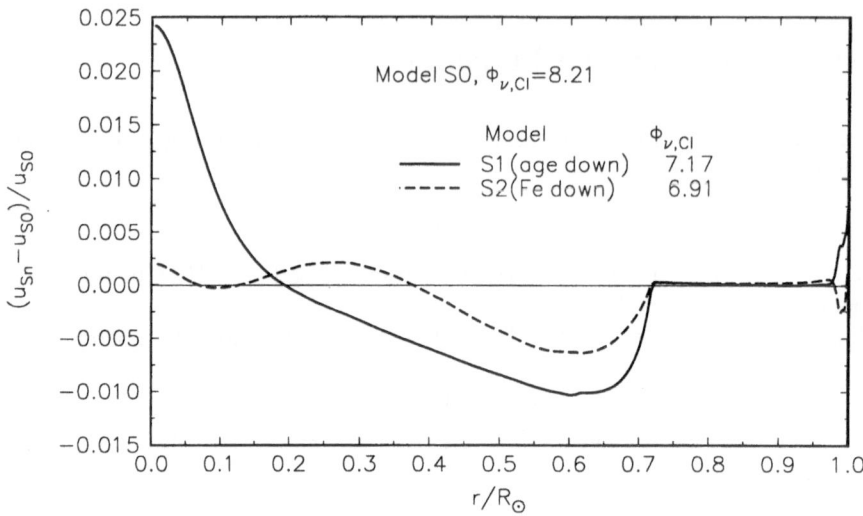

Fig. 3. Relative difference in u between two modified models (1 and 2) and Model S0. In model S1 the meteoritic (instead of photospheric), Fe abundance was assumed. In model S2 the solar age was assumed to be 4 (instead of 4.6) billion years

Changes resulting in an increase of $\Phi_{\nu,Cl}$, such as (i) increase of opacity (inclusion of new absorption effects or increase of metal abundance) (ii) gravitational settling (iii) increasing solar age, cause (i) u to increase in the outer part of the radiative interior and (ii)u to decrease in the core (the effect of the decrease in X dominates!).

In Fig.3 the differences in $u(r)$ between Model S0 and two comparison models are shown. These two models have lower values of $\Phi_{\nu,Cl}$. Thus, u, in these models is lower in the outer part of the radiative interior and higher in the core than in Model S0. All three models were calculated by R. Sienkiewicz in the same way except for the differences described in the caption. Qualitatively, the $\Delta u/u$ behavior is consistent with the general rule. However, the difference in shape of the two functions is very large, which reveals the strength of helioseismology as the method of testing input in the SSM construction procedure.

Model S0 was adopted as the reference model for the inversion of the measured solar p-mode frequencies (Dziembowski *et al*, 1993). The adopted method makes use of an integral formula connecting differences between measured and calculated frequencies, $\Delta\nu_j$, for various modes, j, to differences in $\Delta u(r)$ between the Sun and its model. The functions $\Delta P(r)$ and $\Delta\rho(r)$ may easily obtained from Δu.

Fig.4 shows that the structure of the Sun is very well described by Model S0 and even better by Model BPD. The fact that the difference in the structure of the two models is so small is a result of compensation of the neglect of the helium settling in the S0 model with the effect of adopting the higher Fe abundance implying higher

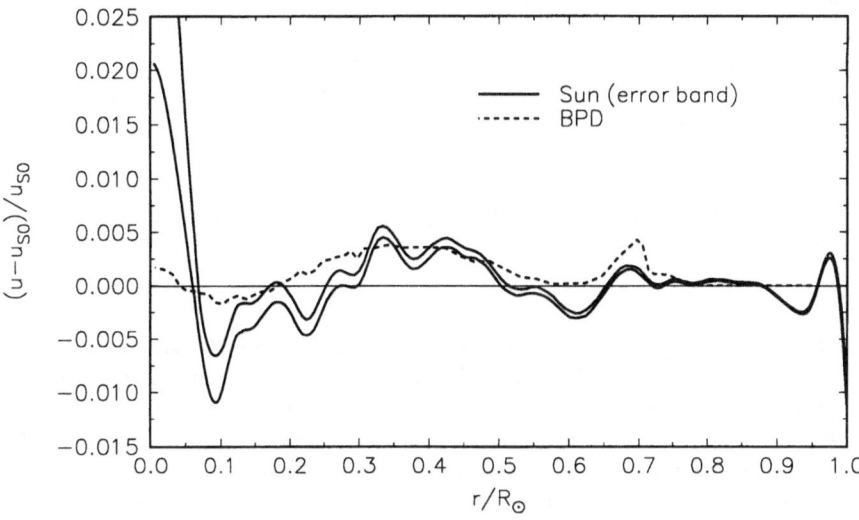

Fig. 4. Difference in u between the Sun and Model S0 inferred by means of inversion of the set of p-mode frequencies containing measurements of Jimenez *et al.* (1988) for $l = 0$ to 3 modes and measurements of Libbrecht, Woddard and Kaufman (1990) for $l = 2$ to 150 modes. The error band was obtained from 1000 determinations of frequencies with a flat distribution 2σ wide. The difference between Models BPD and S0 is shown for $r < 0.95 R_\odot$.

opacity. It is important to notice that the models predict also very similar values of $\Phi_{\nu,\text{Cl}}$, differing only by 0.2 SNU. The surface helium abundance in Model BPD $Y_s = 0.2466$ agrees quite well with the helioseismic range $Y_{s\odot} = (0.2425 - 0.2434)$. In Model S0 we have $Y_s = 0.2827$. The much bigger helioseismic correction is the result of neglecting the He settling, which is the main deficiency of this model.

The only feature in the helioseismic $\Delta u/u$ that cannot be accounted for by small refinements in the input physics is the inward rise below $r \approx 0.1 R_\odot$. The rise near the center in various forms was already seen in earlier inversions (Christensen-Dalsgaard, Gough, and Thompson 1988, Vorontsov 1988, Dziembowski, Pamyat-nykh, and Sienkiewicz 1990, Shibahashi, and Vorontsov 1991). Dziembowski, Pamyat-nykh, and Sienkiewicz found that, depending on the smoothing method, the noisy data might reveal the existence of a small ($r \approx 0.05 R_\odot$) convective core. Mixing in such a small core, however, would have virtually no effect on the calculated value of $\Phi_{\nu,\text{Cl}}$. The minimum value is consistent with their inversion ~ 10 SNU. It should be stressed that helioseismic probing of the inner core is very uncertain. It rests on data of low degree mode, which are less reliable than those of higher degrees. Inversions employing different low degree data sets yield very different $\Delta u/u$ behavior for $r < 0.1 R_\odot$ (Kosovichev and Gough 1993. Dziembowski *et al.*, 1993). The result is also sensitive to the method of frequency inversion. In my opinion, the peculiar feature in the inner core will deserve further attention only after it is confirmed with future data.

It is significant that the most up-to-data solar model of those listed in Table 1. agrees so well with helioseismic data. One effect that still needs to be included is the heavy element settling. We know that it will lead to an increase of u in the core and to a decrease in the outer part of the radiative envelope. We know, as well, that it will cause further increase in the neutrino fluxes.

8 What the Solar Neutrino Experiments Tell us About the Core

Bethe and Bahcall (1991) first noted that greater deficit in the measured neutrino flux in the Homestake experiment than in the Kamiokande makes any astrophysical solution of the solar neutrino problem very unlikely. The latter experiment detects only the ^{8}B neutrinos which are produced in the most temperature sensitive reaction. Thus, if indeed the model temperature is higher than that of the Sun, the deficit in this experiment should be the highest. Langacker (1993) strengthened the argument by making use of the results from the two gallium detectors. He parametrized $\Phi_\nu(pp), \Phi_\nu(^7Be)$ and $\Phi_\nu(^8B)$ in terms of T_c and looked for the best value fitting all the experimental data and obtained the following result.

$$T_c = T_{c,\text{BPD}} \times (0.92 \pm 0.01), \qquad \chi^2 = 20.6$$

Such value of T_c would mean a catastrophe for the stellar evolution theory. The χ^2-value tells us that the fit is unacceptable and demonstrates that the low solar T_c cannot explain the measurements. Thus, if we do not want to question the experimental data, we should abandon further efforts in looking for nonstandard solar models with cooler cores.

Efforts in finding a solution of the neutrino puzzle within the domain of Elementary Particle Physics resulted in the formulation of the MSW mechanism of the neutrino flavor conversion (Miheyev and Smirnov 1985, Wolfenstein 1978). The basic assumption here is that neutrinos have masses and that they undergo conversion as a result of a resonant interaction with matter when travelling through the Sun's interior. Bethe and Bahcall (1991) showed that the MSW theory of neutrino conversions explains both the Homestake and Kamiokande measurements without a revision of the SSM. The theory contains two basic parameters: the mixing angle, θ, and the neutrino mass difference, Δm. Adopting the MSW mechanism Langacker (1993) used the results of the three measured neutrino fluxes to determine simultaneously T_c, Δm, and θ. At the 90% confidence level he found, for the two versions of the mechanism

$$T_c = T_{c,\text{BPD}} \times 1.02^{+0.03}_{-0.05}$$

for an expanded nonadiabatic solution, and

$$T_c = T_{c,\text{BPD}} \times 1.04^{+0.03}_{-0.04}$$

for a large-angle solution. We see there is no conflict with the SSM for either of these two solutions. The values of T_c derived in this way cannot yet be regarded as real constraints on solar central temperature.

9 Conclusions

Recent changes in standard solar models are a consequence of improvement in the input physics, especially better opacities. The main improvement in methodology is an inclusion of the gravitational settling of helium. The changes in the core structure are small. The net result is a greater discrepancy between the predicted and measured neutrino fluxes, but a better agreement with helioseismic data.

The only known physical effect ignored in the models discussed in this review that still needs to be included is the gravitational settling of heavier elements. This a secondary effect which will result in a small increase of the calculated neutrino fluxes. Neither solar neutrino measurements nor helioseismology point to a need for an essential revision of the standard approach to the solar internal structure and evolution.

Results of inversions of solar p-mode frequencies are in a remarkable agreement with the standard solar model of Bahcall and Pinsonneault (1992). The only troublesome discrepancy between the model and helioseismic speed of sound occurs in the inner core ($r < 0.1R_\odot$), but the helioseismic data for this part of the Sun are very unreliable. The precision in frequency measurements for low-l modes must be improved in order to determine the reality of the discrepancy.

The results the neutrino flux measurements cannot be reconciled by modifications in the solar model. On the other hand, the MSW conversion mechanism explains all experimental data without any change of the Sun's model. Within the framework of this mechanism, the present experimental data on neutrino fluxes do not yet constrain solar models. However, the precision of the model has a direct impact on the determination of neutrino mass difference and the mixing angle.

Acknowledgments

The main part of this review was prepared during my stay at Queen Mary and Westfield College, where I was supported by the Perren Fund of the London University. I am grateful to Professor Ian Roxburgh for the invitation as well as for long disscusions on the problems presented here. I thank also the Local Organizing Committee of the Meeting for the financial support.

References

Abazov, A.I., *et al.* 1991, Phys. Rev. Lett., **67**, 3332
Anders, E., and Grevesse, N. 1989, Geochim. Cosmochim. Acta, **53**, 197
Anselmann, P., *et al.*, Phys. Lett. **B285**, 390
Bahcall, J.N. and Pinsonneault, M.H, 1992, *Rev.Mod.Phys.*, **64**, 885.
Bahcall, J.N., and Bethe, H.A. 1990, Phys. Rev. (Letters), **65**, 2233
Bahcall, J.N., and Kumar, P. 1993, ApJ, **409**, L73
Bahcall, J.N., and Ulrich, R.K. 1988, Rev. Mod. Phys., **60, 277**

Berthomieu, G., Provost, J., Morel, P. and LeBreton, Y., 1993, AA, **268**, 775.

Castellani, V., Degl'Innocenti, S., and Fiorentini, G. 1992, preprint INFN-FE-92-06

Christensen-Dalsgaard, J., 1992, Geophys.Astrophys. Fluid Dynamics, **62**, 123

Christensen-Dalsgaard, J., Gough, D.O., and Thompson, M.J. 1988, in Seismology of the Sun and Sun-like Stars, ed. E.J. Rolfe, ESA SP-286, p.493

Cox, A.N., Guzik, J.A., Kidman, R.B. 1989, ApJ, **342**, 1187

Davis, R., Lande, C.K., Lee, C.K., Clevland, B.T., and Ulman, J. 1990, in Inside the Sun, ed. G. Berthomieu and M. Cribier, Cluver, p. 171

De Rújula, A., and Glashow, S., 1992 CERN-Th 66082/92

Dilke, F.W.W., and Gough, D.O. 1972, Nature, **240**, 262

Dziembowski, W.A., Pamyatnykh, A.A., and Sienkiewicz, R. 1990, MNRAS, **244**, 542

Dziembowski, W.A., Goode, P.R, Pamyatnykh, A.A., and Sienkiewicz, R. 1990, Copernicus Astron. Center preprint No. 269

Goode, P.R, and Thompson, M.J, 1992, ApJ, **395**. 307

Gough, D.O. 1991, Ann. NY Acad.Sci,, **647**, 199

Gough, D.O., and Kosovichev, A.G. 1988, in Seismology of the Sun and Sun-like Stars, ed. E.J. Rolfe, ESA SP-286, p.195

Guzik, J.A., and Cox, A.N. 1991, ApJ, **381**, 333

Hirata, K.S., et al. 1989, Phys. Rev. (Letters), **63**, 16

Iglesias, C.A., and Rogers, F.J. 1991, ApJ, **371**, 408

Iglesias, C.A., Rogers, F.J., and Wilson, B.G. 1992, ApJ, **397**, 717

Jiménez, A., Pallé, P.L., Régulo, C., Roca Cortés, T., Isaak, G.R., McLeod, C.P., and van der Raay, H.B. 1988, in Advances in Helio- and Asteroseismology, IAU Symp. 123, ed. J. Christensen-Dalsgaard and S. Frandsen (Dordrecht: Reidel), p.205

Johnson, C.W., Kolbe, E., Koonin, S.E., and Langake, K. 1992, ApJ, **392**, 320.

Langacker, P. 1993, in Unified Symmetry in the Small and in the Large (Proc. of the Conference held in Coral Gables, Florida, January 1993), in press

Libbrecht, K.G., Woodard, M.F., and Kaufman, J.M. 1990, ApJS, **74**, 1129

Loudagh, S. *et al.* 1993, AA, **275**, L25

Mihalas, D., Däppen, W., and Hummer, D.G. 1988, ApJ, **331**, 815

Mikheyev, S.P., and Smirnov, A.Yu. 1985, Sov. J. Nucl. Phys.,**42**, 913

Proffitt, C.R., and Michaud, G. 1991, ApJ, **380**, 238

Rosenbluth, M., and Bahcall, J.N. 1973, ApJ, **184**, 9

Roxburgh, I.W. 1985, Sol. Phys., **100**, 21

Shatzman, E, 1991, in Solar Interior and Atmosphere, eds. A.N. Cox, W.C. Livingston, M.S. Matthews, The University of Arizona Press, p. 192

Shaviv, G., and Salpeter, E. 1971, ApJ, **165**, 171

Turck-Chieze, S., and Lopes, I. 1993, ApJ, **408**, 347

Turck-Chieze, S., Cahen, S., Cassé, M., and Doom, C. 1988, ApJ,**335**, 415

Vorontsov, S.V., and Shibahashi, H. 1991, PASJ,**43**, 739

Wolfenstein, L. 1978, Phys. Rev. D, **17**, 2369

New Sub–Barrier Nuclear Fusion Cross Sections as a Possible Solution to the Solar Neutrino Problem

Lucio Paternò [1], Augusto Scalia [2]

[1] Istituto di Astronomia dell'Università di Catania, Città Universitaria, 95125 Catania, Italy
[2] Dipartimento di Fisica dell'Università di Catania, corso Italia 57, 95129 Catania, Italy

Abstract: The results of the three solar neutrino experiments, Gallex (Gallex coll. 1992a), Homestake (Davis et al. 1990) and Kamiokande-II (Kamiokande-II coll. 1990) indicate the existence of quite tight bounds to the neutrino fluxes produced by the $p(p,\nu_e e^+)D$ main initial reaction, and in the $^3He(\alpha,\gamma)^7Be$ and $^7Be(p,\gamma)^8B$ secondary branches. The Solar Standard Model (Bahcall and Pinsonneault 1992) is still significantly far from the limits of the experiments and the hypotheses on non-conventional neutrino properties are strongly disfavoured (Gallex coll. 1992b), except for the matter neutrino oscillation (Mikheyev and Smirnov 1986; Wolfenstein 1978), this latter surviving within very narrow limits (Gallex coll. 1992b). The present approach is based on the use of new cross sections for the reactions $^3He(^3He,2p)\alpha$, $^3He(\alpha,\gamma)^7Be$ and $^7Be(p,\gamma)^8B$, calculated in the framework of the "shadow" model for the sub-barrier fusion (Scalia 1992; Scalia and Figuera 1992). When the standard reaction rates (Caughlan and Fowler 1988) are replaced by those obtained from the new cross sections, the New Solar Model is consistent with the three neutrino experiments and with the internal stratification as deduced from helioseismology.

1 Solar Neutrino Problem Today

The three solar neutrino experiments, Gallex (Gallex coll. 1992a), Homestake (Davis et al. 1990) and Kamiokande-II (Kamiokande-II coll. 1990) report capture rates $\Phi(Ga) = 83 \pm 19 \pm 8\ SNU$, $\Phi(Ho) = 2.33 \pm 0.25\ SNU$ and $\Phi(Ka) = 0.16 \pm 0.05 \pm 0.03\ SNU$, respectively. The latter comes from the ratio $\Phi(Ka)/\Phi_{ssm}(^8B) = 0.46 \pm 0.13 \pm 0.08$ of the neutrino flux measured, $\Phi(Ka)$, to the ^8B neutrino flux of a Solar Standard Model (SSM) (Kamiokande-II coll. 1990) with $\phi(^8B) = 5.8 \times 10^6\ cm^{-2}\ s^{-1}$, assuming a neutrino electron scattering cross section $\sigma(\nu_e, e^-) = 6.08 \times 10^{-44}\ cm^2$ (Bahcall 1989).

These results have to be compared with the predictions of a series of SSMs recently appeared in the literature (Bahcall and Pinsonneault 1992; Berthomieux et al. 1993; Castellani et al. 1993; Christensen-Dalsgaard 1992; Sackmann et al.

1990; Turck-Chièze et al. 1988). Depending on the opacity and equation of state used, the models give different neutrino fluxes, but all remarkably higher than those measured. The theoretical ranges of variation, relative to the three experiments, are $125\ SNU \leq \Phi_{ssm}(Ga) \leq 134\ SNU$, $5.8\ SNU \leq \Phi_{ssm}(Ho) \leq 8.9\ SNU$ and $0.23\ SNU \leq \Phi_{ssm}(Ka) \leq 0.41\ SNU$. Even when the theoretical and experimental values are stretched to the lowest and highest limits, respectively, the gap between theory and experiment remains, thus giving rise to the solar neutrino problem.

Any attempt to reconcile theory with observations based on exotic solar models, whose internal structure is significantly different from that of the SSMs, is untenable since the inversion of helioseismological data has clearly demonstrated that the "true" solar internal stratification differs only slightly from the SSM stratifications (Dziembowski et al. 1992). On the other hand, very little room is left for non-conventional neutrino properties, even if matter oscillations (MSW effect) may be not completely excluded (Gallex coll. 1992b). Nuclear solutions to the solar neutrino problem have recently been discussed (Castellani et al. 1993; Gallex coll. 1992b; Spiro and Vignaud 1990), and are based on an artificial reduction of the neutrino fluxes. The reduction is obtained either by lowering the solar central temperature, which alters the internal stratification beyond the limits imposed by helioseismology, or by assuming an arbitrary, very low energy resonance in the $^3He(^3He,2p)\alpha$ reaction branch, which does not produce neutrinos.

If we neglect the small contribution to the total neutrino flux from the $p(pe^-,\nu_e)$ D and $^3He(p,\nu_e e^+)\alpha$ reactions and the CNO cycle, more than 90% of the detected neutrinos in the three experiments are produced from the $p(p,\nu_e e^+)D$ reaction (p-p) and in the $^3He(^4He,\gamma)^7Be$ and $^7Be(p,\gamma)^8B$ branches. In the space of the p-p, ^7Be and ^8B neutrino fluxes it is possible to find a volume where the results of the three experiments overlap. Since Gallex measures $\phi(\text{p-p}) + \phi(^7Be) + \phi(^8B)$, Homestake $\phi(^7Be) + \phi(^8B)$ and Kamiokande-II $\phi(^8B)$, it is possible to identify three non-parallel planes, each representing the allowed domain of the individual experiments. The crossing of the planes defines, within the experimental uncertainty, a volume which indicates the the mutual coexistence of the three experiments. These approximately demand that $\phi(\text{p-p}) \simeq 1.07\ \phi_{ssm}(\text{p-p})$, $\phi(^7Be) \simeq 0$ and $\phi(^8B) \simeq 0.34\ \phi_{ssm}(^8B)$. This is shown in Fig. 1, where the intercepts of planes (G = Gallex, H = Homestake, K = Kamiokande-II) with axes (p-p, Be, B) are relative to the neutrino fluxes of an average SSM and the sphere NEC (Neutrino Experiment Convergence) indicates the volume where the experiments are mutually consistent. The distance between the spheres SSM and NEC gives an idea of the present solar neutrino problem.

2 Shadow Model for Sub-Barrier Nuclear Fusion

Our proposed solution is based on the use of a solar model with reaction rates obtained from new cross sections calculated in the framework of the shadow model for the sub-barrier fusion (Scalia 1992; Scalia and Figuera 1992). This semi-phenomenological model considers the fusion as the counterpart of the elastic scattering (Scalia 1987, 1990), and the analytical form of the cross section as a function of energy is determined by a fit to the existing experimental data. The resulting cross sections slightly differ from the standard ones in the range of energies above about 20 keV, but they are sensibly larger at lower energies. The corresponding

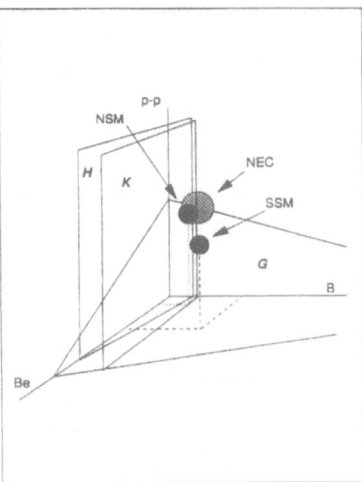

Fig. 1. The three planes G (Gallex), H (Homestake) and K (Kamiokande-II) represent, in the space of the neutrino fluxes from $p(p,\nu_e e^+)D$, $^7Be(e^-,\nu_e)^7Li$ and $^8B(,\nu_e e^+)^8Be^*$ reactions (p-p, Be and B), the allowed domains of the three neutrino experiments. The intercepts of planes with the axes are relative to the neutrino fluxes of an average Solar Standard Model (SSM). The sphere NEC (Neutrino Experiment Convergence) indicates the volume of the mutual coexistence of the three experiments within the experimental uncertainty. The New Solar Model (NSM) produces neutrino fluxes which are in reasonable agreement with the experiments

reaction rates, obtained by numerical integration over a Maxwellian distribution, have been calculated for the reactions $^3He(^3He,2p)\alpha$, $^3He(\alpha,\gamma)^7Be$ and $^7Be(p,\gamma)^8B$, which are the three main branches following the initial p-p reaction.

At the solar central temperature the rate of the first reaction is 7.5 times higher than the standard one, the second 1.5 times lower and the third is about equal. This means that the first reaction, which does not produce neutrinos, is much more efficient with the new reaction rates rather than with the standard ones (Caughlan and Fowler 1988). The consequence is a reduction of the flux of ^7Be and ^8B neutrinos and the increase in the flux of p-p neutrinos.

3 The New Solar Model: Results and Comparison

Our New Solar Model (NSM) has been constructed with the same standard procedures used for constructing a SSM. We only replaced the standard reaction rates (Caughlan and Fowler 1988) of the three above-mentioned reactions with the new ones. We used the FRANEC (Frascati Raphson Newton Evolutionary Code) in its most recent version (Chieffi and Straniero 1989) with OPAL opacities (Rogers and Iglesias 1992) and Anders and Grevesse element mixture (Anders and Grevesse 1989) with $Z/X = 0.0275$. The equation of state is by Straniero (1988) for $T > 10^6$ K, and the code uses Saha equation with ionization pressure according to Ratcliff (1987) for $T < 10^6$ K. The Sun has been evolved for 4.6×10^9 years, reaching the present luminosity (3.846×10^{33} erg s^{-1}) and radius (6.9599×10^{10} cm) with an initial helium abundance $Y_0 = 0.29283$ and a mixing-length parameter $\alpha = 1.8465$. The central temperature, pressure, density, hydrogen abundance and the fractional radius of the convection zone base are $T_c = 1.567 \times 10^7$ K, $P_c = 2.395 \times 10^{17}$ dyne cm^{-2}, $\rho_c = 1.514 \times 10^2$ g cm^{-3}, $X_c = 0.357$ and $x_{cz} = 0.735$, respectively. The NSM neutrino fluxes with the corresponding capture rates for the three experiments are given in Table 1, while the NSM position in the space of the p-p, ^{7}Be and ^{8}B fluxes is shown in Fig. 1.

Table 1. Neutrino fluxes at the Earth and the corresponding capture rates, relative to Gallex, Homestake and Kamiokande-II experiments, predicted by the New Solar Model (NSM)

NEW SOLAR MODEL				
REACTION	FLUX (cm^{-2} s^{-1})	CAPTURE RATES (SNU)		
		Gallex	Homestake	Kamiokande-II
p-p	6.34×10^{10}	74.81	0.00	0.00
^{7}Be	1.20×10^9	8.81	0.29	0.00
^{8}B	2.65×10^6	6.43	2.81	0.16
^{13}N	4.99×10^8	3.08	0.08	0.00
^{15}O	4.33×10^8	5.03	0.29	0.00
^{17}F	5.37×10^6	0.06	0.00	0.00
pep	1.52×10^8	3.22	0.24	0.00
hep	1.39×10^3	0.01	0.01	0.00
TOTAL	6.57×10^{10}	101.46	3.73	0.16

For comparison, we constructed a SSM with the same solar age, using the same code and input physics but the standard reaction rates for the three relevant reactions. In this case we obtain $Y_0 = 0.29397$, $\alpha = 1.8810$, $T_c = 1.575 \times 10^7$ K, $P_c = 2.344 \times 10^{17}$ dyne cm^{-2}, $\rho_c = 1.516 \times 10^2$ g cm^{-3}, $X_c = 0.331$, $x_{cz} = 0.733$. The neutrino fluxes and capture rates of this SSM are given in Table 2. The neutrino capture rates have been computed from the detector cross sections as given by Bahcall (1989). The stratification, namely the behaviour of the ratio $U = P/\rho$, of our two models NSM and SSM has been compared with that of the "true" Sun as deduced from helioseismology (Dziembowski et al. 1992). The relative differences $\Delta U/U(\%)$ are shown in Fig. 2.

The new cross sections based on a semi-phenomenological sub-barrier nuclear fusion model, when applied to a solar model, remarkably reduce the gap between the predicted and measured neutrino flux (Fig. 1) without introducing artificial hypotheses or claiming for non-conventional neutrino properties. The significant reduction of the gap, however, has produced a small alteration in the central stratification of the Sun (Fig. 2).

Table 2. Neutrino fluxes at the Earth and the corresponding capture rates, relative to Gallex, Homestake and Kamiokande-II experiments, predicted by our comparison Solar Standard Model (SSM)

	SOLAR STANDARD MODEL			
REACTION	FLUX $(\text{cm}^{-2}\,\text{s}^{-1})$		CAPTURE RATES (SNU)	
		Gallex	Homestake	Kamiokande-II
p-p	6.00×10^{10}	70.83	0.00	0.00
^7Be	5.11×10^9	37.38	1.23	0.00
^8B	7.11×10^6	17.28	7.54	0.43
^{13}N	4.88×10^8	3.01	0.08	0.00
^{15}O	4.28×10^8	4.96	0.29	0.00
^{17}F	5.32×10^6	0.06	0.00	0.00
pep	1.40×10^8	2.97	0.22	0.00
hep	7.30×10^3	0.05	0.03	0.00
TOTAL	6.62×10^{10}	136.55	9.39	0.43

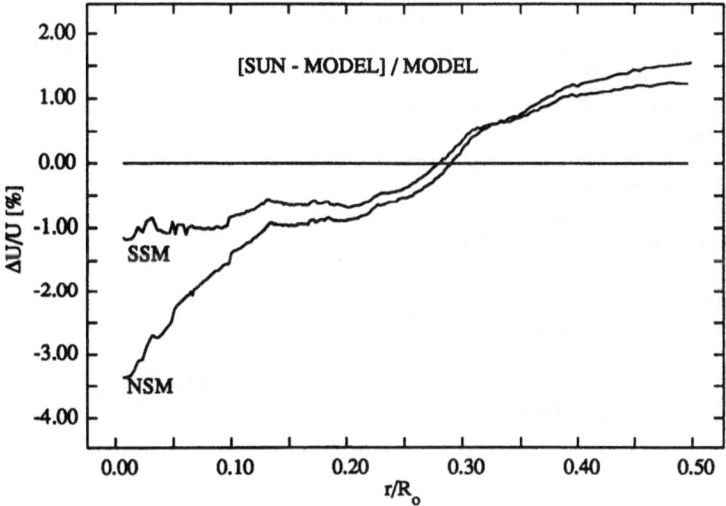

Fig. 2. Relative difference $\Delta U/U(\%) = (U_\odot - U_M)/U_M$, where U is the ratio of pressure to density, between the "true" Sun, $U_\odot$, and a solar model, U_M, as a function of the fractional solar radius. The lines labeled NSM and SSM refer to the New Solar Model and the comparison Solar Standard Model, respectively. The behaviour of $U_\odot$ has been obtained by an inversion of helioseismological data

Acknowledgements

We wish to thank V. Castellani and S. Degl'Innocenti for providing us the new version of the FRANEC evolutionary code and W.A. Dziembowski for providing us the file with the solar stratification as deduced from a helioseismological inversion.

References

Anders, E., Grevesse, N.: 1989, Geochim. Cosmochim. Acta **53**, 197
Bahcall, J.N.: 1989, *Neutrino Astrophysics*, Cambridge University Press, Cambridge
Bahcall, J.N., Pinsonneault, M.H.: 1992, Rev. Mod. Phys. **64**, 885
Berthomieux, G., Provost, J., Morel, P., Lebreton, Y.: 1993, A&A **268**, 775
Castellani, V., Degl'Innocenti, S., Fiorentini, G.: 1993, A&A **271**, 601
Caughlan, G.R., Fowler, W.A.: 1988, *Atomic Data and Nuclear Data Tables* **40**, 284
Chieffi, A., Straniero, O.: 1989, ApJ Suppl **71**, 47
Christensen-Dalsgaard, J.: 1992, Geophys. Astrophys. Fluid Dynamics **62**, 123
Davis, R., Lande, K., Lee, C.K., Cleveland, B.T., Ullman, J., Lehman, H.: 1990, in: *Inside the Sun*, eds. G. Berthomieux and M. Cribier, Kluwer Acad. Publ., Dordrecht, p. 171
Dziembowski, W.A., Pamyatnykh, A.A., Sienkiewicz, R.: 1992, Acta Astron. **42**, 5
Gallex collaboration: 1992a, Phys. Lett. B **285**, 376
Gallex collaboration: 1992b, Phys. Lett. B **285**, 390
Kamiokande-II collaboration: 1990, in *Inside the Sun*, eds. G. Berthomieux and M. Cribier, Kluwer Acad. Publ., Dordrecht, p. 179
Mikheyev, S.P, Smirnov, A.Yu.: 1986, Nuovo Cimento C **9**, 17
Ratcliff, S.J.: 1987, ApJ **318**, 196
Rogers, F.J., Iglesias, C.A.: 1992, ApJ Suppl **79**, 507
Sackmann, I.J., Boothroyd, A.I., Fowler, W.A.: 1990, ApJ **360**, 727
Scalia, A.: 1987, Nuovo Cimento A **98**, 571
Scalia, A.: 1990, Nuovo Cimento A **103**, 85
Scalia, A.: 1992, Nuovo Cimento A **105**, 233
Scalia, A., Figuera, P.: 1992, Phys. Rev. C **46**, 2610
Spiro, M., Vignaud, D.: 1990, Phys. Lett. B **242**, 279
Straniero, O.: 1988, A&A Suppl **76**, 157
Turck-Chièze, S., Cahen, S., Cassé, M., Doom, C.: 1988, ApJ **335**, 415
Wolfenstein, L.: 1978, Phys. Rev. D **17**, 2369

Helioseismic Evidence for Mixing in the Radiative Interior

A.G. Kosovichev

Crimean Astrophysical Observatory, 334413 Nauchny, Crimea, Ukraine[1]
Institute of Astronomy, Madingley Road, Cambridge CB3 0HA, UK
Center for Space Science & Astrophysics, Stanford University, CA 94305

Abstract: Results are presented of a determination of the hydrostatic parameters of the solar structure, namely density, sound speed, and a parameter of convective stability, by direct inversion of solar oscillation frequencies. The analysed data sets include frequencies of acoustic modes of intermediate degree ($l = 4 - 140$), observed by Libbrecht et al. (1990), and those of low degree ($l = 0 - 2$), obtained from the IPHIR space experiment (Toutain and Fröhlich 1992). The low-degree data sets are of particular importance for resolving the structure of the solar core. The inversion results show that the overall structure of the solar interior is consistent with a non-standard solar model constructed by Christensen-Dalsgaard et al. (1993) by increasing opacities beneath the convection zone in accordance with Rogers and Iglesias (1992), and by taking into account gravitational settling of helium. The inversions give also evidence for an overshoot beneath the convection zone and for a moderate localized mixing in the energy-generating core.

1 Introduction

The standard solar model assumes that matter is mixed only in convectively unstable regions. However, there are a number of theoretical arguments and observational indications that mixing actually takes place in the region of radiative energy transport, which is located beneath the convection zone. The radial gradient of specific entropy is positive in the radiative zone, which means that it is convectively stable. However, it is generally believed (e.g. Roxburgh 1985) that convective elements penetrate into the stable region (so-called overshoot effect) resulting in material mixing in the outer part of the radiative zone. The overshoot layer is of importance for understanding the deficit of lithium observed at the solar surface, and the dynamo generation of magnetic fields. It is also the area of rapid transition of solar differential rotation in the convection zone to almost rigid rotation in the radiative interior. The nature and the structure of the overshoot layer are still undetermined. Some mild diffusive mixing in the region of small radial gradient of chemical composition below the convection zone can result from turbulent diffusion generated, for

---

[1] Permanent address

47

instance, by rotationally-driven instabilities. The instabilities, however, are unlikely to cause any mixing in the energy-generating core where a high molecular weight gradient is built up by nuclear fusion. Nevertheless, as was first pointed out by Dilke and Gough (1972), localized mixing in the solar core could be caused by the ^{3}He instability which was particularly strong in the young Sun.

Understanding the mixing processes is one of the most important problems of the theory of solar and stellar evolution, because mixing is likely to cause the most significant corrections to the standard stellar models. Unfortunately, theoretical estimates of the efficiency of the processes are very uncertain since they involve turbulence and other nonlinear effects of the instabilities. Helioseismology provides a tool for direct measurements of the solar structure, and therefore can be used to estimate the effects of possible mixing in the radiative interior.

Perhaps the abundances of hydrogen and heavy elements, obtained as functions of radius by inverting solar oscillation frequencies (Gough and Kosovichev 1990), would give us the most direct measure of mixing. However, this approach requires consideration of the thermal balance in the solar interior, thus involving uncertain microscopic parameters such as the opacity and nuclear reaction rates. It is possible that abundance variations can be partially masked by uncertainties in microscopic physics. A generalized procedure for estimating simultaneously the abundance variations and the uncertainties in physics is currently under development.

Here I present some evidence for mixing from helioseismic inversion of the hydrostatic parameters of the solar structure: density ρ, the ratio of pressure to density, $u = p/\rho$, and the radial gradient of the specific entropy,

$$A = \frac{1}{\gamma}\frac{\mathrm{d}\ln p}{\mathrm{d}r} - \frac{\mathrm{d}\ln\rho}{\mathrm{d}r}.$$

Inversions of this type include only the assumption of hydrostatic equilibrium of the solar interior. We note that the parameter u is proportional to both the ratio of the squared speed of sound c^2 to the adiabatic exponent γ, and the ratio of temperature T to the molecular weight μ: $u = c^2/\gamma = \mathcal{R}T/\mu$. Temperature and molecular weight can be found separately only by applying conditions of the thermal equilibrium in the radiative zone, or by using the equation of state in the convection zone (cf. Kosovichev et al. 1992). The parameter $A(r)$ characterizes the convective stability of the Sun's stratification: it is positive in the convectively stable radiative zone, close to zero in the adiabatic part of the convection zone, and negative in the subsurface area of inefficient convection. It is a particularly sensitive indicator of structure of the transition region between the radiative and convective zones.

2 Inversion Results

A version of an optimal averaging inversion procedure described by Däppen et al. (1991) and Kosovichev (1992) has been used. The inverted data are combinations of 16 frequencies of low-degree modes ($l = 0$, 1 and 2) in the frequency range $2.5 \lesssim \nu \lesssim 3$ mHz, taken from the IPHIR data set (Toutain and Fröhlich 1992), and 598 frequencies of intermediate-degree modes ($l = 4 - 140$, $\nu = 1.5 - 3$ mHz), observed at the BBSO in 1988 (Libbrecht et al. 1990).

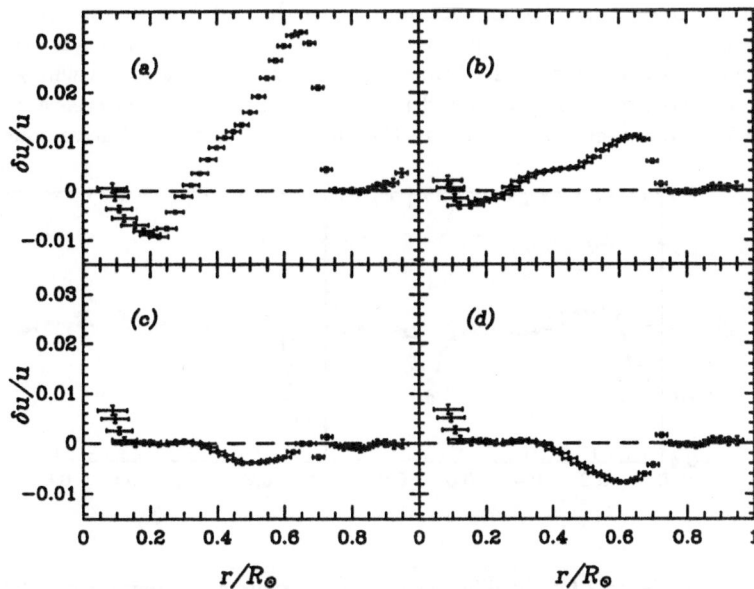

Fig. 1. Relative differences in the parameter $u = p/\rho$ between the inverted solar structure and four solar models: (a) a standard solar model with Los Alamos opacities; (b) a standard solar model with Livermore opacities (OPAL); (c) a model with OPAL and with gravitational settling of helium; (d) a model with OPAL, gravitational settling of helium and turbulent diffusion of hydrogen. The horizontal bars represent the resolution lengths (widths of the optimal averaging kernels); the vertical bars represent standard errors

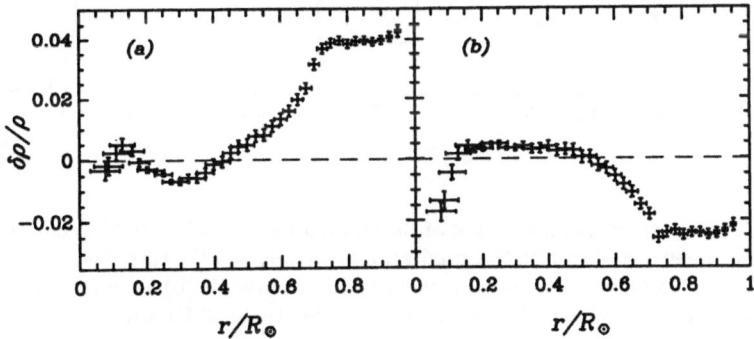

Fig. 2. Relative differences in density between the Sun and the solar models: (a) a standard solar model with Livermore opacities (OPAL); (b) a model with OPAL, gravitational settling of helium and turbulent diffusion of hydrogen

Figure 1 shows relative differences in the parameter u between the Sun and four solar models computed by Christensen-Dalsgaard et al. (1991, 1993). The major improvement of the standard solar model is achieved by using the new Livermore opacity tables (Rogers and Iglesias 1992), in which absorption in iron lines is taken into account more correctly as compared with the old Los Alamos tables (panels *a* and *b* in Fig. 1). Including gravitational settling of helium leads to a further improvement of the solar model (panel *c*). However, the model in which turbulent diffusion of hydrogen is implemented in order to obtain the observed abundance of ^{7}Li shows an increased discrepancy in the outer part of the radiative zone.

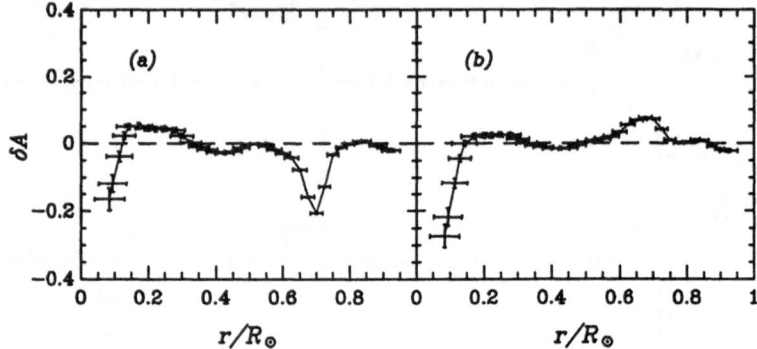

Fig. 3. As Figure 2 for the differences in the parameter of convective stability A

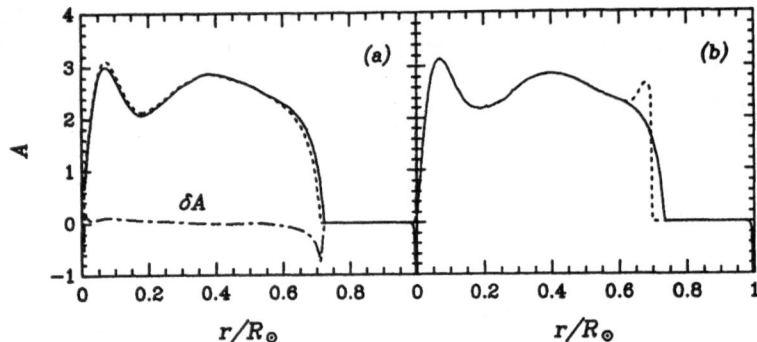

Fig. 4. (a) The structure functions A of the standard model with OPAL (continuous curve) and of the model included both the gravitational settling and the turbulent diffusion (dotted curve), and of the difference between them (dashed curved); (b) as in (a) with the dotted curve representing a model with convective overshoot; the difference is not shown

The inversions for density (Fig. 2) and for the parameter of convective stability (Fig. 3) also show significant large-scale deviations from the Sun. Thus the physics of gravitational settling and turbulent diffusion is not well understood. However, large-scale material mixing processes could be masked by unknown opacity effects.

Along with the large-scale deviations in the hydrostatic parameters, there are small-scale variations localized near the base of the convection zone and in the central region. They are less likely to be represented by opacity variations, and

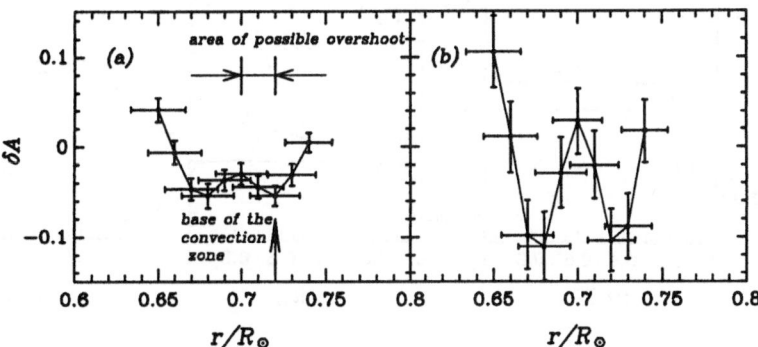

Fig. 5. (a) Differences in the parameter of convective stability A between the Sun and the solar model 13, which has a depth of the convection zone close to that of the Sun. The non-monotonic behavior of δA near the base of the convection zone gives an evidence for convective overshoot. (b) The same as in (a) but with the spatial resolution of inversion increased in expense of errors

could result from convective overshoot and localized mixing in the core. The typical difference in the parameter of convective stability, A, between two solar models displays a sharp peak if the models have convective zones of different depths (Fig. 4a). The overshoot may result in an additional peak of A beneath the convection zone (cf. Skaley and Stix 1991) such as shown in Fig. 4b. Therefore, the combined effect of a difference in the depth of the convection zone and an overshoot layer may result in non-monotonic variations of δA at the base of the convection zone. As the effect of convective overshoot is rather weak, it is more apparent in inversions for δA relative to a solar model with a convection zone depth close enough to that of the Sun. Figure 5 shows δA inverted from the data relative to the solar model 13 of Christensen-Dalsgaard et al. (1991), which was calibrated to an approximately-correct convection zone depth by adjusting opacities in the outer part of the radiative zone. The depth of the overshoot layer estimated from the double-peak structure of δA is approximately $0.2H_\mathrm{p}$, where H_p is the local pressure-scale height.

In the central core, the inversion results show significant deviations from the solar models: a sharp increase of u (Fig. 1c, d), and a decrease of both density (Fig. 2) and convective stability (Fig. 3). These structure variations can be understood in terms of mixing localized in the energy-generating core, where the radial gradient of chemical composition is likely to be flatter than in the standard model. Figure 6 demonstrates typical variations of the solar structure, when the mixing is modeled by a sinusoidal perturbation of the hydrogen abundance X (Kosovichev and Fedorova 1990). It is interesting that such a perturbation produces not only the observed variations in the central region, but also results in perturbations outside the core, which are qualitatively similar to the effect of the gravitational settling and turbulent diffusion, because the model calibration conditions require a lower helium abundance in the outer layer. Comparing the inversion results with the model it is estimated that the localized mixing might cause hydrogen abundance variations $\delta X \simeq 0.01$ at $r \lesssim 0.2R_\odot$. The changes, however, have a small effect on the solar neutrino fluxes.

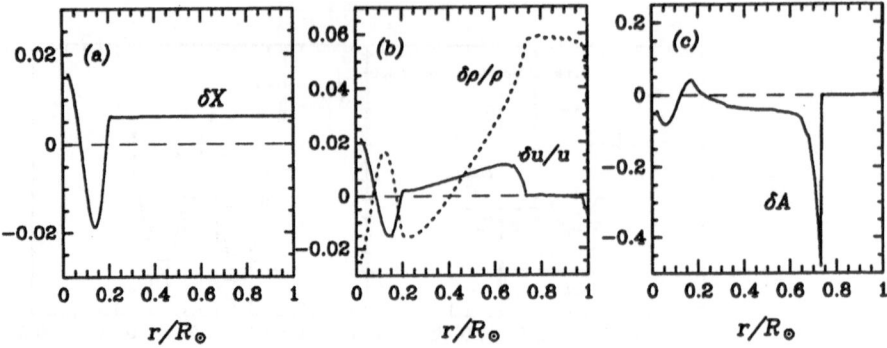

Fig. 6. Effect of localized mixing in the solar core, modeled by perturbing the hydrogen abundance as shown in panel (a), on the structure parameters ρ and u (b), and on A (c)

3 Discussion

Helioseismic structure inversions have shown significant deviations from the standard solar model, suggesting mixing in the radiative interior. Parameters of the mixing effects have been estimated by comparing the inversion results for density, the ratio of pressure to density, and a parameter of convective stability. The next step of this investigation, in which a technique of secondary inversions will be applied (Gough and Kosovichev 1990), will allow us to separate uncertainties in microscopic parameters, and to understand better the physics of the Sun's radiative core.

References

Christensen-Dalsgaard, J., Gough, D.O., Thompson, M.J.: 1991, ApJ 378, 413

Christensen-Dalsgaard, J., Proffitt, C.R., Thompson, M.J.: 1993, ApJ 403, L75

Däppen, W., Gough, D.O., Kosovichev, A.G., Thompson, M.J.: 1991, in: Challenges to theories of the structure of moderate-mass stars, eds. D.O. Gough, J. Toomre, Springer, Heidelberg, p. 111

Dilke, F.W.W., Gough, D.O.: 1972, Nature 240, 262

Gough, D.O., Kosovichev, A.G.: 1990, in: Inside the Sun, eds. G. Berthomieu, M. Cribier, Kluwer, Dordrecht, p. 327

Kosovichev, A.G., Fedorova, A.V.: 1991, SvA 35, 507

Kosovichev, A.G.: 1992, Solar structure inversion package, SOI-TN-084, Stanford Univ.

Kosovichev, A.G., Christensen-Dalsgaard, J., Däppen W., Dziembowski, W., Gough, D.O., Thompson, M.J.: 1992, MNRAS 259, 536

Libbrecht, K.G., Woodard, M.F., Kaufman, J.M.: 1990, ApJ Suppl 74, 1129

Rogers, F.J., Iglesias, C.A.: 1992, ApJ Suppl 79, 507

Roxburgh, I.W.: 1985, Solar Phys. 100, 21

Skaley, D., Stix, M.: 1991, A&A 241, 227

Toutain, T., Fröhlich, C.: 1992, A&A 257, 287

Microscopic Settling and Turbulent Diffusion Induced by Rotation in the Sun

Yves Gaigé and Sylvie Vauclair

Observatoire Midi-Pyrénées, 14 avenue Edouard Belin, F-31400 Toulouse, France.

Abstract: Mixing induced by rotation has been computed together with gravitational settling, using the Geneva stellar evolution code in which an implicit routine for solving the diffusion equation has been introduced. Even in the presence of turbulence, gravitational settling can proceed and lead to μ-gradients which may stabilize the gas : lithium depletion can occur together with helium settling.

Although lithium may be destroyed below the solar convection zone by nuclear reactions even without the help of any macroscopic motions, it seems difficult to reconcile in a consistent way the lithium features observed in galactic clusters and in the Sun, without introducing some transport between the bottom of the convection zone and the nuclear destruction layers. Turbulence induced by rotation has been proved able to provide such a transport process leading to results compatible with the observations (Charbonnel et al. 1993).

On the other hand, helioseismic inversions lead to values of the sound velocity in the Sun which suggest a depletion of helium in the solar outer layers of about 10% by number (Dziembowski 1993). This is the order of magnitude of the helium depletion expected in the case of pure helium diffusion (Proffitt and Michaud 1991; Bahcall and Pinsonneault 1992). Is it possible to reconcile such a value with the macroscopic transport necessary to account for the lithium depletion?

We shall see that the μ-gradient induced by gravitational settling can play a fundamental role in this respect.

1 The Computations

The diffusion processes including turbulence and settling have been described in several papers (e.g. Charbonnel et al. 1992). The evolution of an element concentration c inside the star is obtained by solving the following diffusion equation:

$$\frac{\partial c}{\partial t} = D' \cdot \frac{\partial^2 c}{\partial m_r^2} + \left\{ \frac{\partial D'}{\partial m_r} - V_D' \right\} \frac{\partial c}{\partial m_r} - \left\{ \frac{\partial V_D'}{\partial m_r} + \lambda \right\} c \qquad (1)$$

Here D' is equal to $(4\pi\rho r^2)^2(D_T + D_{12})$ in which D_T is the turbulent diffusivity and D_{12} the atomic diffusion coefficient and V_D' is equal to $(4\pi\rho r^2)V_D$ where V_D is

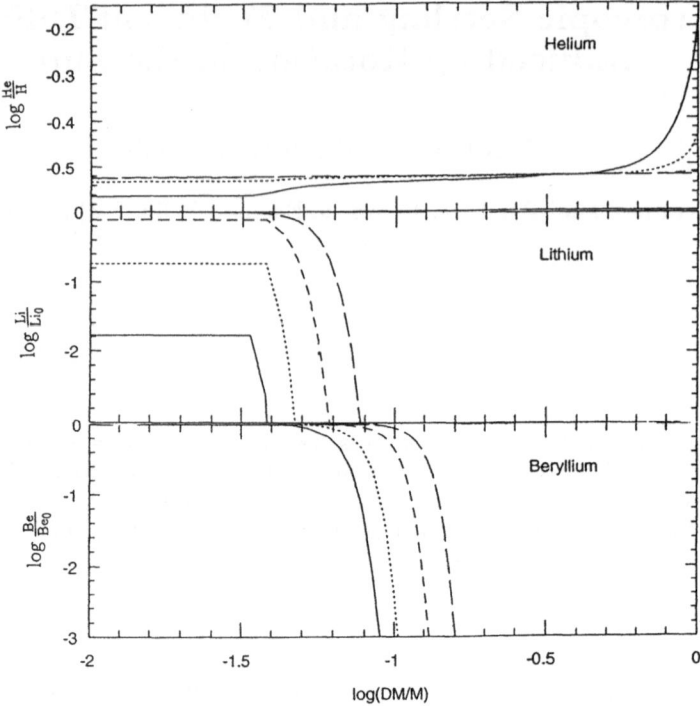

Fig. 1. Helium, lithium and beryllium abundance profiles in the radiative zone of a 1.0 M$_\odot$ star with rotational braking, as a function of the mass fraction (ratio of the external mass to the total mass star of the star). The star is assumed to arrive on the main sequence with a rotation velocity equal to 100 km s^{-1}. Then it slows down according to the 'Skumanich law'. The abundance profiles are shown at four different ages: 0.01 Gyr (*long dashed lines*), 0.1 Gyr (*dashed lines*), 1 Gyr (*dotted lines*) and 4.5 Gyr (*solid lines*)

the settling diffusion velocity:

$$V_{\mathrm{D}} = D_{12} \frac{m_{\mathrm{He}}\, g}{kT} \tag{2}$$

Rotational mixing seems the most promising process proposed so far to transport chemicals such as lithium, down to their nuclear burning layers. Zahn (1992) developped a consistent theory of meridional advection and turbulence induced by rotation, taking into account the feedback effect due to angular momentum transport:

1. In a rotating star, due to centrifugal effects, the gravity equipotentials are no more spherical, which induces a circulation of matter between polar and equatorial regions: the so-called meridional circulation.
2. This circulation itself induces a transport of angular momentum, thereby creating shears which become unstable in the horizontal direction, while the vertical shears are stabilized by the density gradient. This large scale horizontal turbulence decays into small scales and becomes 3D when the turnover rate of the turbulence exceeds the angular velocity.

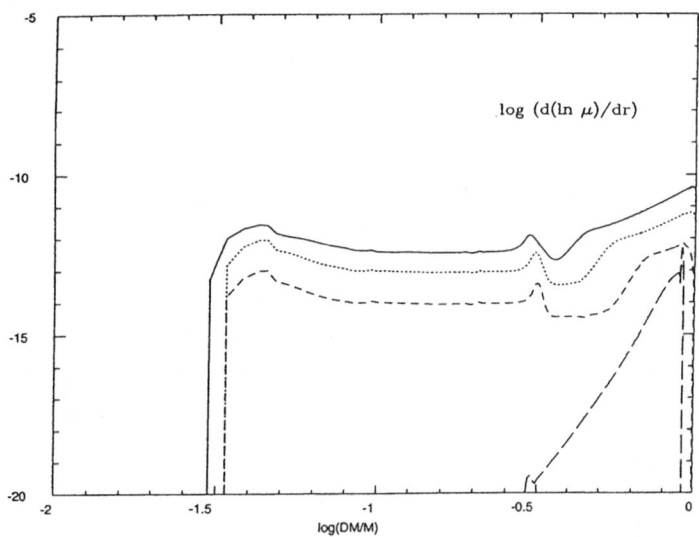

Fig. 2. Molecular weight gradient as a function of the mass fraction for the ages of 0.01, 0.1, 1 and 4.5 Gyr

3. The horizontal turbulence 'cuts down' the effect of advection on the transport of the chemical species, as the elements which go up in the upward flow of matter can be transported into the downward flow by horizontal motions before reaching the top layers. The global effect of advection moderated by horizontal turbulence can also be approximately treated as a diffusion process.
4. Due to this horizontal mixing, the transport of angular momentum is more efficient than the transport of chemicals, as is necessary to explain the observations.
5. The horizontal transport of angular momentum smoothes out the original meridional circulation. Taking this feedback effect into account, Zahn (1992) showed that the whole process is stopped in an Eddington-Sweet time-scale, unless angular momentum is extracted from the star due to a wind. When the rotational braking is taken into account, with a rotational velocity decreasing according to the Skumanich relation, the induced diffusion processes can be treated with an effective diffusion coefficient similar to that used in Charbonnel et al. (1992).

2 The Results

Our aim was to consider the effects of a gradient of molecular weight μ on the circulation and on the mixing. We know that nuclear transformations produce gradients of molecular weight which will tend to inhibit the macroscopic motions and protect the evolving core from rotational mixing. Below the outer convective zone, gravitational settling tends to induce also a small μ-gradient. The more efficient the macroscopic mixing, the smaller the μ-gradient that gravitational settling can build. It is possible, however, that these μ-gradients inhibit the turbulence. So the

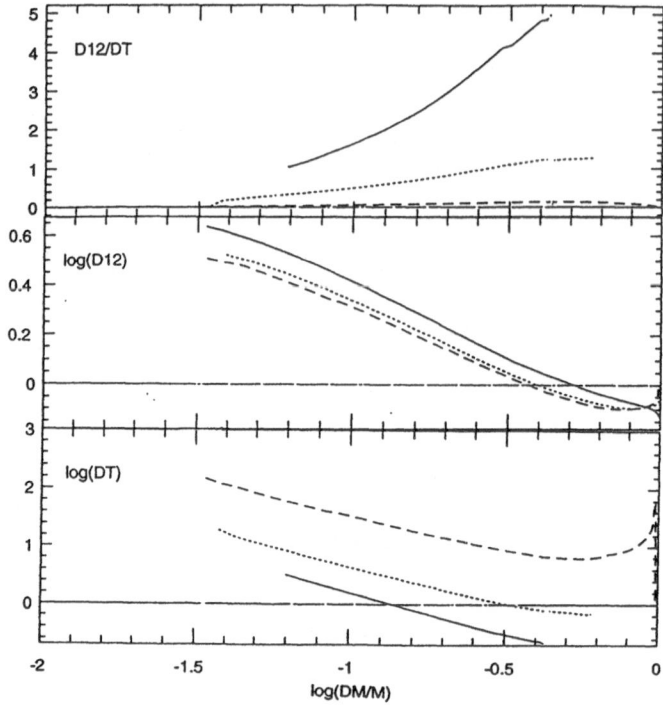

Fig. 3. Effective (D_T) and atomic (D_{12}) diffusion coefficients for different ages: 0.1 Gyr (*dashed lines*), 1 Gyr (*dotted lines*) and 4.5 Gyr (*solid lines*)

following question arises: can gravitational settling below the solar convection zone build a μ-gradient large enough to inhibit turbulence?

We have tested the effect of turbulence on microscopic diffusion in a stellar model of 1.0 $M_\odot$ computed with $\alpha = 1.6$. The evolution of the element abundances was followed during 4.6×10^9 yrs. We assumed that the star began its life on the main sequence with a rotation velocity of 100 km s^{-1}, and was slowed down to the Sun's present value following a Skumanich law. An implicit numerical code was used to solve equation (1) throughout the model (see Charbonnel et al. (1992) for the mathematical method). Mixing was supposed to be efficient only in the zone where the molecular weight gradient was lower than a given critical value; the turbulent diffusivity was set to zero outside this zone.

Figure 1 shows the abundance profiles of helium, lithium and beryllium for a critical value of $\partial \ln \mu / \partial r = 10^{-12}$, for four different ages. Lithium is destroyed by a factor close to 100 at the solar age as observed. Meanwhile helium is depleted by 8.5%. Figure 2 displays the values of molecular weight gradients for the same ages.

Figure 3 shows the diffusion coefficients for the same critical value of 10^{-12} and the same ages. The turbulent diffusivity decreases with age, which is mainly due to the decrease of the rotation velocity. On the other hand the atomic diffusion coefficient increases. Thus the circulation becomes less and less efficient against the microscopic settling. The molecular weight gradient increases with age. It first reaches the critical gradient near the core, and the circulation is progressively ex-

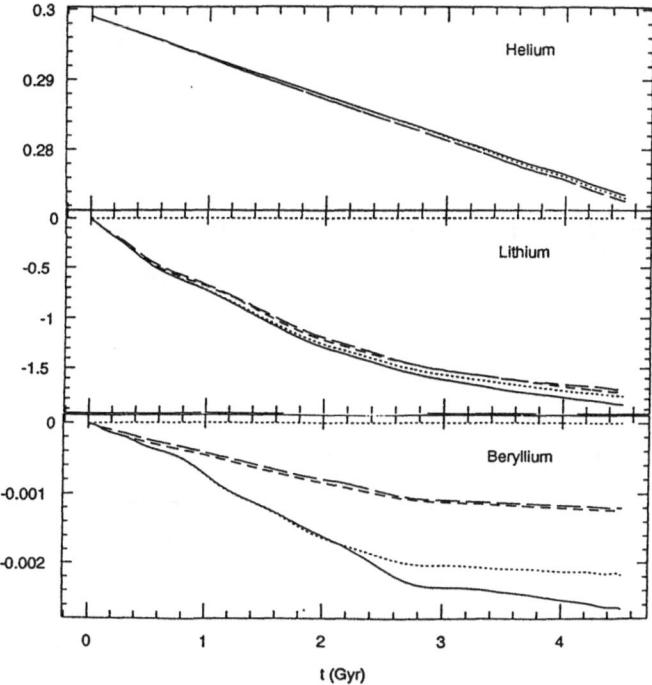

Fig. 4. Time evolution of the surface helium, lithium and beryllium abundances for four different molecular weight gradient limits. The *solid lines* correspond to a critical gradient 10^{-11}; *dotted, dashed* and *long dashed lines* correspond respectively to a limit of 10^{-12}, 10^{-13} and 10^{-14}

pollod towards the external layers. When the gradient due to microscopic settling below the convection zone also reaches the critical value, the external zone is disconnected from the interior. Helium is then depleted at a larger rate because of the atomic diffusion.

Figure 4 summarises the results obtained for different critical values. As the critical value of the molecular weight gradient increases, the period of evolution with mixing increases and thus lithium and beryllium are more depleted whereas helium is less depleted.

In Figure 5, we compare the evolution of a 1.0 $M_\odot$ star with the evolution of 0.9 $M_\odot$ and 1.1 $M_\odot$ stars using the same critical gradient of 10^{-12}. In the three cases the critical gradient is reached during the evolution period. It is reached earlier in the heaviest model because the microscopic diffusion is more efficient. However, even in the lightest model the turbulent diffusion is not efficient enough to inhibit the microscopic diffusion.

In conclusion, the combined effect of turbulence and gravitational settling can lead to a μ-gradient below the convection zone able to inhibit turbulence after about 2 Gyr. Then turbulence stops and gravitational settling proceeds freely, leading to the helium depletion needed to account for the sound velocity measurements.

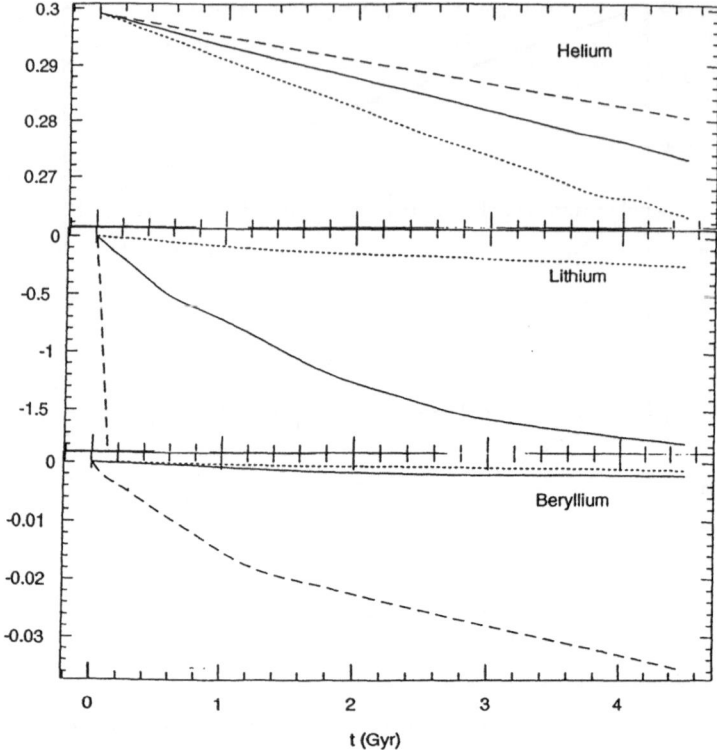

Fig. 5. Time evolution of the surface helium, lithium and beryllium abundances for different masses : 0.9 $M_\odot$ (*dashed lines*), 1.0 $M_\odot$ (*solid lines*) and 1.1 $M_\odot$ (*dotted lines*). The critical molecular weight gradient is in the three cases 10^{-12}

References

Bahcall, J.N., Pinsonneault, M.H.: 1992, *Rev. Mod. Phys.* **64**, 885
Charbonnel, C., Vauclair, S., Zahn, J.P.: 1992, *Astr. Ap.* **255**, 191
Charbonnel, C., Vauclair, S., Maeder, A., Meynet, G., Schaller, G.: 1993, preprint
Dziembowski, W.: 1993, preprint
Proffitt, C.R., Michaud, G.: 1991, *Ap. J.* **380**, 238
Skumanich, A.: 1972, *Ap. J.* **171**, 565
Zahn, J.P.: 1987, in: The Internal Solar Angular Velocity, eds. B.R. Durney and S. Sofia, Reidel, Dordrecht, p. 201
Zahn, J.P.: 1992, *Astr. Ap.* **265**, 115

II

Generation of Large-Scale Magnetic Fields

Mean-Field Theory of the Solar Dynamo

Dieter Schmitt

Univ.-Sternwarte, Geismarlandstraße 11, 37083 Göttingen, Germany

Abstract: The generation of the solar magnetic field is generally ascribed to dynamo processes in the convection zone. The dynamo effects, differential rotation (Ω-effect) and helical turbulence (α-effect) are explained, and the basic properties of the mean-field dynamo equations are discussed in view of the observed properties of the solar cycle. Problems of the classical picture of a dynamo in the convection zone (fibril state of magnetic flux, field strength, magnetic buoyancy, polarity rules, differential rotation and butterfly diagram) are addressed and some alternatives to overcome these problems are presented. A possibility to make up for the missing radial gradient of rotation in the convection zone is an $\alpha^2\Omega$-dynamo with an anisotropic α-tensor. Dynamo solutions then might have the characteristics of the butterfly diagram. Another approach involves meridional circulation as the cause of the migration of a dynamo wave. Another suggestion is that the solar dynamo operates in the overshoot region at the base of the convection zone where strong fields, necessary to explain the polarity rules, can be stored and radial gradients in the angular velocity occur. As an alternative to the turbulent α-effect a dynamic α-effect based on magnetostrophic waves driven by a magnetic buoyancy instability of a magnetic flux layer is introduced. Model calculations which use the internal rotation of the Sun as deduced from helioseismology only show solar cycle behaviour if the turbulent diffusivity is reduced in the layer and the α-effect is concentrated near the equator. Another possibility is a combined model. The non-uniform rotation and most of the azimuthal magnetic flux are confined to a thin layer at the bottom of the convection zone where turbulent diffusion is greatly reduced, with the convective region above containing only weak fields for which the α-effect and turbulent diffusion operate in the conventional manner. The dynamo takes on the character of a surface wave at the interface between the two regions. Another possibility is a fibril field approach where non-axisymmetric flux tube instabilities lead to an α-effect which, together with differential rotation and reconnection of flux tubes, forms the basis of a flux tube dynamo. Furthermore the stochastic excitation of magnetic fields by a fluctuating α-effect is addressed. This contributes to irregularities in the solar cycle and leads to the excitation of a spectrum of dynamo modes which can be compared with decompositions of surface fields into spherical harmonics. An interesting consequence is the generation of a north-south asymmetry in the butterfly diagram.

1 Introduction

The basic observation of the global magnetic field of the Sun is the cyclic variation of the number of sunspots discovered by Schwabe (1844), indicating the 11 yr solar cycle which governs the activity of the Sun. An important additional feature is the latitude migration of the sunspot zones from middle latitudes towards the equator during the course of the cycle (Spörer, around 1860), documented in the famous butterfly diagram (Maunder 1922). Hale (1908) measured the magnetic field in sunspots for the first time and established the polarity rules (Hale 1924), proving that the magnetic cycle is antisymmetric with respect to the equatorial plane and has a period of 22 yr. The systematic behaviour of bipolar sunspot groups is understood in terms of a subsurface toroidal magnetic field which occasionally emerges through the surface to form sunspots. In addition to the dominating toroidal field a weak poloidal magnetic field is observed which is also antisymmetric and shows cyclic reversals which take place at the poles around the maximum of sunspot activity (Babcock 1959). The two field components are almost 180° out of phase.

It is generally believed that these global properties of the magnetic field of the Sun are due to a dynamo process, i.e. the interaction of fluid motions and magnetic fields, an idea first proposed by Larmor (1919). After the frustrating anti-dynamo theorems by Cowling (1934) stating that a dynamo cannot work with completely axisymmetric magnetic fields, and by Bullard and Gellman (1954) saying that the field cannot be maintained by pure rotation, no matter how non-uniform, Parker (1955) suggested that regeneration of the magnetic field is achieved by the cyclonic motions in the convection zone. The effect of small-scale motions on large-scale magnetic fields is systematically described within the framework of mean-field electrodynamics which was established by Steenbeck, Krause and Rädler (1966) and proved useful in discussing the dynamo process.

In mean-field electrodynamics the evolution of a mean magnetic field is investigated. The magnetic field $\boldsymbol{B}$ and the velocity field $\boldsymbol{u}$ are separated into a mean part (indicated by an overbar), understood as a proper average, and a fluctuating part (indicated by a prime). The averaging may be over space or time, or an ensemble average. The mean part of the velocity represents a motion of global scale, notably the solar differential rotation. The fluctuating part describes the irregular or turbulent convective motions. Here we consider the kinematic dynamo problem for which the mean velocity and the statistical properties of the fluctuations are assumed to be given independently of the magnetic field. Substitution of the above decomposition into the induction equation and separation of its mean and fluctuating parts yields

$$\frac{\partial \overline{\boldsymbol{B}}}{\partial t} = \operatorname{curl}\left(\overline{\boldsymbol{u}} \times \overline{\boldsymbol{B}} + \mathcal{E} - \eta \operatorname{curl}\overline{\boldsymbol{B}}\right), \tag{1}$$

$$\frac{\partial \boldsymbol{B}'}{\partial t} = \operatorname{curl}\left(\overline{\boldsymbol{u}} \times \boldsymbol{B}' + \boldsymbol{u}' \times \overline{\boldsymbol{B}} + \mathcal{G} - \eta \operatorname{curl}\boldsymbol{B}'\right). \tag{2}$$

Time is denoted by t and magnetic diffusivity by η. Correlations of the fluctuations lead to the term $\mathcal{E} = \overline{\boldsymbol{u}' \times \boldsymbol{B}'}$ in the equation for the mean field. Since the second equation says that the relationship between $\boldsymbol{B}'$ and $\overline{\boldsymbol{B}}$ is linear, $\mathcal{E}$ can be written as an expanison

$$\mathcal{E}_i = \alpha_{ij}\overline{B}_j + \beta_{ijk}\frac{\partial \overline{B}_j}{\partial x_k} + \cdots \tag{3}$$

The tensors α_{ij} and β_{ijk} are usually calculated neglecting the contribution of $\mathcal{G} = \boldsymbol{u}' \times \boldsymbol{B}' - \overline{\boldsymbol{u}' \times \boldsymbol{B}'}$, which is known as the first order smoothing approximation. This approximation is questionable in the case of the Sun (see e.g. discussion in Stix 1991). In the kinematic approach the tensors are statistical properties of the fluctuating velocity field and independent of $\overline{\boldsymbol{B}}$. In the case of isotropic homogeneous turbulence

$$\mathcal{E} = \alpha\overline{\boldsymbol{B}} - \beta\operatorname{curl}\overline{\boldsymbol{B}} \tag{4}$$

with

$$\alpha = -\frac{1}{3}\int_0^\infty \overline{\boldsymbol{u}'(t)\operatorname{curl}\boldsymbol{u}'(t-\tau)}\,\mathrm{d}\tau\,, \qquad \beta = \frac{1}{3}\int_0^\infty \overline{\boldsymbol{u}'(t)\,\boldsymbol{u}'(t-\tau)}\,\mathrm{d}\tau\,. \tag{5}$$

Substitution yields the mean-field induction equation or dynamo equation

$$\frac{\partial \overline{\boldsymbol{B}}}{\partial t} = \operatorname{curl}(\overline{\boldsymbol{u}} \times \overline{\boldsymbol{B}} + \alpha\overline{\boldsymbol{B}} - \eta_T\operatorname{curl}\overline{\boldsymbol{B}}) \tag{6}$$

where $\eta_T = \eta + \beta$ is the total diffusivity. The dynamo equation differs from the original induction equation in two ways. The first is a substantial increase of the diffusivity, since the turbulent diffusivity β is usually much larger than the molecular diffusivity η. The second and new feature is the term involving α which is crucial for a dynamo. This α-effect ensures that the mean magnetic field is not subject to the anti-dynamo theorems. It occurs if the turbulent flow is helical, i.e. it should be correlated with its own vorticity. This is the case in stratified rotating stellar convection zones. More on mean-field theory can be found in the textbooks of Moffatt (1978), Parker (1979), Krause and Rädler (1980), and Stix (1989).

Axisymmetric mean fields can be decomposed into toroidal and poloidal components $\overline{\boldsymbol{B}} = B\boldsymbol{e}_\varphi + \operatorname{curl}(A\boldsymbol{e}_\varphi)$ where $\boldsymbol{e}_\varphi$ is the unit vector in toroidal direction. With a differential rotation $\overline{\boldsymbol{u}} = u\boldsymbol{e}_\varphi = r\Omega(r,\vartheta)\sin\vartheta\boldsymbol{e}_\varphi$ as the mean velocity, the dynamo equations can be separated into two equations for the toroidal and poloidal field evolution

$$\frac{\partial B}{\partial t} = [\operatorname{curl}\{(u\boldsymbol{e}_\varphi) \times \operatorname{curl}(A\boldsymbol{e}_\varphi) + \alpha\operatorname{curl}(A\boldsymbol{e}_\varphi) - \eta_T\operatorname{curl}(B\boldsymbol{e}_\varphi)\}] \cdot \boldsymbol{e}_\varphi\,, \tag{7}$$

$$\frac{\partial A}{\partial t} = \alpha B - \{\eta_T\operatorname{curl}\operatorname{curl}(A\boldsymbol{e}_\varphi)\} \cdot \boldsymbol{e}_\varphi\,. \tag{8}$$

Here the basic mechanisms of dynamo action become visible. Toroidal field is generated by differential rotation and α-effect from a parent poloidal field, which itself is regenerated by means of the α-effect acting on the toroidal field. Even for rigid rotation, the α-effect is capable of generating a mean magnetic field (α^2-dynamo). In the case of the Sun, however, differential rotation has to be taken into account. If the α-term $\operatorname{curl}\{\alpha\operatorname{curl}(A\boldsymbol{e}_\varphi)\}$ is small compared to differential rotation $\operatorname{curl}\{(u\boldsymbol{e}_\varphi) \times \operatorname{curl}(A\boldsymbol{e}_\varphi)\} = \operatorname{grad}\Omega \times \operatorname{grad}(rA\sin\vartheta)$ it can be neglected in the equation for B. This is named the $\alpha\Omega$-limit.

The essential parameter of the $\alpha\Omega$-dynamo is the dynamo number

$$P = R_\alpha R_\Omega = \frac{\alpha R}{\eta_T}\frac{\Omega R^2}{\eta_T}\,, \tag{9}$$

the product of two Reynolds numbers measuring the strength of the two induction effects relative to diffusion. R is the basic length scale of the system, e.g. the solar radius. A critical value of P must be exceeded for dynamo action. Oscillatory solutions with migrating dynamo waves are typically obtained. The dynamo period is of the order of the time scale of diffusion. The direction of migration is given by $\alpha \operatorname{grad}\Omega \times e_\varphi$, i.e. along the surfaces of constant angular velocity (Yoshimura 1975). With radial shear the migration is towards the equator if $\alpha \, \partial\Omega/\partial r < 0$ in the northern hemisphere. In this case a field antisymmetric with respect to the equatorial plane (dipolar solution) is usually preferred over the symmetric (quadrupolar) solution in spherical dynamos (Roberts 1972).

It has been shown that, by making suitable assumptions about the induction effects and the turbulent diffusivity, many features of the solar cycle could be well represented (e.g. Steenbeck & Krause 1969 and many others, see review by Rädler 1990) such as oscillatory antisymmetric solutions with dominating toroidal field, Maunder's butterfly diagram, Hale's polarity rules and the phase relations between the field components. This general agreement of the calculated fields with the observed patterns provided confidence that the basic ideas are correct. The hope was that minor disagreements could be resolved with a better knowledge of the solar differential rotation and a realistic turbulence model for the mean electromotive force, i.e. the α-effect and turbulent diffusivity.

However, recent observations of the surface magnetic field and the p-mode oscillations as well as theoretical considerations about the field structure within the convection zone have seriously put this picture into question.

2 Difficulties of Conventional Convection Zone Dynamos

A number of recent reviews have dealt with the difficulties of the conventional picture of a mean-field $\alpha\Omega$-dynamo in the convection zone (Brandenburg and Tuominen 1991, Stix 1991, DeLuca and Gilman 1991, Schmitt 1993, Schüssler 1993). I shall therefore only discuss the main arguments briefly and the reader is referred to the mentioned reviews for details and more references.

Investigations of magnetoconvection (reviewed in Galloway and Weiss 1981 and Proctor and Weiss 1982) suggest that the vast majority of the solar magnetic flux in the convection zone is concentrated in small-scale intermittent features such as observed on the solar surface (Stenflo 1989).

These flux concentrations are difficult to store in the convection zone for times comparable to the solar cycle. Several processes, most notably magnetic buoyancy (Parker 1975), transport magnetic flux from the bottom to the top of the convection zone in times of the order of a month (Moreno-Insertis 1992), much too short for the dynamo to generate the field.

Another problem of locating the dynamo within the convection zone is the nearly strict observance of the polarity rules for bipolar active regions. Also their tilt angles show a systematic dependence on latitude. These observations suggest a well-ordered strong toroidal magnetic field and are difficult to explain with a magnetic field originating in the turbulent convection zone (Schüssler 1993).

Helioseismology shows that differential rotation does not at all dominate over convective motions in the convection zone proper. The oscillation data imply that the main convection zone rotates like the solar surface with no significant radial gradient, and that the deep interior rotates almost rigidly at a rate between the equatorial and polar rates on the surface (Dziembowski et al. 1989, Brown et al. 1989, Schou et al. 1992). Thus a radial gradient occurs only in a transition region between the base of the convection zone and the top of the interior.

An $\alpha\Omega$-dynamo in the convection zone with only a latitudinal gradient of angular velocity results in a stationary solution for $\alpha < 0$ or in an oscillatory solution for $\alpha > 0$ in the northern hemisphere. In the latter case the dynamo wave proceeds radially outward from the bottom to the top of the convection zone. The resulting butterfly diagram is not solar-like: most of the flux is concentrated at too high latitudes and there is no clear latitudinal migration (Köhler 1973, Prautzsch 1993); also the period is too short.

3 Alternative Models

3.1 Convection Zone Dynamo

A possibility to make up for the missing vertical gradient of angular velocity in the convection zone is a dynamo with an anisotropic α-tensor. Weisshaar (1982) presented an α^2-dynamo model with a solar-like behaviour (butterfly diagram, period) assuming $\alpha_{rr} < 0, \alpha_{\vartheta\vartheta} = \alpha_{\varphi\varphi} > 0, \alpha_{r\vartheta} = \alpha_{\vartheta r} > 0$ and $|\alpha_{rr}| > |\alpha_{\varphi\varphi}|$. Recent turbulence models yield such an anisotropic α-tensor (Wälder et al. 1980, Brandenburg et al. 1990, Rüdiger and Kichatinov 1993, Ferrière 1993). Rüdiger (1980) and Elstner and Rüdiger (1993) however report that non-axisymmetric ($m = 1$) modes are preferred with such an anisotropic α-tensor. Inclusion of the latitudinal differential rotation may change this to the observed axisymmetric mean solar magnetic field, since differential rotation favours axisymmetric modes (Rädler 1986).

3.2 Overshoot Layer Dynamo

Some of the problems mentioned above can be alleviated if the source region of the magnetic flux which emerges in active regions and the site of the dynamo is the slightly subadiabatic region of overshooting convection between the convection zone proper and the radiative interior (e.g. Spiegel and Weiss 1980, Schüssler 1983, Hughes 1991). In an overshoot layer with a depth of some 10^4 km (van Ballegooijen 1982, Schmitt et al. 1984, Pidatella and Stix 1986, Skaley and Stix 1991) strong fields of 10^4 G to 10^5 G can be stored for times of the order of the cycle (Moreno-Insertis et al. 1992). Sunspots are formed by erupting flux loops while other parts of the flux tubes are still firmly anchored in this layer (Moreno-Insertis 1986). This layer is also a favourable site for the solar dynamo: the radial gradient of angular velocity yields dynamo waves migrating in latitudinal direction, a dynamic α-effect occurs which is acting on a strong toroidal field, and turbulent diffusivity may be reduced so that the cycle period increases. These points are taken up below.

The results from helioseismology indicate that the radial gradient of angular velocity at the base of the convection zone changes its sign at a latitude of about 30°. The gradient is positive near the equator and negative near the pole. At the pole the gradient seems to be steeper approximately by a factor of two. There is still a latitudinal gradient which is somewhat smaller than at the surface.

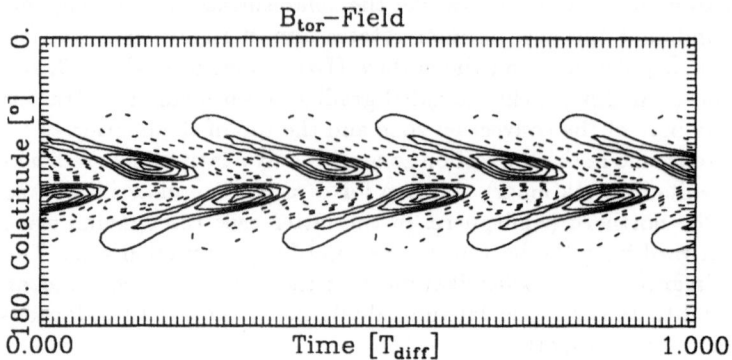

Fig. 1. Butterfly diagram of the toroidal field component of an overshoot layer $\alpha^2\Omega$-dynamo with differential rotation as deduced from helioseismology. α is concentrated in the overshoot region within $\pm 30°$ around the equator and has a negative sign in the northern hemisphere. The magnetic diffusivity is reduced in the overshoot region

The second and most important ingredient of a dynamo is the α-effect. As an alternative to overshooting convection which gives a negative α in the northern hemisphere (Yoshimura 1972, Glatzmaier 1985), an intense toroidal magnetic field layer in the overshoot region provides an α-effect by itself (Schmitt 1984, 1985, 1987). The top parts of the layer, where the magnetic field decreases rapidly with height, are unstable due to the magnetic Rayleigh-Taylor instability (Cattaneo and Hughes 1988). Because of the solar rotation the instability takes the form of growing magnetostrophic waves (Acheson and Gibbons 1978). These are helical and are therefore capable of inducing an electromotive force parallel to the toroidal field. This electric field drives a current which regenerates the poloidal field. This represents a dynamic α-effect because the velocity is not prescribed but follows from the actual forces and the interaction of the magnetic field with the velocity field is taken into account. The effect is independent of convective velocity fields.

By superimposing the most unstable magnetostrophic waves, an α-effect is derived which is concentrated mainly at lower latitudes. α is antisymmetric with respect to the equator. In the northern hemisphere it is negative from the equator to a latitude of approximately 30° where it changes sign and becomes slightly positive in higher latitudes.

With differential rotation as deduced from helioseismology realistic butterfly diagrams (Fig. 1) are only obtained if the α-effect is concentrated near the equator and negative in the northern hemisphere (Prautzsch 1993, Schmitt 1993, Prautzsch et al. 1993). Magnetostrophic waves thus naturally provide the right kind of induction effect for the poloidal field. The reduced, but still turbulent, diffusivity of the overshoot region leads to a period of the order of the solar cycle. The magnetic field is only generated in a small region near the equator at the bottom of the convection zone.

A problem of overshoot layer dynamos is the phase relation between the toroidal and radial field components (Stix 1976, 1987). Observationally they show a phase difference of 180°. With $\partial\Omega/\partial r > 0$ near the equator an equatorward migration of the dynamo wave is obtained for $\alpha < 0$. Then, however, the field components are in phase. The observed phase relation can be obtained if the signs of both differential rotation and α are reversed.

For a dynamo operating in a thin layer Choudhuri (1990) estimated the magnitudes of the induction effects needed in order to reproduce the period and latitudinal wavelength of the solar cycle. He found that the radial shear should be of the order of $G = \partial v_\varphi / \partial r \approx 10^{-5}\,\mathrm{s}^{-1} ... 10^{-6}\,\mathrm{s}^{-1}$, values consistent with the results from acoustic oscillations, the α-effect of magnitude $\alpha \approx 10\,\mathrm{cm\,s}^{-1}$, and the magnetic diffusivity of $\eta \approx 10^{10}\,\mathrm{cm^2 s^{-1}}$.

3.3 Interface Wave Dynamo

Recently Parker (1993) suggested a two component model which is a combination of the convection zone containing only weak fields in which the α-effect and turbulent diffusion are operative in the conventional manner, and a thin shear layer below with greatly reduced diffusion and strong toroidal fields. The concentration of the vertical shear to a thin layer below the bottom of the convection zone is imposed by helioseismology. It follows from dynamo theory that the toroidal magnetic field is concentrated in and around the same thin layer. Parker states that the confinement of the field to such a thin layer requires such strong fields that the turbulent diffusivity of the field is completely suppressed (Cattaneo and Vainshtein 1991, Vainshtein and Cattaneo 1992). A turbulent diffusivity is however essential for the operation of the dynamo. Therefore the convection zone is needed in the dynamo process. The downward penetration of the poloidal field produced by the α-effect in the convection zone into the shear layer as well as the escape of toroidal field out of the shear layer into the convection zone are essential parts of the dynamo which is then a surface wave at the interface between convection and non-uniform rotation. Such a dynamo is not as effective as a dynamo with overlapping induction effects. Parker shows that his picture is still effective enough and selfconsistent if the suppression of the turbulent diffusivity scales inversely with the magnetic energy density. But he also stresses that there is today no compelling knowledge about the suppression of turbulent diffusion by a magnetic field.

Brandenburg (1993, this volume) suggested a similar picture based on magnetoconvection simulations of a rotating stratified compressible fluid (Nordlund et al. 1992). The simulations show that the field regeneration occurs in the convection zone, but the magnetic field is pushed downward by concentrated spinning downdrafts into the overshoot layer below the convection zone where the field accumulates.

3.4 Flux Tube Dynamo

Schüssler (1980, 1993) discusses a picture of the dynamo based on magnetic flux tubes. He conjectures that the magnetic flux in the convection zone is in an intermittent, fragmentated state (Schüssler 1984) which may be described by an ensemble of magnetic flux tubes. The dynamics of flux tubes can strongly differ from the behaviour of a passive, diffuse field which is often assumed in conventional mean-field dynamo theory. His view is based on observed properties of active regions like emergence in low latitudes, Hale's polarity rules, the tilt angles and the way sunspots form, and on theoretical considerations of the dynamics of buoyant flux tubes. He argues that the magnetic structures which erupt in an emerging active region are not passive to convection and originate in a source region with field strengths of at least 10^5 G. This is much larger than equipartition values with respect to convective kinetic energy densities. To avoid the magnetic buoyancy problem the source region is presumably the overshoot region below the convection zone.

Ferriz-Mas and Schüssler (1993a,b) studied the equilibrium and stability of thin toroidal flux tubes in the stably stratified overshoot region of a differentially rotating star. They identified two instability mechanisms, one of which takes place in a small strip of width of approximately 20° in latitude, which starts at middle latitudes for $B \approx 10^5$ G and proceeds towards the equator for increasing field strengths of up to $B \approx 1.5 \cdot 10^5$ G. The growth rate of this instability is rather small. Ferriz-Mas et al. (1993) derived an α-effect of this non-axisymmetric flux tube instability under the influence of rotation (very similar to the α-effect of magnetostrophic waves mentioned above), which together with differential rotation could form the basis of a flux tube dynamo of $\alpha\Omega$-type in the overshoot layer. The other instability domain occurs at larger field strengths and extends over almost all latitudes exept close to the poles. It has larger growth rates and is thought to be responsible for the rapid eruption of flux loops toward the surface.

Schüssler (1993) sketched a flux tube dynamo where a toroidal flux tube forms a loop by an undulatory instability which is twisted by Coriolis force and reconnects. Differential rotation stretches the loop until the ends meet again and reconnect to create a pair of alternately directed flux tubes in the vicinity of the original toroidal tube. Diffusion is ascribed to fluctuations of the tube location, and tubes of opposite polarities at the same location annihilate. A time delay is introduced for the above described dynamo process. He found an oscillatory dynamo with propagating waves if excitation exceeds a critical value.

4 Random Forcing and the Mode Structure of the Solar Magnetic Field

Various kinds of averages in mean-field theory can be used; ensemble, spatial or temporal averaging is possible. Consider for example averages over longitude. This means that mean values can fluctuate in the other spatial directions and in time. The mean electromotive force, described by the parameters α and β, then also fluctuates. Such stochastic fluctuations excite magnetic fields (Hoyng 1987, 1988). Neglecting fluctuations $\delta\beta$ in the turbulent diffusivity β, Hoyng et al. (1993) investigated the effect of random fluctuations of the α-effect on the solutions of a linear mean-field dynamo. They used a one-dimensional (latitude-dependent) solar dynamo model of $\alpha\Omega$-type with constant radial shear $\partial\Omega/\partial r$ and constant turbulent diffusion η. Radial turbulent diffusion is modeled by a prescribed factor $\exp(\mathrm{i}kr)/r$ in the field variables A and B: $A = A(\vartheta, t)\exp(\mathrm{i}kr)/r$, B analogous. The radius r is then a free parameter for which $r = R$, the radius of the dynamo shell, was choosen; ϑ is the colatitude.

The α-effect may fluctuate statistically in latitude and time around a mean value which is marginally critical for excitation of the first mode, an oscillatory dipole: $\alpha = \alpha_0 \cos \vartheta + \delta\alpha(\vartheta, t)$. With no fluctuating α-effect, $\delta\alpha = 0$, the other dynamo modes, oscillatory dipolar (antisymmetric or odd) as well as monotonic and oscillatory quadrupolar (symmetric or even) modes, are damped. Growth rates and frequencies of the dynamo mode spectrum are denoted by γ and ν, respectively.

A fluctuating α-effect, $\delta\alpha \neq 0$, represents a random forcing term in the dynamo equation. The butterfly diagram acquires a random component due to the α-fluctuations which excite the fundamental dipolar mode and the higher dynamo modes to a fluctuating amplitude. The mean amplitude ratios of the various dynamo modes scale with the inverse square root of their damping rates: $a_k/a_0 \sim |\gamma_k|^{-1/2}$. The resulting superposition of dynamo modes is decomposed into spherical harmonics. The amplitudes for odd spherical harmonic degree l are primarily determined by the fundamental dipolar mode because the next dipolar mode has already a very small relative amplitude. The curve for even l is mainly determined by the first and second quadrupolar modes. Higher modes contribute only to the power at higher values of l. The ratio between odd and even amplitudes depends on the parameter $(\delta\alpha/\alpha_0)\lambda_c(\tau_c)^{1/2}$ where λ_c is the correlation length and τ_c the correlation time. The fundamental period, corresponding to the 22 yr period of the solar cycle, remains dominant for odd l. There is only little power at other frequencies. The weaker power at even l is widely distributed over frequency. The frequency spectrum of each mode is a Lorentzian with central frequency ν and width $|\gamma|$. Since $|\gamma|$ increases much more rapidly than ν the frequency distribution becomes broad and featureless.

Hoyng et al. compared the results with recently published harmonic decompositions of the observed surface magnetic field of the Sun (Stenflo 1988). The relative amplitudes and phase differences of the various harmonics are remarkably well reproduced. Also the 22 yr resonance of the odd harmonics and the broadening of the frequency spectrum of the even harmonics is found. The reported discrete higher frequencies in the even harmonics are however not seen. They are also absent in a similar analysis of sunspot data (Gokhale et al. 1992).

In addition to a random component the fluctuations also generate a north-south asymmetry in the butterfly diagram which can be directly compared with the properties of the solar cycle. It is sensitive to the structure of the first two or three modes of the dynamo model. The dominating mode is always an oscillatory dipole. Depending on the radial wave number k the second mode is a monotonic quadrupole for $kR = 1$ or an oscillatory quadrupole for $kR = 4$. Higher modes are more strongly damped and are unimportant for the asymmetry. The coherence time is proportional to the inverse of the damping rate $|\gamma|^{-1}$ and the second modes can be followed for several periods of the fundamental mode. In the first case the superposition of the dominating oscillatory dipole and the weaker monotonic quadrupole is an asymmetry with one polarity dominating such that asymmetry changes from north to south and vice versa after each cycle. In the other case the superposition of the two oscillatory modes, which have approximately the same frequency, leads to a dominating hemisphere and asymmetry does not alternate. Only in the case $kR = 2$ the second and third modes, monotonic and oscillatory quadrupoles, have approximately the same damping rate. The result is an irregularly changing north-south asymmetry (Fig. 2) not unlike that observed on the Sun (White and Trotter 1977, Vizoso and Ballester 1990, Carbonnell et al. 1993).

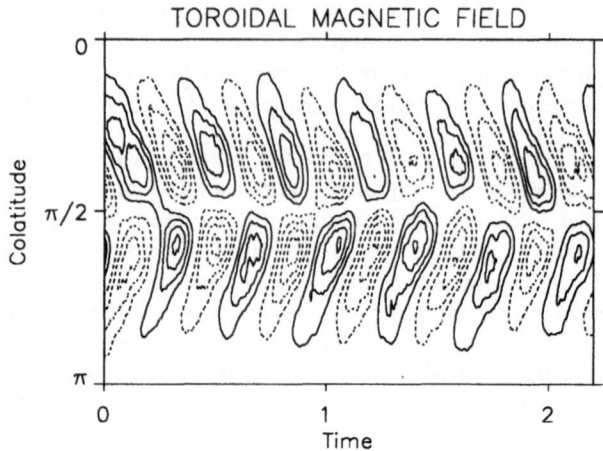

Fig. 2. Butterfly diagram of the toroidal field component $B(\vartheta, t)$ for $kR = 2$, $P = -685$, $\delta\alpha/\alpha \approx 4$, $\tau_c \approx 3$ months and $\lambda_c \approx 30$ Mm. An irregularly changing north-south asymmetry can be seen in addition to the fluctuations

Another interesting result of mean helicity fluctuations is a simple explanation for the fact that weaker solar cycles last longer than stronger ones (Hoyng 1993).

Acknowledgements

I would like to thank W. Deinzer, P. Hoyng, T. Prautzsch, G. Rüdiger and M. Schüssler for valuable discussions.

References

Acheson, D.J., Gibbons, M.P. (1978): *J. Fluid Mech.* **85**, 743
Babcock, H.D. (1959): *Astrophys. J.* **130**, 364
Brandenburg, A. (1993): "Hydromagnetic Simulations of the Solar Dynamo", this volume
Brandenburg, A., Tuominen, I. (1991): "The Solar Dynamo", in *The Sun and Cool Stars: activity, magnetism, dynamos*, Tuominen, I., Moss, D., Rüdiger, G. (eds.), *IAU-Coll.* **130**, *Lecture Notes in Physics* **380**, 223
Brandenburg, A., Nordlund, Å., Pulkinnen, P., Stein, R.F., Tuominen, I. (1990): *Astron. Astrophys.* **232**, 277
Brown, T.M., Christensen-Dalsgaard, J., Dziembowski, W.A., Goode, P., Gough, D.O., Morrow, C.A. (1989): *Astrophys. J.* **343**, 526
Bullard E., Gellman, H. (1954): *Phil. Trans. R. Soc. London* **A247**, 213
Carbonnell M., Oliver, R., Ballester J.L. (1993): *Astron. Astrophys.* **274**, 497
Cattaneo, F., Hughes, D.W. (1988): *J. Fluid Mech.* **196**, 323
Cattaneo, F., Vainshtein, S.I. (1991): *Astrophys. J.* **376**, L21
Choudhuri, A.R. (1990): *Astrophys. J.* **355**, 733
Cowling, T.G. (1934): *Mon. Not. R. Astron. Soc.* **94**, 39
DeLuca, E.E., Gilman, P.A. (1991): "The Solar Dynamo", in *The Solar Atmosphere and Interior*, Cox, A.N. (ed.), Univ. of Arizona Press, Tucson

Dziembowksi, W.A., Goode, P.R., Libbrecht, K.G. (1989): *Astrophys. J.* **337**, L53

Elstner, D., Rüdiger, G. (1993): "Stellar dynamos with anisotropic α-tensor", Krause, F., Rädler, K.-H., Rüdiger, G. (eds)., *The Cosmic Dynamo, IAU-Symp.* **157**, in press

Ferrière, K. (1993): *Astrophys. J.* **404**, 162

Ferriz-Mas, A., Schüssler, M. (1993a): "Instabilites of magnetic flux tubes in a stellar convection zone. I: Toroidal flux tubes in differentially roatating stars", *Geophys. Astrophys. Fluid Dyn.*, in press

Ferrix-Mas, A., Schüssler, M. (1993b): "Instabilites of magnetic flux tubes in a stellar convection zone. II: Flux rings outside the equatorial plane", preprint

Ferriz-Mas, A., Schmitt, D., Schüssler, M. (1993): "A dynamo effect due to instabilities of magnetic flux tubes", *Astron. Astrophys.*, submitted

Galloway, D.J., Weiss, N.O. (1981): *Astrophys. J.* **243**, 945

Glatzmaier, G.A. (1985): *Astrophys. J.* **291**, 300

Gokhale, M.H., Javaraiah, J., Narayanan Kutty, K., Varghese, B.A. (1992): *Solar Phys.* **138**, 35

Hale, G.E. (1908): *Astrophys. J.* **28**, 315

Hale, G.E. (1924): *Nature* **113**, 105

Hoyng, P. (1987): *Astron. Astrophys.* **171**, 348 and 357

Hoyng, P. (1988): *Astrophys. J.* **332**, 857

Hoyng, P. (1993): *Astron. Astrophys.* **272**, 321

Hoyng, P., Schmitt, D., Teuben, L.J.W. (1993): "The effect of random α-fluctuations and the global properties of the solar magnetic field", *Astron. Astrophys.*, submitted

Hughes, D.W. (1991): "Magnetic Buoyancy", in *Advances in Solar System Magnetohydrodynamics*, Priest, E.R., Wood, A.W. (eds.), Cambridge University Press, p. 77

Köhler, H. (1973): *Astron. Astrophys.* **25**, 467

Krause, F., Rädler, K.-H. (1980): *Mean-Field Magnetohydrodynmics and Dynamo Theory*, Pergamon Press, Oxford

Larmor, J. (1919): Rep. Brit. Assoc. Adv. Sc., 159

Maunder, W. (1922): *Mon. Not. R. Astron. Soc.* **82**, 534

Moffatt, H.K. (1978): *Magnetic Field Generation in Electrically Conducting Fluids*, Cambridge University Press

Moreno-Insertis, F. (1986): *Astron. Astrophys.* **166**, 291

Moreno-Insertis, F. (1992): "The motion of magnetic flux tubes in the convection zone and the subsurface origin of active regions", in *Sunspots, Theory and Oberservations*, Thomas, J.H., Weiss, N.O. (eds.), Kluwer, Dordrecht, p. 385

Moreno-Insertis, F., Ferriz-Mas, A., Schüssler, M. (1992): *Astron. Astrophys.* **264**, 686

Nordlund, A., Brandburg, A., Jennings, R.L., Rieutord, M., Ruokolainen, J., Stein, R.F., Tuominen, I. (1992): *Astrophys. J.* **392**, 647

Parker, E.N. (1955): *Astrophys. J.* **122**, 293

Parker, E.N. (1975): *Astrophys. J.* **198**, 205

Parker, E.N. (1979): *Cosmical Magnetic Fields*, Clarendon Press, Oxford

Parker, E.N. (1993): *Astrophys. J.* **408**, 707

Pidatella, R.M., Stix, M. (1986): *Astron. Astrophys.* **157**, 338

Prautzsch, T. (1993): "The Dynamo Mechanism in the Deep Convection Zone of the Sun", in NATO ASI *Theory of Solar and Planetary Dynamos*, Proceedings of the Isaac Newton Institute, Cambridge

Prautzsch, T., Schmitt, D., Schüssler, M (1993): "Non-linear dynamos. II. Two-dimensional model of an overshoot layer dynamo", in preparation

Proctor, M.R.E., Weiss, N.O. (1982): *Rep. Prog. Phys.* **45**, 1317

Rädler, K.-H. (1986): "On the effect of differential rotation on axisymmetric and non-axisymmetric magnetic field of cosmical bodies", in *Plasma-Astrophysics*, ESA SP-251, p. 569

Rädler, K.-H. (1990): "The Solar Dynamo", in *Inside the Sun*, Berthomieu G., Cribier M. (eds.), Kluwer, Dordrecht, *IAU-Coll.* **121**, 385

Roberts, P.H. (1972): *Phil. Trans. R. Soc. London* **A272**, 663

Rüdiger, G. (1980): *Astron. Nachr.* **301**, 181

Rüdiger, G., Kichatinov, L.L. (1993): *Astron. Astrophys.* **269**, 581

Schmitt, D. (1984): "Dynamo Action of Magnetostrophic Waves", in *The Hydromagnetics of the Sun*, Guyenne, T.D., Hunt, J.J. (eds.), ESA SP-220, 223

Schmitt, D. (1985): "Dynamowirkung magnetostrophischer Wellen", Thesis, Universität Göttingen

Schmitt, D. (1987): *Astron. Astrophys.* **174**, 281

Schmitt, D. (1993): "The Solar Dynamo", in *The Cosmic Dynamo*, Krause, F., Rädler, K.-H., Rüdiger, G. (eds.), *IAU-Symp.* **157**, in press

Schmitt, J.H.M.M., Rosner, R., Bohn, H.U. (1984): *Astrophys. J.* **282**, 316

Schou, J., Christensen-Dalsgaard, J., Thompson, M.J. (1992): *Astrophys. J.* **385**, L59

Schüssler, M. (1980): *Nature* **288**, 150

Schüssler, M. (1983): "Stellar Dynamo Theory", in *Solar and Stellar Magnetic Fields: Origins and Coronal Effects*, Stenflo, J.O. (ed.), *IAU-Symp.* **102**, 213

Schüssler, M. (1984): "On the Structure of Magnetic Fields in the Solar Convection Zone", in *The Hydromagnetics of the Sun* Guyenne, T.D., Hunt, J.J. (eds.), ESA SP-220, 67

Schüssler, M. (1993): "Flux Tubes and Dynamos", in *The Cosmic Dynamo*, Krause, F., Rädler, K.-H., Rüdiger, G. (eds.), *IAU-Symp.* **157**, in press

Schwabe, H. (1844): *Astron. Nachr.* **21**, 233

Skaley, D., Stix, M. (1991): *Astron. Astrophys.* **241**, 227

Spiegel, E.A., Weiss, N.O. (1980): *Nature* **287**, 616

Steenbeck, M., Krause, F. (1969): *Astron. Nachr.* **291**, 49

Steenbeck, M., Krause, F., Rädler, K.-H. (1966): *Z. Naturforsch.* **21a**, 369

Stenflo, J.O. (1988): *Astrophys. Space Sci.* **144**, 321

Stenflo, J.O. (1989): *Astron. Astrophys. Rev.* **1**, 3

Stix, M. (1976): *Astron. Astrophys.* **47**, 243

Stix, M. (1987): "On the Origin of Stellar Magnetism", in *Solar and Stellar Physics*, Schröter, E.-H., Schüssler, M. (eds.), *Lecture Notes in Physics* **292**, 15

Stix, M. (1989): *The Sun, An Introduction*, Springer, Berlin

Stix, M. (1991): *Geophys. Astrophys. Fluid Dyn.* **62**, 211

Vainshtein, S.I., Cattaneo, F. (1992): *Astrophys. J.* **393**, 165

van Ballegooijen, A.A. (1982): *Astron. Astrophys.* **113**, 99

Vizoso, G., Ballester, J.L. (1990): *Astron. Astrophys.* **229**, 540

Wälder, M., Deinzer, W., Stix, M. (1980): *J. Fluid Mech.* **96**, 207

Weisshaar, E. (1982): *Geophys. Astrophys. Fluid Dyn.* **21**, 285

Whight, O.R., Trotter, D.E. (1977): *Astrophys. J. Suppl.* **33**, 391

Yoshimura, H. (1972): *Astrophys. J.* **178**, 863

Yoshimura, H. (1975): *Astrophys. J.* **201**, 740

Hydrodynamical Simulations of the Solar Dynamo

Axel Brandenburg

HAO/NCAR*, P.O. Box 3000, Boulder, CO 80307-3000, USA

Abstract: Hydrodynamic simulations of the solar convection zone can be used to model the generation of differential rotation and magnetic fields, and to determine mean-field transport coefficients that are needed in mean-field models. The importance of the overshoot layer beneath the solar convection zone is discussed: it is the place where the magnetic field accumulates, although most of the field regeneration can still occur in the convection zone proper. We also discuss how systematically oriented bipolar regions can emerge from the convection zone where the magnetic field is highly intermittent.

1 Introduction

The engine driving solar and stellar activity is the dynamo. In theories of the solar corona and solar wind the dynamo magnetic fields are an important input quantity. In order to compute the loss of angular momentum during the evolution of the Sun we need to know the magnetic field strength as a function of the angular velocity. Properties of differential rotation and magnetic field geometry are bound to change during this complicated evolutionary process which can only be understood using detailed and realistic dynamo models. Thus, a better understanding of the solar dynamo is essential.

At present it is not feasible to compute realistically in a direct simulation the evolution of magnetic fields and fluid turbulence, because of the large range of different time and length scales that are important. Therefore, one expects the mean-field approach to be well-suited to address certain questions of solar and stellar magnetism (e.g. Schmitt 1993). In this theory, nondiffusive contributions to the turbulent electromotive force and the Reynolds stress tensor are described by α- and Λ-effects, respectively. These effects are responsible for generating large scale magnetic fields and driving differential rotation (Krause & Rädler 1980, Rüdiger 1989). Progress has recently been made to derive the Rossby number dependence of α- and Λ-effects, as well as the turbulent magnetic diffusivity, the eddy viscosity, and the eddy conductivity (e.g. Rüdiger & Kitchatinov 1993, Küker et al. 1993).

---

* The National Center for Atmospheric Research is sponsored by the National Science Foundation

In the mean-field approach solutions are only found for long time and length scales. In spite of such simplifications, this approach is actually rather complicated compared to direct three-dimensional simulations, because there are so many different turbulent transport coefficients, and because they are nonlinear in the magnetic field strength and the angular velocity. However, since our knowledge of these dependencies is based on uncontrolled approximations, it is essential to confirm such results using direct simulations. The problem here is that the conditions under which the mean-field approach applies usually do not overlap with those accessible to direct simulations. This includes in particular the requirement of scale separation which is not satisfied in our simulations (and not even in the Sun!).

We first discuss some key issues of dynamo simulations, such as magnetic buoyancy and the formation of large scale fields and bipolar regions. Such simulations are used to evaluate α and its dependence on various parameters. Some recent progress in constructing solar mean-field dynamos is reported and finally the question of the seat of the dynamo is discussed.

2 Numerical Simulations

Using a numerical simulation of idealised turbulent compressible convection in a small box at reasonably high magnetic Reynolds number it has been possible to study properties of the dynamo process, the formation of magnetic flux tubes and the magnetic buoyancy associated with such flux tubes; see Nordlund et al. (1992) and Brandenburg et al. (1993a). In these simulations the magnetic field grows on a dynamical (turnover) time scale until saturation sets in and a statistically steady state is reached approximately. By splitting the Lorentz force $\boldsymbol{J} \times \boldsymbol{B}$ in the simulation into its various components, it has been demonstrated that both the magnetic pressure gradient force (magnetic buoyancy) and the tension force (component of $\boldsymbol{B} \cdot \nabla \boldsymbol{B}$ in the direction of $\boldsymbol{B}$) are unimportant for saturation (Nordlund et al. 1992). Thus, saturation is accomplished mainly by the curvature force that prevents the flux tubes from bending beyond a certain point.

There is a strong tendency for the magnetic field to be sucked by the concentrated convective downdrafts and subsequently advected downwards to the bottom of the convection zone (Brandenburg et al. 1991a). This raises the question whether it makes sense to consider magnetic flux tubes as passive objects subjected to the influence of magnetic buoyancy. On the other hand, the magnetic field in these simulations is not yet strong enough to produce highly buoyant flux tubes. Nevertheless, in the simulations mentioned above, the maximum magnetic field strength in the overshoot layer can be 2 to 8 times larger than the local value of the equipartition field strength, $B_{\mathrm{eq}} = u_t(\mu_0 \rho)^{1/2}$, where μ_0 is the permeability, ρ the density, and u_t the rms-velocity of the turbulent motions.

What needs to be changed in the simulations to make the flux tubes more intense? First of all, magnetic flux tubes are rather small objects (at least in the simulations) and they eventually disappear due to dissipation. In the Sun, dissipation is much smaller than in simulations with a finite number of mesh points. Thus, flux tubes would live longer and there would then be more time for them to gain maximal field strength. Secondly, in the simulations the Mach number at the bottom of the convection zone is still unrealistically large and the magnetic field energy is only 3-10% of the kinetic energy density, i.e.

$$v_A \ll u_t \ll c \quad \text{(in simulations)}, \tag{1}$$

where $v_A = \langle B^2/\mu_0\rho \rangle^{1/2}$ is the Alfvén velocity, and c the speed of sound. At the bottom of the solar convection zone, the situation is probably more like

$$u_t \lesssim v_A \ll c \quad \text{(solar overshoot layer)}. \tag{2}$$

Larger sound velocities and smaller Mach numbers (Ma) are automatically obtained by choosing the Rayleigh number (Ra) large enough. This may be demonstrated by solving the standard mixing length equations for the same setup used in numerical simulations of convection in an unstable layer with a stable layer beneath (Table 1). Here we define $\mathrm{Ra}\,\mathrm{Pr} = (\mathrm{gd}^3/\overline{\chi}^2)(1 - \nabla_{\mathrm{ad}}/\nabla_{\mathrm{rad}})$, where g is gravity ($5\,10^4$ cm/s^2), d the thickness of the unstable layer in the model (100 Mm), and $\overline{\chi}$ the mean radiative diffusion coefficient, which is varied in order to vary Ra. $\mathrm{Pr} = \nu/\overline{\chi}$ is the Prandtl number.

Table 1. Turbulent velocity and Mach number at the bottom of the unstable layer for mixing length models with different Rayleigh number, but solar values for temperature and density. For $\mathrm{Ra}\,\mathrm{Pr} = 10^{24}$, the solar luminosity $L = L_{\odot}$ is reached

Ra Pr	u_t	Ma	$L/L_{\odot}$
10^6	20 km/s	10^{-1}	10^9
10^{12}	2 km/s	10^{-2}	10^6
10^{24}	20 m/s	10^{-4}	1

The turbulent velocity u_t is a slowly *decreasing* function of Ra. Realistic models of the lower part of the solar convection zone can only be obtained when $\mathrm{Ra}\,\mathrm{Pr}$ is of the order 10^{24} (Ra of the order 10^{30}). If Ra was too small, $\overline{\chi}$ would be too large, and the radiative flux that enters from below would be too large for a realistic temperature gradient. Consequently, in order to model realistically the deep solar convection zone, subgrid-scale diffusivities have to be employed to stabilise the scheme whilst reducing $\overline{\chi}$ to realistically small values.

3 Magnetic Field Loops and Bipolar Regions

Even though in our simulations the magnetic field is unrealistically weak, there is a clear tendency for the formation of bipolar regions. An example of such an event is shown in Fig. 1, where we plot magnetic field vectors for a simulation with an imposed horizontal magnetic field with $B_0 = 0.06B_{\mathrm{eq}}$, in the presence of rotation (the inverse Rossby number is $\mathrm{Ro}^{-1} \equiv 2\Omega d/u_t = 1.6$), the magnetic Reynolds number $R_m \equiv u_t d/\eta = 120$, and the resolution $63 \times 63 \times 37$ mesh points (Brandenburg et al. 1993b).

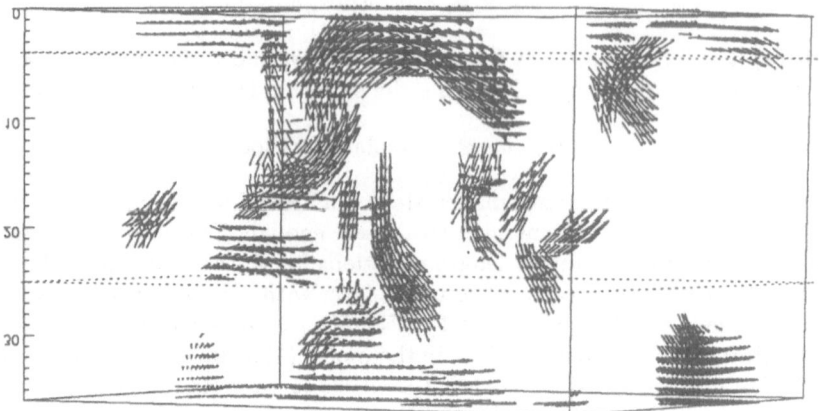

Fig. 1. Magnetic field vectors in a simulation with an imposed magnetic field B_0 in the x-direction (from the left to the right on the paper). Note the formation of a big loop producing a bipolar region at the upper surface. The boundaries of the unstable region are marked by dotted lines

It turns out that bipolar regions are often aligned with the direction of the mean magnetic field, even though it is rather weak compared with the field in typical flux tubes. The large scale magnetic field is usually strongest in the lower overshoot layer. Due to persistent pumping of magnetic field into this layer, the magnetic field is able to accumulate here (Brandenburg et al. 1991a, Rüdiger & Kitchatinov 1992, Petrovay & Szakály 1993). In the present simulations the field is still rather irregular in the overshoot layer, but it is conceivable that under more realistic circumstances the turbulent time scales are much longer in this layer, thus giving enough time for individual flux tubes to line up with the large scale field.

4 Development of Large Scale Fields

Following the pioneering work of Frisch et al. (1975) and Pouquet et al. (1976), a large scale magnetic field is generated in the presence of either magnetic or kinetic helicity by an inverse cascade of the magnetic helicity which, in turn, gives rise to an inverse cascade of the magnetic energy. This leads to a build-up of the magnetic field at large scales. This effect is also seen in simulations of a convective dynamo action. In Fig. 2 we plot the magnetic energy spectrum $M(k)$ for two different times (approximately 10 turnover times apart) for a run with $R_m = 1000$, $\mathrm{Ro}^{-1} = 1$, at a resolution of 63^3 mesh points (Nordlund et al. 1992).

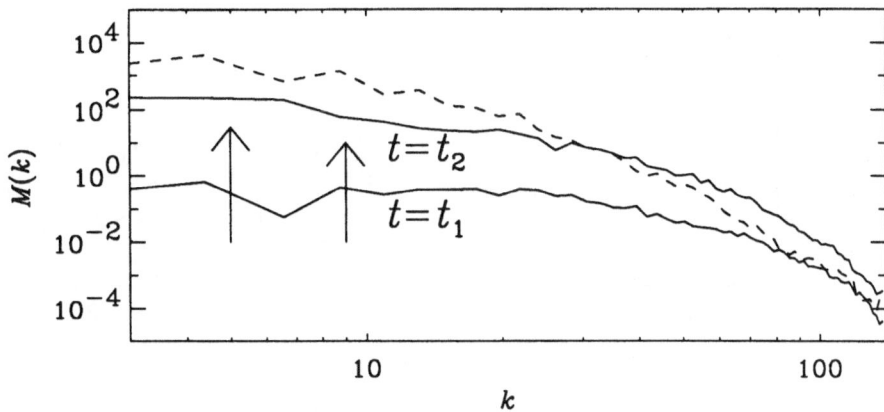

Fig. 2. Magnetic energy spectra during the growth phase ($t = t_1$) and the saturated phase ($t = t_2$) of a dynamo (lower and upper solid curve). For comparison, the kinetic energy spectrum at $t = t_2$ is also shown (dotted line)

The α-effect formalism of Steenbeck, Krause & Rädler (1966) may be considered as a linearised version of the fully nonlinear approach mentioned above. In Fig. 3 we show that a simple, one-dimensional, α^2-dynamo gives rise to an inverse cascade, similar to the inverse cascade in MHD turbulence. This is illustrated in the second panel of Fig. 3, where we show a sequence of magnetic energy spectra from a simple cascade model of convective MHD turbulence (Brandenburg 1992). The main difference is that in the α-effect dynamo the growth of magnetic energy at small scales is not described.

The difference in the form of the magnetic energy spectra during the growth phase of the dynamo and during the saturated phase is an important property. The development of small scale structures is an inherently kinematic process: as time goes on, larger and larger structures develop. This is also seen by comparing snapshots of simulations: during the growth phase of the dynamo the flux tubes are thinner than at later times when the flux tubes become more clearly defined (Brandenburg et al. 1991b).

5 Intermittency and Small Scale Fields

It has been argued that magnetic fluctuations that are very strong compared to the large scale magnetic field can lead to a severe quenching of the α-effect and the turbulent magnetic diffusivity (Vainshtein & Cattaneo 1992). The fluctuating magnetic field is considered strong enough when the ratio $q \equiv \langle B^2 \rangle / \langle B \rangle^2$ is of the order of R_m. This would be the case if the magnetic energy spectrum had an inertial

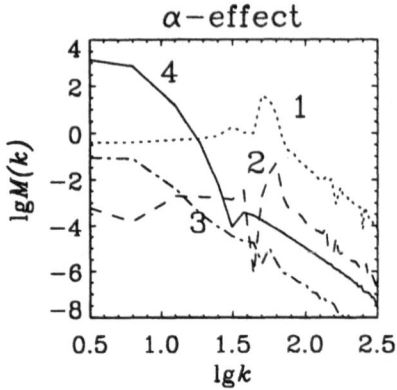
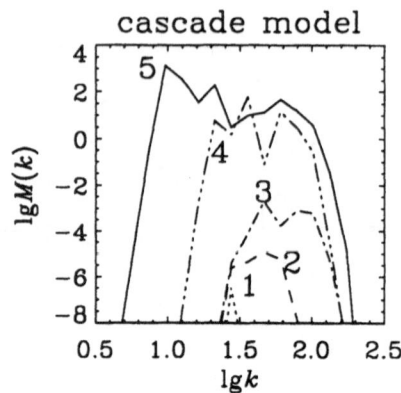

Fig. 3. Magnetic energy spectra of a one-dimensional α-effect dynamo (left panel) and a simple cascade model of MHD turbulence (right panel). The numbers on the curves indicate increasing time

range that increases like $k^{+1/3}$ with wave number k (Moffatt 1961). By contrast, if the magnetic energy had for example a k^{-1} spectrum, q would asymptotically only increase like $\ln R_m$ (Zeldovich et al. 1983, Kleeorin et al. 1990). Furthermore, as the mean magnetic field strength increases, the quantity q is quenched and becomes of order unity as $|\langle B \rangle| \to B_{\mathrm{eq}}$ (Kleeorin et al. 1990, Brandenburg et al. 1993b). In other words, it is possible that the magnetic fluctuations are of comparable order of magnitude to the mean magnetic field.

It is important to note that the quantity $\langle B^2 \rangle$ can be significantly underestimated if the magnetic field continues to be intermittent and nonsmooth down to the smallest scales resolved. This is a particular problem when observational data are analysed. It is sometimes possible to extrapolate to the limit of perfect resolution by measuring the moments of the average magnetic field at different resolution,

$$\mathcal{B}_n(r) = \langle |\langle \boldsymbol{B} \rangle_r|^n \rangle. \tag{3}$$

Here, $\langle ... \rangle_r$ denotes an average over a box of scale r, and $\langle ... \rangle$ is an average over the entire computational volume. The function $\mathcal{B}_2(r)$ is closely related to the magnetic energy spectrum. As an example we show in Fig. 4 the scaling of $\mathcal{B}_n(r)$ for data from a numerical simulation with $\mathrm{Ra} = 10^6$, $\mathrm{Pr} \equiv \nu/\overline{\chi} = 0.2$, and $\mathrm{Pr}_M \equiv \nu/\eta = 4$ (Nordlund et al. 1992).

Note that in the lin-log plot (Fig. 4) the curves are almost straight lines at small scales of r, which is related to an exponential power spectrum of the magnetic energy in the dissipation range. This suggests that it is possible to extrapolate to $r = 0$. For example, at the smallest resolved scale, $r_0 = 1\,\mathrm{mesh}$ size, $\mathcal{B}_2(r_0)$ is $4.7\,10^{-4}$, but the extrapolated value $\mathcal{B}_2(0)$ increases to about $6.4\,10^{-4}$. In the present case we have $q \approx \mathcal{B}_2(0)/\mathcal{B}_2(32 r_0) = 350$. Since the average magnetic field over the entire box vanishes due to boundary and initial conditions, this ratio depends strongly on

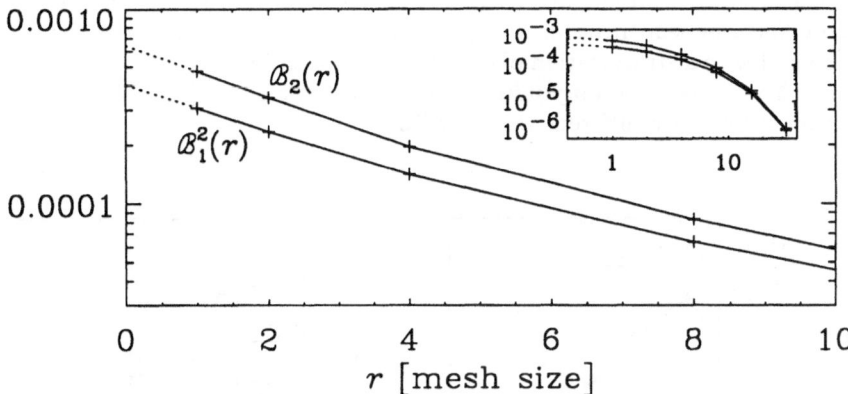

Fig. 4. $\mathcal{B}_2(r)$ and $\mathcal{B}_1^2(r)$ for data of a numerical simulation with dynamo effect. (In the plot, the square of $\mathcal{B}_1(r)$ is shown to allow comparison with $\mathcal{B}_2(r)$.) Ra = 10^6, Pr = 0.2, and $\mathrm{Pr}_M = 4$. The inset shows a log-log plot of the same data

the scale at which the denominator is evaluated. For example at half the scale this ratio is much smaller: $\mathcal{B}_2(0)/\mathcal{B}_2(16r_0) = 32$.

The slope $\kappa = d\ln\mathcal{B}_1/d\ln r$, evaluated at the skin layer scale $r \sim LR_m^{-1/2}$, is the cancellation exponent (Ott et al. (1992, Bertozzi et al. 1993). In the present case, where there is a small scale dynamo, this exponent is around 0.2. A nonvanishing exponent indicates that the field is still "rough" down to the smallest scale resolved. There is however no power law behaviour, but instead $\mathcal{B}_n(r)$ is proportional to $\exp(-r/r_d)$ with $r_d \approx 3r_0$ (for $n = 2$).

6 The α-Effect Evaluated from Simulations

The α-effect can in principle be evaluated from numerical simulations of convection (e.g. Brandenburg et al. 1990). Using such simulations one can compute α as a function of $|\langle B \rangle|$. In the simulations of Brandenburg et al. (1993b), α seems to be surprisingly insensitive to $|\langle B \rangle|$. Unfortunately it is not (yet?) possible to attain sufficiently large values of R_m, and it therefore remains open whether α is a sensitive function of the magnetic Reynolds number R_m. For example, $\alpha \sim R_m^{-1/2}$ has been obtained by Childress (1979) and Perkins & Zweibel (1989) for a model with a steady flow, and Vainshtein & Cattaneo (1992) argue that the onset of α-quenching occurs for progressively weaker fields as R_m increases. However, their argument requires $q = \mathcal{O}(R_m)$, implying a magnetic energy spectrum that increases towards larger k. This issue is still controversial, but it should be noted that there is evidence that in the inertial range the magnetic spectrum does indeed decrease like k^{-1} (see discussion in Brandenburg et al. 1993b).

Another important application of such simulations is to determine the latitudinal dependence of α. Current theories (e.g. Rüdiger & Kitchatinov 1993) are restricted to only linear dependencies of α on $\hat{\boldsymbol{g}} \cdot \hat{\boldsymbol{\Omega}} \propto \cos \theta$, where $\hat{\boldsymbol{g}}$ symbolises the preferred direction due to a stratification of density or turbulent intensity, and θ is colatitude. In Fig. 5 we show the longitudinal (ϕ-) component of α, $\langle \boldsymbol{u}' \times \boldsymbol{B}' \rangle_\phi / \langle B_\phi \rangle$, as a function of θ for two different values of Ro^{-1}.

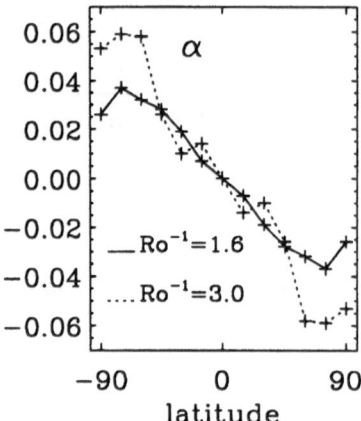 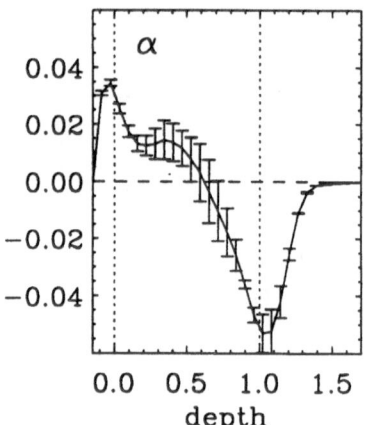

Fig. 5. *Left panel:* The latitudinal dependence of α for a simulation with $\mathrm{Ra} = 3\,10^5$ ($\mathrm{Ro}^{-1} = 1.6$) and $\mathrm{Ra} = 10^5$ ($\mathrm{Ro}^{-1} = 3.0$). In all cases, $\mathrm{Ta} = 3\,10^4$, $\mathrm{Pr} = 0.5$, and $\mathrm{Pr}_M = 0.5$. *Right panel:* The depth dependence of α for the same parameters as before, but with $\mathrm{Ro}^{-1} = 3.0$ and $\theta = 90°$. The dotted vertical lines denote the boundaries of the convection zone

Note that α deviates from a simple $\cos \theta$ law. Indeed, there is no reason to expect such a simple dependence to be valid for strong stratification and rapid rotation, in which case higher powers in $\hat{\boldsymbol{g}} \cdot \hat{\boldsymbol{\Omega}}$ must occur in the α expressions. Rüdiger & Brandenburg (1993) used the empirical formula

$$\alpha(\theta) = \alpha_0 \cos \theta (1 - \alpha_U \cos^2 \theta). \qquad (4)$$

with $\alpha_U \approx 1$ to model the solar dynamo in the overshoot layer. A similar latitudinal dependence of α has previously been suggested in order to explain a concentration of the sunspot activity maximum at low latitudes (Yoshimura 1975, Belvedere et al. 1991). Schmitt (1987) finds a similar dependence of an α-effect that is based on magnetostrophic waves.

There is another effect that can be seen in simulations in a spherical shell: inside the cylinder tangent to the inner radius of the shell the kinetic helicity is virtually zero for rapid rotation (e.g. Rieutord et al. 1993). It is not obvious whether this effect operates in the Sun, or whether it is an artifact of the simulations not being sufficiently turbulent.

In contrast to earlier estimates of α from convection simulations with weak stratification (Brandenburg et al. 1990), α now tends to be reduced in the bulk of the convection zone and concentrated towards the boundaries of the convection zone (Fig. 5). In any case, α is rather small and only a few percent of u_t. Whether or not mean-field dynamo action is possible with such a weak and localised α-effect depends crucially on the value of the turbulent magnetic diffusion, η_t. Theory predicts a decrease of η_t for large inverse Rossby numbers (Kitchatinov & Rüdiger 1993), and simulations also give values of η_t that are smaller than standard estimates.

The effect of η_t-quenching by the magnetic field has long been recognised (Roberts & Soward 1975). A severe suppression of η_t has been found in a special two-dimensional case (Cattaneo & Vainshtein 1991), but this does not seem to carry over into the three-dimensional regime (Nordlund et al. 1993). A suppression of η_t can be inferred from simulations by monitoring the decay rate of the large scale magnetic field component. For the convection simulations of Brandenburg et al. (1993b) we estimate $\eta_t \approx 0.2\eta_0$ for strong fields ($B \approx 0.2B_{eq}$), and $\eta_t \approx 0.4\eta_0$ for weak fields ($B \approx 0.006B_{eq}$), where $\eta_0 = \frac{1}{3}u_t d$ is a rough estimate for reference.

7 Mean-Field Dynamos

Using an α-effect of the form (4) with $\alpha_U = 1$, Rüdiger & Brandenburg (1993) computed mean-field dynamos for the overshoot layer beneath the solar convection zone, taking the full Rossby number dependence, the full α-tensor and the turbulent diffusivity into account (Rüdiger & Kitchatinov 1993). We already know from the Krause formula for α (Krause 1967) that α becomes negative at the bottom of the convection zone, because of the sharp gradient in the turbulent velocity. Beneath the interface of the convection zone and the radiative interior, the turbulent magnetic diffusivity gradually goes to zero. The magnetic field tends to accumulate in this interface; see Fig. 6.

Magnetic buoyancy acts mostly in the upper part of the convection zone, but it turns out that this effect can drastically increase the cycle period. Furthermore, due to the intermittent nature of the magnetic field, the effective electromotive force is reduced by a factor ϵ. For $\epsilon = 0.2 - 0.5$ the correct cycle period can be obtained.

All dynamo models with solar-like differential rotation ($\partial\Omega/\partial r > 0$ in the equatorial plane) have the common problem that at low latitudes poloidal and toroidal fields are in phase, which is in contrast to the observations. However, the indicators of these two field components probe different depths in the convection zone, and it is therefore plausible that the observed phase relation strongly depends on the depths where poloidal and toroidal fields are measured (B_r somewhere close to the surface and B_ϕ at the bottom of the convection zone). At intermediate and high latitudes, the poloidal and toroidal fields are in phase – in agreement with the solar field.

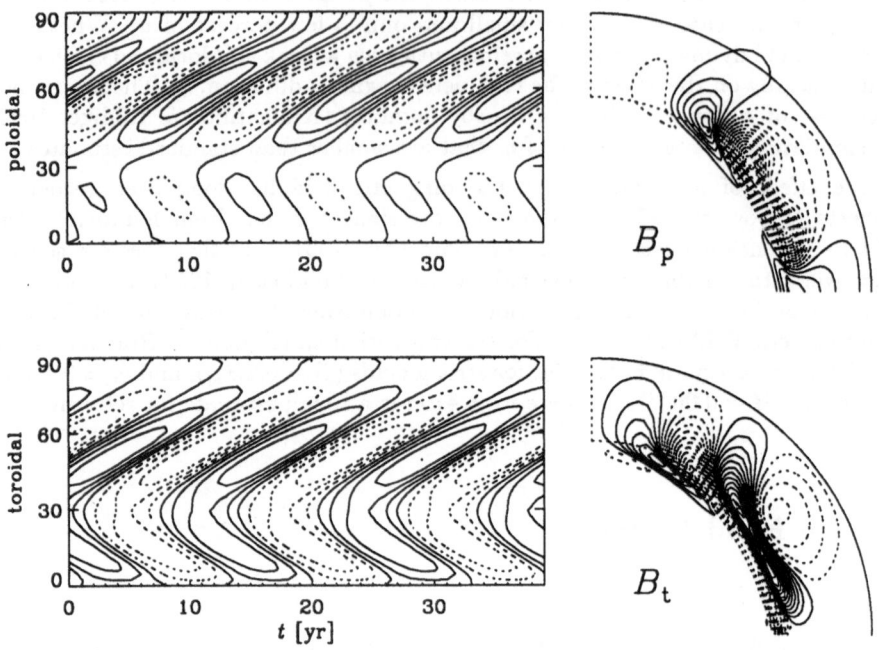

Fig. 6. Butterfly diagram for the B_r and the B_ϕ components of the mean magnetic field from a $\alpha\Omega$-dynamo model for the Sun together with meridional cross-sections showing poloidal field lines and contours of the toroidal field

8 Discussion

Numerical simulations of MHD convection can help to improve significantly our understanding of solar and stellar magnetism. Such simulations suggest the possibility of dynamo action in the entire convection zone with field advection down to the bottom. Thus, whilst the magnetic field turns out to be strongest at this interface, the actual generation of the magnetic field in this layer is perhaps relatively unimportant.

This picture seems to be in contrast to the other possibility that most of the dynamo generation happens in the overshoot layer itself. The general problem with this approach is that in the overshoot layer the kinetic energy of the fluid motions is probably relatively weak and of the order of, or less than, the magnetic energy in that layer. Indeed, convective dynamo models presented so far typically generate magnetic fields whose strength does not significantly exceed the kinetic energy density of the turbulent motions that are generating this field. It seems therefore more natural to generate magnetic field in the convection zone where the kinetic energy of the turbulent motions is large. Turbulent pumping and suction of the magnetic field by the intense downdrafts leads to an accumulation of the magnetic field at the interface, where the magnetic energy may then easily exceed the kinetic energy of the motions.

A systematic large scale magnetic field is expected to occur in deeper regions where the motions are slow enough. The dynamo process seen in numerical simulations generates a small scale magnetic field that consists of a number of intense flux tubes. Future mean-field models of the solar magnetic field should therefore incorporate such small scale fields which, in principle, may play an active role in the formation of large scale fields via inverse cascade type mechanisms. Future simulations, on the other hand, should be carried out in larger boxes that include effects of the spherical geometry of the Sun.

References

Belvedere, G., Proctor, M. R. E., Lanzafame, G. (1991): "The latitude belts of solar activity as a consequence of a boundary-layer dynamo", *Nature* **350**, 481–483

Bertozzi, A. L., Chhabra, A. B., Kadanoff, L. P., Ott, E., Vainshtein, S. I. (1993): On dynamo generation of magnetic flux and fractal properties of the field. (preprint).

Brandenburg, A. (1992): "Energy spectra in a model for convective turbulence", *Phys. Rev. Lett.* **69**, 605–608

Brandenburg, A., Nordlund, Å., Pulkkinen, P., Stein, R.F., Tuominen, I. (1990): "3-D Simulation of turbulent cyclonic magneto-convection", *Astron. Astrophys.* **232**, 277–291

Brandenburg, A., Jennings, R. L., Nordlund, Å., Stein, R. F., Tuominen, I. (1991a): The role of overshoot in solar activity: A direct simulation of the dynamo in The Sun and cool stars: activity, magnetism, dynamos, ed. by I. Tuominen, D. Moss & G. Rüdiger (Lecture Notes in Physics 380, Springer-Verlag) pp.86-88

Brandenburg, A., Jennings, R. L., Nordlund, Å., Stein, R.F. (1991b): Magnetic flux tubes as coherent structures in Spontaneous formation of space-time structures and criticality, ed. by T. Riste & D. Sherrington (Nato ASI Series) pp.371-374

Brandenburg, A., Jennings, R. L., Nordlund, Å., Rieutord, M., Stein, R. F., Tuominen, I. 1993a Magnetic structures in a dynamo simulation. *J. Fluid Mech.* (submitted).

Brandenburg, A., Krause, F., Nordlund, Å., Ruzmaikin, A. A., Stein, R.F., Tuominen, I. 1993b On the magnetic fluctuations produced by a large scale magnetic field. *Astrophys. J.* (submitted).

Cattaneo, F., Vainshtein, S. I. (1991): "Suppression of turbulent transport by a weak magnetic field", *Astrophys. J.* **376**, L21–L24

Childress, S. (1979): "Alpha-effect in flux ropes and sheets", *Phys. Earth Planet. Int.* **20**, 172-180

Frisch, U., Pouquet, A., Léorat, J., Mazure, A. (1975): "Possibility of an inverse cascade of magnetic helicity in hydrodynamic turbulence", *J. Fluid Mech.* **68**, 769–778

Kitchatinov, L. L., Rüdiger, G. (1992): "Magnetic-field advection in inhomogeneous turbulence", *Astron. Astrophys.* **260**, 494–498

Kitchatinov, L. L. & Rüdiger, G. (1993): (in preparation)

Krause, F. (1967): *Eine Lösung des Dynamoproblems auf der Grundlage einer linearen Theorie der magnetohydrodynamischen Turbulenz*, (Habilitationsschrift, University of Jena)

Krause, F., Rädler, K.-H. (1980): *Mean-Field Magnetohydrodynamics and Dynamo Theory*, (Akademie-Verlag, Berlin)

Kleeorin, N. I., Rogachevskii, I. V., Ruzmaikin, A. A. (1990): "Magnetic force reversal and instability in a plasma with advanced magnetohydrodynamic turbulence", *Sov. Phys. JETP* **70**, 878–883

Küker, M., Rüdiger, G., Kitchatinov, L. L. 1993 An $\alpha\Omega$-model of the solar differential rotation. *Astron. Astrophys.* (in press).

Moffatt, H. K. (1961): "The amplification of a weak magnetic applied magnetic field by turbulence in fluids of moderate conductivity", *J. Fluid Mech.* **11**, 625–635

Nordlund, Å., Brandenburg, A., Jennings, R. L., Rieutord, M., Ruokolainen, J., Stein, R. F., Tuominen, I. (1992): "Dynamo action in stratified convection with overshoot", *Astrophys. J.* **392**, 647–652

Nordlund, Å. et al. (1993): (in preparation)

Ott, E., Du, Y., Sreenivasan, K. R., Juneja, A. & Suri, A. K. (1992): "Sign-singular measures: fast magnetic dynamos, and high-Reynolds-number fluid turbulence", *Phys. Rev. Lett.* **69**, 2654–2657

Perkins, F. W., Zweibel, E. G. (1987): "A high magnetic Reynolds number dynamo", *Phys. Fluids* **30**, 1079–1084

Petrovay, K., Szakály, G. (1993): "The origin of intranetwork fields: a small-scale solar dynamo", *Astron. Astrophys.* **274**, 543–554

Pouquet, A., Frisch, U., Léorat, J. (1976): "Strong MHD helical turbulence and the non-linear dynamo effect", *J. Fluid Mech.* **77**, 321–354

Rieutord, M. et al. 1993 Reynolds stress and differential rotation in Boussinesq convection in a rotating spherical shell. *Astron. Astrophys.* (to be submitted).

Roberts, P. H., Soward, A. M. (1975): "A unified approach to mean field electrodynamics", *Astron. Nachr.* **296**, 49–64

Rüdiger, G., Kitchatinov, L. L. (1993): "Alpha-effect and alpha-quenching", *Astron. Astrophys.* **269**, 581–588

Rüdiger, G., Brandenburg, A. 1993 A solar dynamo in the overshoot layer: cycle period and butterfly diagram. *Astrophys. J.* (submitted).

Schmitt, D. (1987): "An $\alpha\omega$-dynamo with an α-effect due to magnetostrophic waves", *Astron. Astrophys.* **174**, 281–287

Schmitt, D. (1993): (this volume)

Steenbeck, M., Krause, F., Rädler, K.-H. (1966): "Berechnung der mittleren Lorentz-Feldstärke $\overline{v \times B}$ für ein elektrisch leitendendes Medium in turbulenter, durch Coriolis-Kräfte beeinflußter Bewegung", *Z. Naturforsch.* **21a**, 369-376

Vainshtein, S. I., Cattaneo, F. (1992): "Nonlinear restrictions on dynamo action", *Astrophys. J.* **393**, 165–171

Yoshimura, H. (1975): "A model of the solar cycle driven by the dynamo action of the global convection in the solar convection zone", *Astrophys. J. Suppl.* **29**, 467–494

Zeldovich, Ya. B., Ruzmaikin, A. A., Sokoloff, D. D. (1983): *Magnetic fields in Astrophysics*, (Gordon & Breach, New York)

The Asymmetric Behaviour of Solar Activity

M. Carbonell, R. Oliver, J.L. Ballester

Departament de Fisica, Universitat de les Illes Balears, E - 07071 Palma de
Mallorca, Spain

Abstract: We present a thorough study of the North-South asymmetry of solar activity
made with the daily sunspot areas since they are good indicators of magnetic activity.
To perform the study, we have constructed an asymmetry time series and analyzed it by
different methods. First of all, its significance and statistical behaviour have been studied.
Secondly, using a Monte Carlo simulation we have generated synthetic asymmetries and
compared them to the real one and, thirdly, we have used modern techniques, developed
to study the chaotic behaviour of time series, to search for low-dimensional chaos in the
asymmetry time series. Our results show that in most cases the north-south asymmetry is
statistically highly significant and that it cannot be obtained from a distribution of sunspots
areas generated, in a random way, from a binomial or uniform distribution of probability
between hemispheres. Moreover, the real asymmetry time series can be represented by a
means of a multi-component model made up of a long term trend, a sinusoidal component
with a period of 12.1 yr and a dominant purely random component; while chaotic analysis
does not reveal the existence of a strange attractor in the time series.

1 Introduction

The existence of asymmetries in the spatial distribution of solar activity has been
claimed for a long time. This claim is mainly based on various studies made with
partial records of sunspots and solar flares, since these are the most important
manifestations of solar activity, the north-south asymmetry being the most studied
one.

One of the first studies of the N-S asymmetry of sunspots was made by Newton
and Milson (1955) who analysed the yearly values of sunspot areas from 1874 to
1954, comprising seven sunspot cycles. They concluded that, on the one hand, the
fluctuations in the values of the N-S asymmetry of sunspot areas are real and, on
the other hand, that the relative behaviour of the two hemispheres is maintained
during successive cycles without any indication of an alternation related to the 22
year cycle of magnetic polarities of sunspots. Waldmeier (1957, 1971) investigated
the N-S asymmetry of sunspot areas between 1874 and 1954 and the N-S asymmetry of sunspot number between 1955-1969, arguing that the real asymmetry is
strengthened by a phase difference between hemispheres. Roy (1977) studied the

N-S asymmetry of sunspot magnetic classes and areas of large sunspots from 1955-1974, and found that in spots with complex magnetic configurations the asymmetry is more pronounced, that the asymmetry in favour of the northern hemisphere is less pronounced for large sunspots than for major flare events, and that the degree of the asymmetry does not depend on the area of sunspot groups. Swinson et al. (1986), using combined data of sunspot areas and sunspot number from White and Trotter (1977) and Koyama (1985), showed that, in general, the activity in the northern hemisphere peaks about two years after the sunspot minimum and that the peak is greater during even cycles. Hence, they suggested a 22-year periodicity in the N-S asymmetry of solar activity connected with the solar magnetic cycle. Related to sunspots, an interesting study was performed by Howard (1974) who analyzed magnetic flux data over the interval 1967-1973, and found that the total flux in the northern hemisphere was greater than in the southern hemisphere by about 7% over this period, which suggests a higher level of activity in the northern hemisphere. Also, Mouradian and Soru-Escaut (1991) have analysed the solar activity cycle, from the point of view of large-scale magnetic fields, using the synoptic charts for cycles 20, 21 and 22 and they found that the activity cycle is asymmetrical, such that the north polar background predominates at the minimum and the south polar background at the maximum.

Thus, although the existence of a north-south asymmetry in solar activity is generally accepted, most of its characteristic features, such as its secular trend and the fact that it peaks at the sunspot minimum, have been noticed by means of qualitative studies based on limited data sets of sunspots or solar flares. Taking this into account, our main aim has been to apply classical and modern quantitative techniques to the best available record of solar activity, given by daily sunspot areas (1874-1989), in order to characterize clearly the existence and behaviour of the north-south asymmetry of solar activity.

2 The Analysis of the Asymmetry Time Series

The World Data Center of The National Oceanic and Atmospheric Administration provided us with the daily sunspot areas; the data came from the Greenwich Photoheliographic Results (1874-1982) and the United States Air Force (1983-1989). To perform the statistical analysis we computed the asymmetry (A) between hemispheres by means of:

$$A = \frac{N - S}{N + S} \tag{1}$$

where N and S stand for north and south sunspot areas. In the case of daily data we obtain an asymmetry time series composed of 42248 values, while if the data are grouped by Carrington rotation, from rotation 276 till 1839, they generate 1563 values.

The statistical significance of the asymmetry values has been assessed by computing the actual probability of obtaining these results or one having a larger difference due to chance. The results indicate that in most of the cases, the asymmetry in the distribution of daily and rotational sunspot areas between hemispheres is statistically significant, suggesting that it is a real feature of solar activity.

To obtain further information about the behaviour of the time series, we have performed power spectra, for the rotational and daily asymmetry, using the Maximum Entropy Method (MEM) and a significant peak at a period of 163.93 rotations (12.1 years) is found. Also, significant power at low frequencies appears suggesting the existence of an underlying long-term trend in the time series whose presence can be confirmed by means of the Cox-Stuart test. To determine the form of the trend we have performed the first, second, third and fourth differencing of the rotational asymmetry time series and computed the MEM power spectrum of those differenced time series. The power spectrum indicates that with a third or fourth order differencing the power at low frequencies is negligible; also, the chi-square test of goodness of fit indicates that the best fit is a cubic polynomial, but there is a small difference with respect to the first and second order polynomials. Thus, it seems that there is a cubic trend in the asymmetry time series which plays an important role in its behaviour, being responsible for the shift from south dominant asymmetry, at the end of last century and the beginning of the present one, to north dominant asymmetry during the most recent solar cycles.

From the Wold decomposition theorem (see Cox and Miller, 1968) we know that any discrete stationary process can be expressed as the sum of two uncorrelated processes, one purely deterministic (i.e. that the past of the process completely determines its future forever), and one which is stochastic. In our case, the trend and the sinusoidal component, with a period of 163.93 rotations, represent the deterministic process since the sinusoid is a deterministic cycle. This can be seen by computing the power spectrum for different sizes of the sample and by observing that the height of the peak increases with the sample size (Gottman, 1981). Once the trend and the sinusoid have been subtracted from the time series, the remaining signal represents the stochastic process. Inspection of the autocorrelation function for this remaining signal suggests that a tentative model to fit to it is an autoregressive process with an order p to be determined. This fitting has been performed by means of the Burg algorithm (Marple, 1987) which also provides the variance of the residual white noise. The variance of the original time series was 0.34 and it decreases to 0.32 after the subtraction of the trend and the sinusoid. However, the fitting of AR models of different orders does not decrease this variance substantially, and we obtain 0.27 as the variance of the residual white noise, which indicates that most of the variance of the signal can be accounted for by the white noise, i.e., by a purely random process. Therefore, we conclude that the asymmetry time series can be modelled, basically, as a multi-component model consisting of the addition of a trend plus a sinusoidal component, with a period of 163.93 rotations, describing the deterministic process, and a purely random (white noise) component.

On the other hand, Roy (1977) and Yau (1988) suggested that the north-south asymmetry is anticorrelated with solar activity, while Swinson et al. (1986) claimed that some cyclic behaviour can be found in the asymmetry since it peaks two years after the minimum, this peak being more important during even cycles, and that there is probably a 22-year periodicity in it. However, these conclusions are mainly based on a visual comparison of the shapes of the curves of solar activity and asymmetry versus time. In the power spectrum the most prominent and clearly significant peak is at 163.93 rotations (12.1 years), which suggests that the asymmetry has a cyclic behaviour with a period very similar to that of the solar cycle. To establish or refute the reality of this suspected anti-correlation in a quantitative way, we have applied two standard rank statistical tests, namely, the Kendall tau and the Spearman rank-order correlation coefficients. The computations have been performed using the rotational asymmetry time series and the results indicate that it presents a very poor correlation with the sunspot areas time series. However, if we remember the existence of a deterministic cycle inside the asymmetry time series, we can try to compare this cycle with the solar cycle. To this end, we have fitted sinusoids to both time series and, using the same tests as above, we have obtained anticorrelation between the deterministic cycles inside the asymmetry and the solar activity. However, our feeling is that this result is due to the fact that the sine waves have different periods, which produces epochs of correlation and anticorrelation and then, for the interval of time considered, the net result is anticorrelation since they are not completely balanced. In conclusion, statistical tests do not support the existence of an anticorrelation between the asymmetry of solar activity and solar activity itself.

To gain still further knowledge about the behaviour of solar activity we have developed Monte Carlo simulations of the north-south distribution of sunspot areas. The scheme was as follows. Firstly, for each rotation, and assuming a binomial distribution of probability between hemispheres with $p = 0.5$, we have randomly distributed the sunspot areas within it and computed the north-south asymmetry of the synthetic distribution. In order to know if the real asymmetry may be based on a binomial distribution of probability, we have performed a chi-square test of goodness of fit by taking four cells between -1 and $+1$ and distributing among them the observed and expected (synthetic) values of the asymmetry. In our case, we have performed 10000 simulations and in all of them the assumed null hypothesis can be rejected. Secondly, we have assumed a uniform distribution of probability between hemispheres and generated an integer random number between plus and minus the total sunspot area. If this number is positive this area is assigned to the northern hemisphere and the remnant area, up to the total, to the southern hemisphere, and visa-versa. As above we have computed the synthetic asymmetry and performed the chi-square test of goodness of fit using ten cells and the same significance level. In this case, after 10000 simulations, only in 3% of the cases the null hypothesis can be accepted. For the daily asymmetry, the results obtained from the simulations are similar.

The irregular and unpredictable behaviour of a time series can be due to chaos, noise or both, and in recent years techniques have been developed to search for the presence of low-dimensional deterministic chaos in a measured time series. In general, the origin of low-dimensional chaos is associated with the presence of a strange attractor in the space phase of the system. This strange attractor is characterized by its fractal dimension D and the larger the dimension, the more chaotic the dynamics of the system. The fact that the dynamics of a deterministic system with an underlying strange attractor of fractal dimension D can in principle be modelled by $2D + 1$ coupled nonlinear ordinary differential equations, has led to the development of methods for reconstructing the space phase of the system and for calculating the fractal dimension of the strange attractor from a time series (Takens, 1981; Packard et al., 1980). The most widely used method to search for chaotic behaviour in a time series is the correlation integral method of Grassberger and Procaccia (1983a,b). A detailed description of the application of this method to an astrophysical time series can be found in Voges et al. (1987). The convergence of the correlation dimension, at increasing embedding dimensions, is a rigorous result for systems which are dominated by a low-dimensional strange attractor. In our case, there is no convergence of the correlation dimension and this result is in agreement with the dominant stochastic behaviour obtained before.

3 Conclusions

We have studied the north-south asymmetry of sunspot areas, from 1874 till 1989, using classical and modern quantitative techniques, in order to confirm or refute the reality of some features pointed out by different authors from a qualitative point of view. This study has produced the following conclusions:

1) The statistical significance of the asymmetry time series indicates that in most of the cases the asymmetry is highly significant, that is, the asymmetry is a real feature in the north-south distribution of sunspot areas.

2) The statistical analysis and the power spectrum of the asymmetry confirm the existence of an underlying cubic trend which accounts for the slow shift from south dominant asymmetry, at the end of last century and beginning of the present one, to north dominant during most of this century.

3) The statistical analysis also reveals that the asymmetry time series can be modelled by means of a multi-component model made by a trend, a deterministic cycle with a period of 163.93 rotations and a dominant purely random component which accounts for most of the variance of the time series.

4) Statistical rank correlation tests do not confirm the existence of anticorrelation between the asymmetry and the solar cycle. Surely, this anticorrelation is only present during some epochs and is due to the different period of the deterministic cycles inside the asymmetry and the solar cycle.

5) Using Monte Carlo simulations we have generated synthetic asymmetries, based on either the null hypothesis that the real distribution of sunspot areas can be obtained randomly from a binomial distribution, or the uniform distribution of probability between hemispheres. These synthetic asymmetries have been compared to the real one by means of an chi-square test of goodness of fit and the results indicate that, practically, in all cases the assumed null hypothesis can be rejected.

6) With the help of modern techniques developed to search for the presence of low-dimensional deterministic chaos in time series, we have analyzed the asymmetry time series. Our results suggest that this time series has a stochastic behaviour, in agreement with what we had previously found in the statistical analysis.

One might expect that the features of the asymmetry should be explained using theoretical models for the solar dynamo. However, while Weiss (1988) argues that the behaviour of the solar cycle is an example of deterministic chaos, and evidence for the presence of a strange attractor in it has been given by Kurths (1987), Gizzatullina et al. (1990), Morfill et al. (1991), Ostryakov and Usoskin (1991) and Mundt et al. (1991), our study indicates that the behaviour of the asymmetry of solar activity is dominated by a stochastic process. Moreover, it has been shown (Brandenburg et al. 1989) that from some nonlinear dynamo models, with adequate parameters, it is possible to obtain a very regular asymmetry signal which for example peaks at the minimum of solar activity but does not show other features of the real asymmetry.

As a final conclusion, then, we can say that at present neither an empirical reproduction of the solar cycle such as performed by Morfill et al. (1991), nor solar dynamo models are able to account for the observed features and behaviour present in the asymmetry of solar activity.

More extensive information about this research can be found in Carbonell et al. (1993).

References

Brandenburg, A., Krause, F. and Tuominen, I., 1989, in: Turbulence and nonlinear Dynamics in MHD Flows, eds. M. Meneguzzi, A. Pouquet and P.L. Sulem, North-Holland, Amsterdam

Carbonell, M., Oliver, R. and Ballester, J.L., 1993, A&A **274**, 497

Cox, D.R. and Miller, H.D., 1968, The Theory of Stochastic Processes, Chapman and Hall, London

Gizzatullina, S.M., Rukavishnikov, V.D., Ruzmaikin, A.A. and Tavastsherna, K.S., 1990, Solar Phys. **127**, 281

Grassberger, P. and Procaccia, I., 1983a, Phys. Rev. Lett. **50**, 346

Grassberger, P. and Procaccia, I., 1983b, Physica **9D**, 189

Gottman, J.M., 1981, Time-series Analysis, Cambridge University Press, Cambridge

Howard, R., 1974, Solar Phys. **38**, 59

Koyama, H., 1985, Observations of sunspots 1947-1984, Kawadeshoboshinsha, Tokyo

Kurths, J., 1987, Pre-ZIAP, N0 02

Marple, S.L., 1987, Digital Spectral Analysis with Applications, Prentice-Hall, New Jersey

Morfill, G.E., Scheingraber, H., Voges, W. and Sonnett, C.P., 1991, in: The Sun in Time, eds. C.P. Sonnett, M.S. Giampapa and M.S. Mattews, University of Arizona Press, Tucson

Mouradian, Z. and Soru-Escaut, I., 1991, A&A **251**, 649

Mundt, M.D., Maguire II, W.B. and Chase, R.R.P., 1991, J. Geophys. Res. **96**, 1705

Newton, H.W. and Milson, A.S., 1955, M.N.R.A.S. **115**, 398

Ostryakov, V.M. and Usoskin, I.G., 1990, Solar Phys. **127**, 405

Packard, N.H., Crutchfield, Farmer, J.D. and Shaw, R.S., 1980, Phys. Rev. Lett. **45**, 712

Roy, J.R., 1977, Solar Phys. **52**, 53

Swinson, D.B., Koyama, H. and Saito, T., 1986, Solar Phys. **106**, 35

Takens, F., 1981, in: Dynamical Systems and Turbulence, eds. D.A. Rand and L.S. Young, Springer-Verlag, Berlin

Voges, W., Atmanspacher, H. and Scheingraber, H., 1987, Ap. J. **320**, 794

Waldmeier, M., 1957, Z. f. Astrophysik **43**, 149

Waldmeier, M., 1971, Solar Phys. **29**, 232

Weiss, N., 1988, in: Secular Solar and Geomagnetic Variations in the last 10000 years, eds. F.R. Stephenson and A.W. Wolfendale, Kluwer Acad. Publishers, Dordrecht

White, O.R. and Trotter, D.E., 1977, Ap. J. Suppl. **33**, 391

Wilson, R.M., 1987, Statistical Aspects of Solar Flares, NASA Technical paper 2714

Yau, K.K.C., 1988, in: Secular Solar and Geomagnetic Variations in the last 10000 years, eds. F.R. Stephenson and A.W. Wolfendale, Kluwer Acad. Publishers, Dordrecht

On the Possibility of Supergiant Stable Flows in the Convection Zone of the Sun

Evgeniy Tikhomolov

Institute of Solar Terrestrial Physics, 664033 Irkutsk, P.O.Box 4026, Russia

Abstract: We present a numerical model of convection in a rotating fluid layer with a deformable upper free surface, which exhibits a solution in the form of a large-scale dissipative structure.

1 The Convection Zone of the Sun is an Open Dissipative System

According to direct observations and numerical simulations (Gilman and Miller 1986), convection on the various scales is nonstationary. The existence of long-lived phenomena, such as active longitudes, complexes of activity, and large-scale structures of background magnetic fields with clear-cut rotation periods, under chaotic conditions of convective motions, is a non-trivial fact. A theoretical interpretation of these phenomena as a result of self-organization of the magnetic field under turbulent convection conditions on the Sun (non-axisymmetric dynamo) requires rather strong suppositions (Makarov et al. 1989).

In this paper an approach is developed with which it is possible to avoid eventual problems. Large-scale long-lived formations on the Sun are supposed to be the result of the large-scale structuring of giant convection. The magnetic field has only the role of a tracer that reflects the characteristics in the distribution of hydrodynamical fields. One of the main necessary (but not sufficient) formation conditions for a stationary large-scale structure is mutual feedback between convection and large-scale flow. One of the examples of the appearance of feedback between convection in a rotating layer and an ordered mean flow is provided by numerical experiments that simulate the differential rotation on the Sun (Gilman and Miller 1986). The action of feedback leads to a change in the shape of the convective rolls and the establishment of a certain stationary profile of differential rotation. What must be the form of the interaction between convection and large-scale flows to enable a stationary large-scale structure to be sustained?

This paper is concerned with one of the possible variants of the interaction between large-scale flows (Rossby waves) and convection which leads to structuring of convection. The effect is associated with a large-scale modulation of the thickness of the giant convection zone. A large-scale disturbance in this case may not be observed directly because it is hidden by the upper layer where granule and supergranule cells exist.

The problem of investigating the possible effects giving rise to structuring is rather complicated. Therefore, in theoretical physics particular attention is at present devoted to very simple models (so-called base models) that have the required properties of the physical system being simulated. In this paper the simplest model is constructed and investigated. It has the main properties of convection in a rotating layer and it exhibits the possibility of large-scale structuring.

2 Structuring of Convection in a Two-Dimensional Numerical Model

Convection in the Boussinesq approximation is a classical model of an open dissipative system which has been and is being used to investigate fundamental questions of the theory of bifurcations, the transition to chaos, and the formation of structures. In changing over to a Cartesian co-ordinate system to model convection in a rotating spherical layer the beta-plane approximation is used. One of the solutions of the equations in view of the sphericity is a Rossby wave. In the absence of convection, for a layer with a free upper surface the excitation of a barotropic Rossby wave leads to surface deformation. Usually this deformation is neglected. In this paper the deformation of a free upper surface is taken into account for modelling the modulation of the thickness of the giant convection zone. A rather fast rotation is considered when it becomes possible to use a quasi-geostrophic approximation. We have considered a two-dimensional variant of the system of equations which can be applied to describing convection in the form of rolls (there is no dependence on latitude). As a result of the approximations made, we obtain the following system of equations in dimensionless units:

$$w_z - \frac{1}{D^2}p_{xxt} - \frac{\beta}{D}p_x + \frac{1}{D^2}\Delta p_{xx} = 0 \tag{1}$$

$$w_t + uw_x + ww_z = -p_z + RT + \Delta w \tag{2}$$

$$T_t + uT_x + w(T_z - 1) = \Delta T \tag{3}$$

$$u = -\frac{1}{D^2}p_{zt} + \frac{1}{D^2}\Delta p_x \quad \Delta = \frac{\partial^2}{\partial x^2} + \frac{\partial^2}{\partial z^2} \tag{4}$$

where u and w are the horizontal and vertical velocity components, respectively; p is the deviation of pressure from an undisturbed hydrostatic profile, and T is the deviation of the temperature from a linear profile; $D = 2\Omega H^2/\nu$; Taylor number $Ta = D^2$; $R = g\alpha\Theta H^3/(\nu\kappa)$ is the Rayleigh number; the Prandtl number $P = \nu/\kappa$ is assumed to be unity. Here ν is viscosity, Ω is rotation frequency, g is gravity, α is the coefficient of thermal expansion; Θ is a temperature difference between the upper and lower surfaces, H is an undisturbed depth, and κ is the thermometric conductivity. The vertical co-ordinate is measured from the lower boundary. The parameter β appears as a result of differentiation of the angular velocity with respect

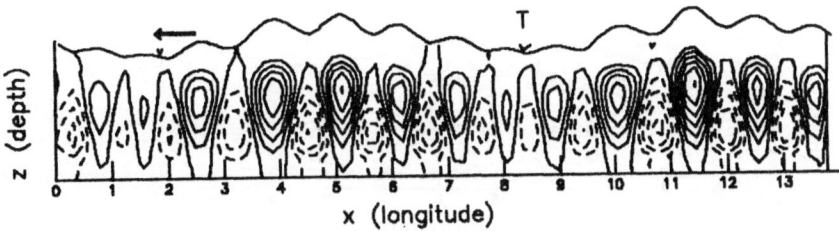

Fig. 1. Convection, modulated by a Rossby wave. Steady stationary picture. Model parameters: $R = 10^4$, $Ta = 10^4$, $\beta = 0.2$. Positive (solid) and negative (dashed) temperature isolines are shown

to latitude (beta-plane approximation). For the upper boundary we introduced the variable h – the deviation of the surface from the undisturbed state.

The analysis of the linearized system of equations shows the existence of two classes of solutions: large-scale weakly damped barotropic Rossby waves travelling opposite to the direction of rotation, and convective modes travelling along the direction of rotation, which agrees with the results of three-dimensional calculations (Busse and Cuong 1977; Gilman 1975).

Numerical calculations of the nonlinear system demonstrated a dependence of the convection intensity on large-scale deformation of the upper free surface. One can clearly see in Fig. 1 a temperature rise above elevations of the upper free surface and a temperature decline below depressions.

The interaction between a large-scale flow and convection, resulting from this effect, leads to a sustained, stationary large-scale structure. The presence of the beta-effect leads to a drift of the large-scale structure opposite to the rotation direction (indicated by the arrow).

To obtain the spatial distribution of the large-scale component of physical fields, spatial filtering of the stationary picture was carried out. An intensification of convection leads to the formation of a large-scale temperature profile, which is shown in Fig. 2.

3 The Result of the Interaction is Negative Diffusion

The results obtained in numerical calculations make it possible to understand qualitatively the way in which it becomes possible that the large-scale structure is sustained. The convection modulation caused by deformation of the upper free surface is associated with the fact that fluid particles from the lower to upper boundary traverse a different path. The convection effect with respect to the large-scale component is equivalent to introducing heat sources that adjust themselves to a change of the upper free surface. Feedbacks between convection and a large-scale disturbance are realized in such a manner. The equation of hydrostatic equilibrium for

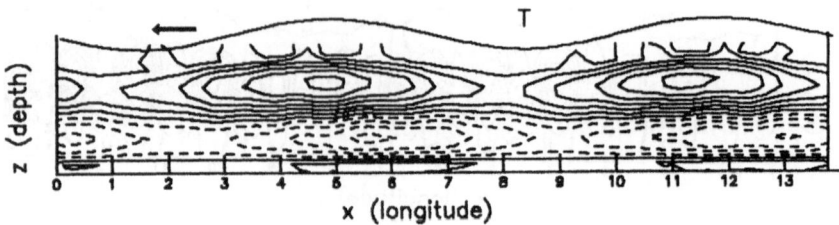

Fig. 2. The picture after filtering for a steady state large-scale temperature field

the large-scale component of the flows has the form $p_z = RT(z, h(x))$. Integration of the continuity equation over the co-ordinate z leads to the equation for deformation of the upper free surface:

$$h_t - \frac{q}{D^2} h_{xxt} - \frac{q}{D} \beta(1 + h)h_x + \frac{q}{D^2} h_{xxxx} + c(h)h_{xx} = 0 , \qquad (5)$$

with $q = gH^3/(\nu\kappa)$. The functional $c(h(x))$ always assumes a positive value. A diffusion term with a negative diffusion coefficient appears in equation (5) for the deformation of the upper free surface. This term ensures the pumping of a large-scale disturbance at the expense of an infinite energy reservoir, namely the heating of the convection zone from below. This equation is a generalization, on the one hand, of a regularized long-wave equation (Peregrine 1966) and, on the other hand, of the Kuramoto-Sivashinsky equation (Kuramoto and Tsuzuki 1976; Sivashinsky 1983). A fundamental property of this equation is the existence of a soliton-like stationary solution with some fixed amplitude (Tikhomolov 1993). Possibly, this property would become important when explaining theoretically three-dimensional long-lived large-scale features seen on the solar surface.

This work was supported by the Russian Fund of Fundamental Investigations, project number 93-02-17014

References

Busse, F.H. and Cuong, P.G.: 1977, Geophys. Astrophys. Fluid Dyn. 8, 17
Gilman, P.A.: 1975, J. Atmos. Sci. 32, 1331
Gilman, P.A. and Miller, J.: 1986, Astrophys. J. Suppl. 61, 585
Kuramoto, Y., Tsuzuki, T.: 1976, Prog. Theor. Phys. 55, 356
Makarov, V.I., Ruzmaikin, A.A., Starchenko, S.V., Tavastsherna, K.S.: 1989, Solnechnyie
    Dann. 10, 104
Peregrine, D.H.: 1966, J. Fluid Mech. 25, 321
Sivashinsky, G.I.: 1983, Ann. Rev. Fluid Mech. 15, 179
Tikhomolov E.M.: 1993, in press

Stellar Dynamos

Gaetano Belvedere

Istituto di Astronomia dell'Università di Catania, Città Universitaria, 95125 Catania, Italy

Abstract: The observational constraints on stellar dynamos are discussed in the framework of the solar-stellar connection. The non-linear theoretical approach to stellar activity is then outlined and some results of the boundary layer solar dynamo are pointed out, which suggest that future observations from space of activity features on distant stars may give some insight on their internal rotation.

1 The Solar-Stellar Connection

The principle of solar-stellar connection states that:

- the phenomenology of solar and stellar magnetic activity is substantially the same, despite the differences in strength, topology and timescales of active phenomena occurring in the Sun and stars;
- it can be interpreted in terms of the same basic mechanism, the dynamo, operating in stellar convective envelopes, even if remarkable differences between various dynamo operation modes are expected, depending on stellar fundamental parameters like mass, luminosity, rate of rotation and age. This is also reflected in the dependence of solutions of non-linear dynamo equations on the dynamo number D, a parameter which characterizes the strength of dynamo action.

Thus we have to view solar and stellar observations as complementary in order to highlight the basic phenomena and global mechanisms of stellar magnetic activity.

Unity in multiplicity is therefore the basic concept of the solar-stellar connection: unity as for the mechanism, multiplicity as for its operation modes. However some caution is needed when making use of the analogy to the Sun to understand activity in stars. For instance:

- extrapolating from slow rotators (hypo-active stars) to fast rotators (hyperactive stars) is very dangerous since the convection pattern in the presence of rotation is radically different;
- the topology and strength of activity are different in single and close binary stars, since in the latter case the physical interaction between the two companions affects differential rotation and leads to a more complex magnetic field topology than in the former case;

- different activity signatures are likely for dwarfs and giants of the same effective temperature, since the onset of convection in the outer layers starts at different spectral types (F0 in dwarfs, G0 in giants);
- the dynamo operation modes may be different as to both the location (the whole convection zone, its bottom, or the transition layer (boundary or over-shooting) just beneath the convective layer) and the driving mechanism (radial or latitudinal gradient of angular velocity);
- significant differences in activity signatures of very similar stars are expected when the non-linear hydromagnetic regime is analysed theoretically (Weiss et al. 1984; Belvedere et al. 1990). This is confirmed by observations (Rodonò 1987).

2 Basic Observational Characteristics of Stellar Dynamos

The basic observational characteristics of stellar dynamos are the following:

- the dynamo operates in dwarfs (later than F0) and in giants (later than G0);
- the dynamo efficiency, which is a function of the magnetic field intensity B and the filling factor f (i.e. the fraction of the area covered with active regions), increases with the rotation rate;
- the dynamo efficiency increases with the fractional thickness of the convection zone;
- the two previous points may be synthetically expressed in a single sentence which states that the dynamo efficiency decreases with the Rossby parameter, P/τ, where P is the rotational period and τ is the convective turnover time. The Rossby parameter equals the inverse of the square root of the dynamo number D, so that the level of dynamo action increases when D increases, as predicted on theoretical grounds. Notice that the Rossby parameter is a hybrid, since P is an observational datum while τ is deduced from theory. The dynamo efficiency is conventionally related to the chromospheric flux measured in the CaII H and K lines (Noyes et al. 1984; Soderblom et al. 1992). However the matter is controversial, since the Utrecht group argues for an explicit (B–V) or mass dependence of the chromospheric flux (once the "basal" flux or "quiet chromosphere" contribution is subtracted) on the Rossby parameter (Rutten and Schrijver 1986; Schrijver 1993);
- the dynamo action has as an ultimate consequence the loss of angular momentum via a magnetic torque, which leads to stellar rotation braking (Skumanich's (1972) law: $P \sim Mt^{1/2}$, where M is the mass and t the age of a star);
- the dynamo may operate in a cyclic or non-cyclic way, even in the same star. This has been proved in the Sun after the discovery of the Maunder and similar minima (Eddy 1976). Fairly well-defined stellar cycles, as well as chaotic activity in non-cyclic form, have been discovered in a sample of about a hundred stars whose chromospheric activity has been monitored by the Mount Wilson group in a number of years (Baliunas and Vaughan 1985; Saar and Baliunas 1992). In the observed sample 85% of stars show variability (60% smooth and solar-like, 25% chaotic) while the remaining 15% show no variability of the chromospheric flux (Maunder state?);

- cycle periods seem to increase with the fractional thickness of the convection zone;
- no correlation seems to exist between cycle periods and suitable parameters such as mass, rate of rotation, Rossby parameter and age;
- direct measurements of magnetic fields in stellar active regions, made possible after the Robinson (1980) method and successive improvements, show magnetic field strengths up to some kGauss and correlations such as $\Phi \sim T_{\text{eff}}^{2.8} V_{\text{rot}}^{0.55}$, where $\Phi = fB$ is the magnetic flux, and $\Phi \sim \tau\omega$, where ω is the angular velocity. For a review see Saar (1991). Future improvement of the observational techniques will check the previous results. In this regard we look with interest at the recent Zeeman-Doppler Imaging Method (Donati et al. 1989), which can give intensity and cartography of the magnetic field on stellar surfaces simultaneously, by measuring the Stokes parameters I and V. In principle, this is an extension of the Doppler Imaging Method (Vogt 1983; Vogt et al. 1987), which allows us to reveal spotted regions on distant stars.

3 Non-Linear Modeling of Stellar Dynamos

Global hydromagnetic dynamos in a rotating spherical shell, with a fully dynamical prescription of the convection zone, have been investigated in classical works by Gilman (1983, 1986), Gilman and Miller (1981), Glatzmaier (1985a,b). Although such models have illustrated the difficulties of modeling the solar dynamo, nevertheless they have shown some interesting results, for instance the various regimes that appear when feedback from the Lorentz force on differential rotation is introduced: "no dynamos", "dynamos with cycles", "dynamos without cycles". However these models require a large amount of computer time and a straightforward interpretation of results is sometimes not easy.

Thus it still makes sense to parametrize the dynamics of the convection zone by the α-effect and to try to solve simpler models, albeit somewhat idealized, which include the essential physics of the dynamo action together with a source of non-linearity such as the Lorentz force, the magnetic buoyancy of toroidal flux tubes, the parametrized quenching of the α- and ω- effects or the so called α-Λ effect, which is essentially a dynamo driven by Reynolds stresses, investigated in recent years by the Helsinki and Potsdam groups. For recent reviews on the subject see Brandenburg (1994) and Schmitt (1994).

The effect of the non-linearity introduced by the Lorentz force has been investigated in a classical work by Ruzmaikin (1981), who has shown, by discussing a simple system of non-linear differential equations, that the solar cycle may be interpreted in terms of a "strange attractor". Then, Weiss et al. (1984) have contructed a low order system of ordinary differential equations describing stellar cycles, adopting a severely truncated representation of the spatial structure. They have found, in various parameter ranges, cyclical behaviour, quasi-periodic oscillations, and aperiodic solutions resembling the Maunder minima. More recently, Belvedere et al. (1990) have improved the Weiss et al. model, adopting only a radial truncation, thus allowing full latitudinal dependence of fields and flows, that results in the construction of butterfly diagrams. The model equations have been solved in a spherical shell representing the stellar convection zone. A large variety of stable solutions have been found in different ranges of the dynamo number $D = \alpha \omega_o d^3 / \eta^2$, where ω_o is the equatorial rotation rate, d is the thickness of the convection zone and η is the turbulent viscosity. These are: periodic symmetric, periodic asymmetric, quasi-periodic and pulsed, the last resembling the intermittent activity characteristic of the Grand Minima. A notable result is that two or three different solutions can coexist in suitable ranges of D, depending on the initial conditions. This suggests that large differences in the activity regimes and signatures of very similar stars might exist, as shown by observational evidence. Thus solutions of the dynamo equations in the non-linear regime do indeed predict a complex time modulation, which indicates that stellar activity must be interpreted in terms of deterministic chaos.

4 Internal Rotation, Boundary Layer Dynamo and the Latitudinal Belts of Stellar Activity

The location of dynamo action in the Sun has been debated vigorously in recent years. Three possible locations have been suggested: (a) the whole convection zone; (b) the bottom of the convection zone; (c) the boundary (overshoot) layer just beneath the convection zone. The last possibility appears to be the most realistic, on the basis of toroidal flux tube stability arguments (Parker 1979; Spiegel and Weiss 1980; Spruit and Van Ballegooijen 1982) and recent results of helioseismology (Belvedere 1990).

Assuming the internal rotation profile inferred by the most recent helioseismological data (Brown et al. 1989; Libbrecht 1989), Belvedere et al. (1991) carried out a non-linear α-ω dynamo model in the boundary layer betweeen the convective and radiative zones (about $0.65\,R_\odot$ to $0.70\,R_\odot$), reproducing the essential features of the solar activity cycle: an equatorward migration of sunspots and most faculae, which belong to the equatorial activity belt (latitude $\lambda < 35°$) and a poleward migration of polar faculae, filaments and large scale magnetic flux, which are characteristic of the polar activity belt ($\lambda = 40°$ - $50°$). This means that the two latitudinal belts of solar activity, which are distinguished by opposite latitudinal migration of tracers, would appear as a natural and direct consequence of the internal rotation profile, via the operation of a dynamo in the boundary layer. Specifically, it is a consequence of the rigid rotation rate $\omega_o = 2.7 \times 10^{-6}$ rad s^{-1} below the convection zone ($r = 0.65\,R_\odot$) being precisely the value at the surface at the critical latitude $\lambda_o = 35°$, that defines the boundary between the two latitudinal belts. Since no radial variation of the surface rotation (a function of latitude) has been found in the convection zone, we should have in the boundary layer $\partial \omega / \partial r > 0$ for $\lambda < \lambda_o$ and $\partial \omega / \partial r < 0$ for $\lambda > \lambda_o$. Accordingly, the α-ω dynamo criterion (Parker 1955; Yoshimura 1975a), based on the sign of the product $\Gamma = \alpha \partial \omega / \partial r$ allows equatorward propagation of dynamo waves for $\lambda < \lambda_o$ and poleward propagation for $\lambda > \lambda_o$, since α is negative in the boundary layer in the northern hemisphere (Yoshimura 1975b; Glatzmaier 1985a,b; Rüdiger and Kichatinov 1993).

Indeed, this picture has received a substantial support by the non-linear α-ω dynamo calculations performed by Belvedere et al. (1991) in a thin spherical shell

($d = 0.05\ R_\odot$) representing the boundary layer, with full latitude-time resolution. In fact, they have shown the existence of periodic (dynamo wave-like) stable solutions with both equatorward and poleward migrating branches, for dynamo numbers of the order of unity. This is a basic result: the simulation shows that horizontally propagating dynamo waves can exist even in a very thin layer.

Although many uncertainties still exist even in the case of the Sun, we are nevertheless tempted to suggest that, conversely, observations of latitudinal distribution and migration of active regions in late main sequence (G, K, M) slowly rotating stars, by photometric and spectroscopic methods, may in principle allow us to infer the internal rotation profile and angular momentum distribution. For this we need accurate measurements of: (i) the angular velocity at the equator or some fixed latitude; (ii) the latitudinal differential rotation profile at the surface, over a suitable latitude interval; (iii) the latitude drift of activity tracers over a suitable time span. All necessary data may in principle be given by photometric and spectroscopic observations (Baliunas and Vaughan 1985; Rodonò 1987).

Thus it should be possible to determine the surface latitude (if any) at which the direction of migration changes its sign and the corresponding value of the surface angular velocity, and these are the essential data needed to deduce the internal rotation profile. The angular momentum profile would be consequently computed taking into account the density profile given by theory of stellar structure. Unfortunately, the present available photometric and spectroscopic data of surface distribution of stellar active regions (Rodonò 1986; Vogt et al. 1987; Vogt and Hatzes 1991) only refer to fast rotating, hyperactive binary stars of the RS CVn and BY Dra type – whereas the analogy to the Sun requires single, slowly rotating, mildly active main sequence stars – and they show a definite tendency of active regions to be concentrated at high latitudes. Moreover, the sensitivity and resolving power of present observational capabilities are far from making it possible to perform high precision measurements, such as those we suggest here, not only for solar type single stars with relatively weak activity, but even for very active RS CVn and BY Dra type stars.

However, future improvement of observational techniques based on the analysis of both light curves and line profiles (in this regard we mention the Zeeman-Doppler Imaging Method that is an extension of the Doppler Imaging Method and allows surface magnetic cartography), as well as observation from space with large instruments, may make it possible in the near future.

Further, comparison between surface activity data and acoustic oscillations data for a suitable sample of stars may offer the opportunity to test the validity of the internal rotation probing method, whose basic principles are outlined here, and eventually the reliability of the boundary layer dynamo concept.

References

Baliunas, S.L., Vaughan, A.H.: 1985, *Ann. Rev. Astron. Astrophys.* **23**, 379

Belvedere, G.: 1990, in: Inside the Sun, *Proc. IAU Coll.* **121**, eds. G. Berthomieu, M. Cribier, Kluwer, Dordrecht, p. 371

Belvedere, G., Proctor, M.R.E., Pidatella, R.M.: 1990, *Geophys. Astrophys. Fluid Dyn.* **51**, 263

Belvedere, G., Proctor, M.R.E., Lanzafame, G.: 1991, *Nature* **350**, 481

Brandenburg, A.: 1994, in: Advances in Solar Physics, *Proc. 7th European Meeting on Solar Physics*, eds. G. Belvedere, M. Rodonò, G.M. Simnett, in press

Brown, T.M., Christensen-Dalsgaard, J., Dziembowski, W., Goode, P.R., Gough, D.O., Morrow, C.A.: 1989, *Astrophys. J.* **343**, 526

Donati, J.F., Semel, M., Praderie, F.: 1989, *Astron. Astrophys.* **225**, 467

Eddy, J.A.: 1976, *Science*, **286**, 1189

Gilman, P.A.: 1983, *Astrophys. J. Suppl.* **53**, 243

Gilman, P.A.: 1986, *Geophys. Astrophys. Fluid Dyn.* **31**, 137

Gilman, P.A., Miller, J.: 1981, *Astrophys. J. Suppl.* **46**, 211

Glatzmaier, G.A.: 1985a, *Geophys. Astrophys. Fluid Dyn.* **31**, 137

Glatzmaier, G.A.: 1985b, *Astrophys. J.* **291**, 300

Libbrecht, K.G.: 1989, *Astrophys. J.* **336**, 1092

Noyes, R.V., Hartmann, L.W., Baliunas, S.L., Duncan, D.K., Vaughan, A.H.: 1984, *Astrophys. J.* **279**, 763

Parker, E.N.: 1955, *Astrophys. J.* **122**, 293

Rodonò, M.: 1987, in: Solar and Stellar Physics, eds. E.H. Schröter, M. Schüssler, Lecture Notes in Physics **292**, 39

Rodonò, M. et al.: 1986, *Astron. Astrophys.* **165**, 135

Robinson, R.D.: 1980, *Astrophys. J.* **239**, 261

Rüdiger, G., Kichatinov, L.L.: 1993, *Astron. Astrophys.* **269**, 581

Rutten, R.G.M., Schrijver, C.J.: 1986, in: Cool Stars, Stellar Systems and the Sun, eds. M. Zeilik, D. Gibson, Springer, Berlin, p. 19

Ruzmaikin, A.A.: 1981, *Comments on Astrophys.* **9**, 85

Saar, S.H.: 1991, in: The Sun and Cool Stars: Activity, Magnetism, Dynamos, *Proc. IAU Coll.* **130**, eds. I. Tuominen, D. Moss, G. Rüdiger, Lecture Notes in Physics **380**, 389

Saar, S.H., Baliunas, S.L.: 1992, in: The Solar Cycle, ed. K.L. Harvey, Astronomical Society of the Pacific Conference Series **27**, 150

Schmitt, D.: 1994, in: Advances in Solar Physics, *Proc. 7th European Meeting on Solar Physics*, eds. G. Belvedere, M. Rodonò, G.M. Simnett, in press

Schrijver, C.J.: 1993, in: Inside the Stars, *Proc. IAU Coll.* **137**, eds. W.W. Weiss, A. Baglin, Astronomical Society of the Pacific Conference Series **40**, 591

Skumanich, A.: 1972, *Astrophys. J.* **171**, 565

Soderblom, D.R., Stauffer, J.R., Hudon J.D.: 1993, *Astrophys. J. Suppl.* **85**, 315

Vogt, S.S.: 1983, in: Activity in Red Dwarf Stars, *Proc. IAU Coll.* **71**, eds. P.B. Byrne, M. Rodonò, Reidel, Dordrecht, p. 137

Vogt, S.S., Hatzes, A.P.: 1991, in: The Sun and Cool Stars: Activity, Magnetism, Dynamos, *Proc. IAU Coll.* **130**, eds. I. Tuominen, D. Moss, G. Rüdiger, Lecture Notes in Physics **380**, 297

Vogt, S.S., Penrod, G.D., Hatzes, A.P.: 1987, *Astrophys. J.* **321**, 496

Weiss, N.O., Cattaneo, F., Jones, C.A.: 1984, *Geophys. Astrophys. Fluid Dyn.* **30**, 305

Yoshimura, H.: 1975a *Astrophys. J.* **201**, 740

Yoshimura, H.: 1975b *Astrophys. J. Suppl.* **29**, 467

On Magnetic Fields, Rossby Numbers and Dynamo Action in Late-Type Stars

B. Montesinos [1], C. Jordan [2]

[1] Laboratory for Space Astrophysics and Fundamental Physics, ESA–IUE
Satellite Tracking Station, P.O. Box 50727, 28080 Madrid, Spain
[2] Department of Physics, Theoretical Physics, University of Oxford,
1 Keble Road, Oxford OX1 3NP, UK

Abstract: In this paper we give a brief account of our comparisons of predicted and observed trends in the dependence of the magnetic flux, $B_s f_s$, and the filling factor, f_s, on the Rossby number, Ro, in main-sequence late-type stars. A sample of stars with reliable measurements of B_s and f_s has been selected. The surface average magnetic field, $B_s f_s$, and the fraction of the star covered by this field, the filling factor, f_s, were computed by using a simple dynamo model plus some assumptions on how the field generated at the bottom of the convection zone emerges at the photosphere. The best fits between $B_s f_s$ and f_s with Ro are shown and a possible theoretical basis for the saturation observed at small values of Ro, both in $B_s f_s$ and f_s is discussed. Full details of the work summarised here can be found in Montesinos and Jordan (1993).

1 Introduction

The presence of magnetic fields on the solar surface has been known since the beginning of the century, when fields of the order of 1 - 2 kG were found to be associated with sunspots. However, the detection and measurement of magnetic fields and filling factors in late-type stars was an elusive task until about fifteen years ago. The low spatial resolution and the intrinsic observational difficulty in detecting magnetic fields via the Zeeman effect delayed progress in this area.

Spatially averaged magnetic fields can now be measured in the photospheres of cool main-sequence stars (see e.g. Saar 1990) and the fields and filling factors found from modelling. Structures analogous to sunspots have been detected in some stars (e.g. Hall 1976; Bopp and Stencel 1981) and the presence of activity cycles is well established through rotational modulation of emission in the Ca II H and K lines (e.g. Baliunas and Vaughan 1985). It is therefore reasonable to assume that the features observed in the solar chromosphere, transition region and corona have their analogues on other main-sequence stars.

2 Magnetic Fields and Filling Factors

We have adopted measurements of magnetic fields and filling factors from the compilation by Saar (1990). We have added two more measurements by Saar (1991), for HD 17925 and V833 Tau, and one by Valenti (1991), for ξ Boo A. For the Sun, Tarbell, Title and Schoolman (1979) found that about 0.83% of the quiet Sun, and about 8.5% of a plage is filled with fields of 1200 G. Sheeley (1966) estimated that $\leq 20\%$ of the Sun is covered by plage regions. Schrijver (1987) found similar results for active regions, i.e. around 5 - 10% of the area covered by fields of 1000 - 2000 G. These numbers give a filling factor of $f_s \simeq 2\%$. Saar and Schrijver (1987) use a smaller value for the area covered by active regions, but this is compensated by a larger network value at high activity, and their maximum value of f_s is the same. Regarding $f_s = 1\%$ as a typical low activity value we adopt a mean of 1.5%.

We have re-examined the relations between the magnetic parameters and $P_{\rm rot}$ and Ro. Of the possible forms of correlation (i.e. $\log B_s f_s$ with $P_{\rm rot}$, Ro, $\log P_{\rm rot}$, $\log Ro$) we find that $\log B_s f_s$ is best fitted with a linear dependence on Ro, such that

$$\log B_s f_s = 3.27(\pm 0.08) - 0.97(\pm 0.08)Ro . \qquad (1)$$

The observed filling factors can be fitted in a similar form

$$\log f_s = -0.01(\pm 0.07) - 0.86(\pm 0.07)Ro , . \qquad (2)$$

In Figure 1 we show the observed values of $B_s f_s$ and f_s plotted against Ro (empty circles) and results of both fits as solid lines.

3 Theoretical Predictions

The method for estimating the surface magnetic flux and the filling factor consists of two steps. First, the convection zone magnetic field is computed from a kinematic dynamo model for each star in the sample, and then some assumptions are made concerning the way in which the magnetic field escapes from the generation region through the convection zone and reaches the photosphere.

The calculation of the magnetic field at the base of the convection zone has been computed based on the method of Durney and Robinson (1982) which considers a toroidal configuration of tubes around the equator and uses the Parker (1979) formalism to relate the several local values of convection zone quantities. Some modifications have been introduced to the original Durney and Robinson approach (see Montesinos and Jordan 1993 for details). The dependence of the angular velocity with depth and latitude has been scaled for stars other than the Sun from the solar law fitted by Durney (1985) to the results of Duvall *et al.* (1984). The shear in the angular velocity as one moves down in the radial coordinate depends on the angular velocity itself, i.e. $d\omega/dr \propto \omega$.

Estimates of the filling factor are based on two approximations following the solar analogy. The magnetic structures in the solar photosphere can be divided into large-scale polarity patterns and small-scale structures. The most conspicuous examples of the former kind are the active regions which represent the largest concentrations

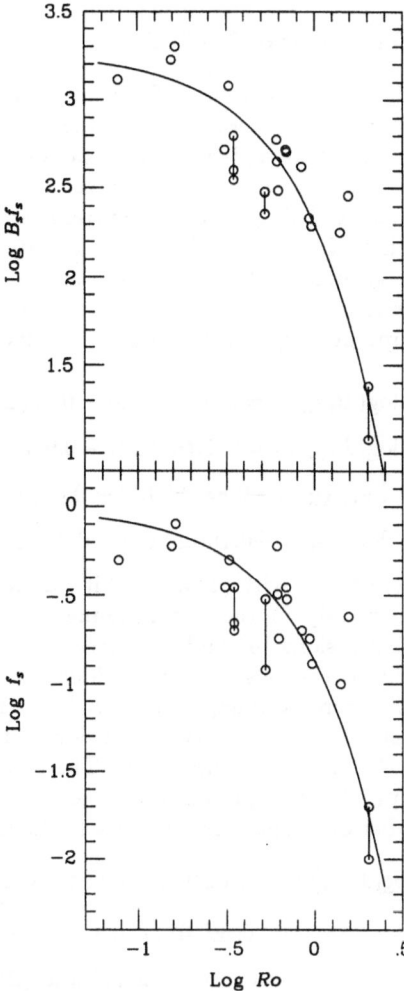

Fig. 1. Observed values of $\log B_s f_s$ (upper panel) and $\log f_s$ (lower panel) (solid circles) plotted against the Rossby number $\log Ro$. The solid lines are the fits given by equations (1) and (2)

of magnetic flux. They emerge from the solar convection zone as compact, bipolar regions, which contain sunspots in some cases, and decay over periods of weeks to months. On a smaller scale there is a background field which is not smoothly distributed but occurs in a pattern forming a network whose characteristic cell diameter is about 30,000 km, the so-called supergranulation network. The lifetime of these cells is of the order of one day. It has been shown that more than 90% of the magnetic flux reaching the solar photosphere, outside sunspots, is in the form of kGauss flux elements whose typical dimensions are of the order of 100 km, i.e., below the resolution limits of the currently available telescopes (Stenflo 1989), hence one may regard the supergranulation network to be composed of bundles of such magnetic elements. Our theoretical predictions of the filling factor take into account both types of structure.

3.1 Results of the Theoretical Predictions

As far as the magnetic field at the base of the convection zone is concerned, the general trend is that, for a given spectral type, the field strength increases as the rotation period decreases, and for a given rotation period, the field strength increases towards later spectral types, i.e. with deeper convection zones.

For the calculation of the filling factor two extreme approaches have been taken. It has been assumed that *all* magnetic field appears *either (a)* as active regions *or (b)* in the form of supergranulation network. The calculations have been normalized to a value of 0.015 for the solar filling factor. Subscripts *'AR'* and *'SN'* stand for 'active regions' and 'supergranulation network' respectively. The resulting relations are

$$(a) \quad \log(B_s f_{AR}) = 2.55(\pm 0.08) - 0.69(\pm 0.11)Ro \tag{3}$$

$$(b) \quad \log(B_s f_{SN}) = 3.21(\pm 0.10) - 0.99(\pm 0.12)Ro \tag{4}$$

$$(a) \quad \log f_{AR} = -0.83(\pm 0.07) - 0.60(\pm 0.10)Ro \tag{5}$$

$$(b) \quad \log f_{SN} = -0.15(\pm 0.08) - 0.93(\pm 0.10)Ro \tag{6}$$

Comparing equations (3) to (6) with (1) and (2) it can be seen that the assumption that the field emerges mostly in the supergranulation network gives a closer fit to the observations. Expressions (4) and (6) agree with (1) and (2) to within the deviations. The best fit to the observations is found for a combination of active regions and supergranulation assuming that $\sim$ 75% of the strong field appears in the latter kind of structures for *all* the stars. Obviously this combination may well be a function of both Ro and the spectral type. The results of the calculations are shown in Figure 2. Empty circles represent the observational results and solid circles are the theoretical predictions. The regression fits to the theoretical predictions give

$$\log(B_s f_s) = 3.11(\pm 0.10) - 0.94(\pm 0.12)Ro \tag{7}$$

$$\log f_s = -0.26(\pm 0.08) - 0.88(\pm 0.10)Ro \tag{8}$$

which are in close agreement with expressions (1) and (2).

4 Modeling the Saturation

Not only the observations of $B_s f_s$ and f_s, but also those of other activity indicators, such as ΔR_{HK} appear to show a 'saturation' at small values of Ro (Noyes et al. 1984; Vilhu 1984). Stauffer and Hartmann (1987) report observations of rapid and slow rotating Pleiades and α Per stars and find modest differences in chromospheric emission for stars ranging in $v \sin i$ from 10 to 100 km s^{-1}. Also, to explain successfully the distribution in rotational velocities observed in the Pleiades, these authors suggest the existence of a different braking mechanism, based on angular momentum loss independent of the rotational velocity for stars with $v_{eq} \geq 10$ km s^{-1}. This difference must also be reflected in the dependence of ω on r. We have therefore calculated the locus of the values of $B_s f_s$ for stars of a fixed spectral type (G2 V), using $d\omega/dr \propto \omega$ for the angular velocity with depth for stars with $P_{rot} \geq 5$ days, which corresponds to $Ro \geq 0.4$ ($v_{eq} < 10$ km s^{-1}) but keeping a *constant* shear

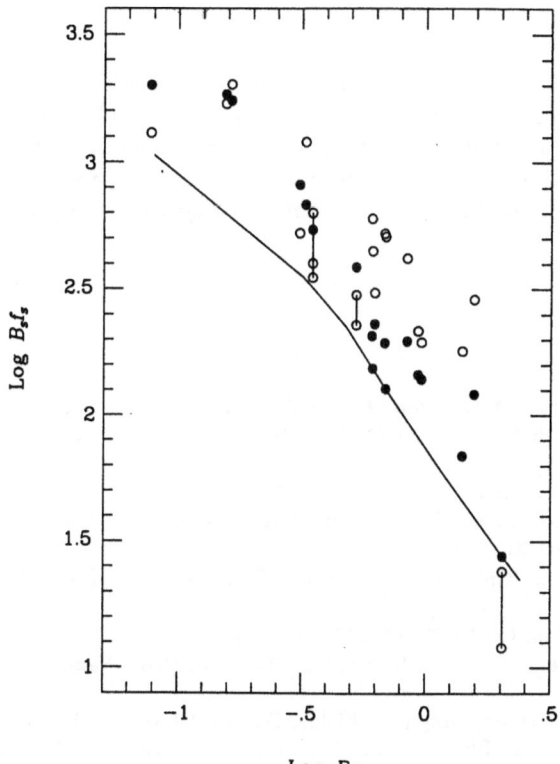

Fig. 2. Measured values of $\log(B_s f_s)$ (open circles) and theoretical estimates (solid circles) plotted against the Rossby number $\log Ro$. The theoretical points have been derived using for all the stars a similar dependence of the angular velocity on depth as that given for the sun by Durney (1985) and a combination 75%/25% for supergranulation and active regions. The solid line shows the theoretical prediction of $B_s f_s$ for stars of spectral type G2 V, using the above angular velocity law ($d\omega/dr \propto \omega$) for $Ro \geq 0.4$ days but a constant shear in ω for $Ro < 0.4$

in ω for stars with $Ro < 0.4$ ($v_{eq} > 10$ km s^{-1}), which gives a less steep gradient. The locus is shown in Figure 2 as a solid line. Thus the observed log-linear form of the whole sample of stars, and the tendency to saturate at small values of Ro could arise naturally from a change in the dependence of $d\omega/dr$ on $\omega(r)$ at $Ro \simeq 0.4$. This is of course only one possible cause of the 'saturation' effect but it is encouraging that simple physical assumptions lead to theoretical predictions that go in the same directions as the observed trends.

5 Conclusions

We have shown that the observed surface magnetic fields and filling factors can be fitted with a log-linear dependence on the Rossby number Ro. The agreement between the theory and the present sample of observations is encouraging given the simplifying assumptions in the model we have used. The apparent saturation of $B_s f_s$ and f_s at small values of Ro may arise from a change in the dependence of $\omega(r)$ on r at around $Ro = 0.4$.

References

Baliunas, S.L., Vaughan, A.H., 1985, ARA&A 23, 379

Bopp, B.W., Stencel, R.E., 1981, ApJ 247, L131

Durney, B.R., 1985, ApJ 297, 787

Durney, B.R., Robinson, R.D., 1982, ApJ 253, 290

Duvall, P.L. Jr., Dziembowski, W.A., Goode, P.R., Gough, D.O., Harvey, J.W., Leibacher, J.W., 1984, Nature 310, 22

Hall, D.S., 1976, in: Multiple Periodic Variable Stars, ed. W.S. Fitch, Reidel, Boston, p. 287

Montesinos, B., Jordan, C., 1993, MNRAS 264, 900

Noyes, R.W., Hartmann, L.W., Baliunas, S.L., Duncan, D.K., Vaughan, A.H., 1984, ApJ 279, 763

Parker, E.N., 1979, Cosmical Magnetic Fields, Oxford, Clarendon Press, p. 145

Saar, S.H., 1990, in: Solar Photosphere: Structure, Convection and Magnetic Fields, IAU Symp. No. 138, ed. J.O. Stenflo, Kluwer, Dordrecht, p. 427

Saar, S.H., 1991, in: Mechanisms of Chromospheric and Coronal Heating, eds. P. Ulmschneider, E.R. Priest, R. Rosner, Springer-Verlag, p. 273

Saar, S.H., Schrijver, C.J., 1987, in: Proc. of the Fifth Cambridge Workshop on Cool Stars, Stellar Systems and the Sun, eds. J.L. Linsky, R.E. Stencel, Springer-Verlag, p. 38

Schrijver, C.J., 1987, A&A 180, 241

Sheeley, N.R. Jr., 1966, ApJ 147, 1106

Stauffer, J.R., Hartmann, L.W., 1987, ApJ 318, 337

Stenflo, J.O., 1989, A&AR 1, 3

Tarbell, T.D., Title, A.M., Schoolman, S.A., 1979, ApJ 229, 387

Valenti, J.A., 1991, in: The Sun and Cool Stars: activity, magnetism and dynamos (Proc. IAU Coll. 130), eds. I. Tuominen, D. Moss, G. Rüdiger, Springer-Verlag, p. 411

Vilhu, O., 1984, A&A 133, 117

III

Coupling Between Interior and Corona

III

The Magnetic Field of the Solar Corona

C. E. Alissandrakis

Section of Astrophysics, Astronomy and Mechanics
Department of Physics, University of Athens
GR-15784 Athens, Greece

Abstract: The structure of the solar corona, on all observable scales, is intimately controlled by the magnetic field. Although direct measurements of the magnetic field are from difficult to impossible, its presence is evident in all spectral ranges where the coróna is observable. This review discusses the measurement of coronal magnetic fields and the indirect information provided by radio, white light and soft X-ray observations.

1 Introduction

The Sun is made up of plasma and magnetic field. The latter enters in practically all solar phenomena, in all layers of the solar atmosphere. In particular the structure of the upper atmospheric layers is the result of the interaction of the plasma with the magnetic field. Contrary to the photosphere, the magnetic energy density in the corona is much higher than the internal energy density of the plasma; consequently, it is the magnetic field that gives the corona its highly structured appearance. Plasma, electric currents, heat, all flow along channels provided by the lines of force of the magnetic field.

Important as it may be for the structure of the corona, the magnetic field is shaped much lower; formed in the solar interior, it is rearranged by the photospheric motions and extends into the corona in a nearly force-free configuration. Thus, in addition to shaping the corona, the magnetic field plays the role of the principal physical link with the lower solar atmosphere. With the exception of eruptive phenomena, the corona itself does little to change the magnetic field. At large heights it is the solar wind flow that stretches the magnetic field lines and gives them an almost radial orientation. The field extends into interplanetary space and acts as a link with the solar atmosphere.

Although the magnetic field is everywhere present in the corona, its measurement is much more difficult there than in the photosphere or even in the chromosphere. This is primarily due to the low intensity of the coronal emission. In order to get quantitative information about the magnetic field one has to go through a careful and detailed analysis of emission processes and, even then, the information is not complete. This review will give a brief account of the various methods that have been developed for the direct measurement of the coronal magnetic field and will discuss the qualitative information that the coronal structure itself provides.

2 Quantitative Information

The strength and/or the direction of the magnetic field enter directly into a number of emission mechanisms, mainly in the radio part of the spectrum. In principle, the entire corona is accessible to radio observations, the effective height of observation increasing with wavelength. Difficulties arise from the scarcity of high spatial resolution radio observations at short radio wavelengths and the relatively low spatial resolution at long wavelengths. In addition, the magnetic field cannot be measured everywhere but only in structures that are both resolved and bright, and show measurable circular polarization. In spite of these difficulties the radio band gives the most reliable quantitative information about coronal magnetic fields.

2.1 Zeeman and Hanlé Effect Measurements

The short wavelength of coronal lines in the EUV makes it very difficult to apply the Zeeman effect in the measurement of the magnetic field. Circular polarization in the wings of the C IV line (formed in the transition region at $T \sim 10^5$ K) was observed with the UVSP above sunspots (Henze *et al.*, 1982; Hagyard *et al.*, 1983); they gave magnetic field strengths of ~1100-1400 G and, combined with photospheric measurements, a gradient of ~0.2 G/cm.

The Hanlé effect, which requires measurement of linear polarization, has been used to determine the magnetic field in prominences only. The extensive work at Pic du Midi (Leroy *et al.*, 1977; Sahal-Bréchot, 1977; Leroy *et al.*, 1984) has given values of a few Gauss.

2.2 Thermal Bremsstrahlung

The two electromagnetic waves present in a magnetized plasma (extraordinary and ordinary) have different properties: the thermal bremsstrahlung absorption coefficient for the e-mode is greater than in the unmagnetized plasma and that of o-mode is lower; thus a magnetized region appears with (slightly) higher total intensity and is circularly polarized.

Thermal bremsstrahlung is a useful diagnostic when the magnetic field is not strong enough to give rise to gyroresonance emission (see Sect *2.3*), i.e. in the quiet Sun and in plage regions. High spatial resolution observations of the quite Sun at short cm wavelengths show increased emission associated with the chromospheric network (Kundu *et al.*, 1979) and with the photospheric magnetic field (Gary and Zirin, 1988); this is a consequence of higher density and temperature. We should note, however that this emission comes from the transition region rather than from the corona, where the network is no longer visible. Measurements of the magnetic field above a plage region were made by Bogod and Gelfreikh (1980). They used total intensity and circular polarization data from RATAN-600 and derived values of ~ 25 G.

There are no reported measurements of circular polarization from quiet regions at long wavelengths. The expected polarization is low; for an isothermal, optically thin source (e.g. a coronal loop) the degree of polarization is:

$$p \approx 2\omega_H/\omega \tag{1}$$

where ω_H is the gyrofrequency and ω is the frequency of observation. Thus a field of 5 G will give a polarization of ~18% at 160 MHz and only 7% at 410 MHz.

2.3 Thermal Gyroresonance

In the presence of a magnetic field, thermal electrons emit electromagnetic radiation at the harmonics of the gyrofrequency:

$$\omega = s\,\omega_H = s\frac{eB}{mc} \tag{2}$$

where ω is the emission frequency, s is the harmonic number and B is the magnetic field. Under coronal conditions, it is mainly the second and the third harmonics that contribute to the emission (Zheleznyakov, 1962; Kakinuma and Swarup, 1962). At a wavelength λ, the emission from the harmonic s comes from a region where the magnetic field is:

$$B = \frac{10800}{s\lambda} = \frac{360f}{s} \tag{3}$$

where the field is in Gauss, the wavelength in cm and the frequency in GHz.

Thus at 6 cm a field of 900 G is required for emission at the second harmonic and 600 G for emission at the third. Consequently thermal gyroresonance is a useful diagnostic of the magnetic field mainly above sunspots and in the microwave range. At shorter wavelengths the required magnetic field is too high, at longer wavelengths thermal bremsstrahlung masks the gyroresonance emission; at still longer wavelengths the harmonic layers are hidden from the observer by refraction effects.

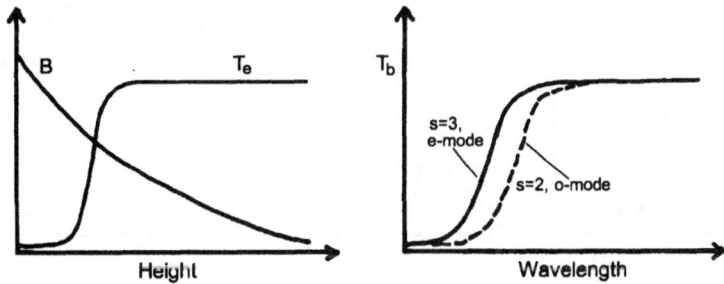

Fig. 1. Variation of the electron temperature and the magnetic field with height (left) and the corresponding brightness temperature of gyroresonance emission at the second and third harmonic as a function of wavelength (right)

A given harmonic layer radiates strongly if it is optically thick and if the associated magnetic field is located in a high temperature region. The spectrum of the emission depends on the variation of the temperature and the magnetic field with height. In the standard source (Zlotnik, 1968a, 1968b) the temperature increases and the magnetic field decreases monotonically with height (Fig. 1). As a consequence, the lower harmonics which are associated with stronger magnetic field are located lower in the atmosphere and their height increases with wavelength. At short wavelengths both the second and third harmonic layers are located below the transition region (TR) and the emission is weak. As the wavelength increases the third harmonic layer enters into the transition region while the second is still in the chromosphere. The third harmonic is opaque in the extraordinary and transparent in the ordinary mode; thus the brightness temperature increases and the radiation is almost 100% circularly polarized. The brightness rises sharply with wavelength, which reflects the fast increase of temperature with height. At longer wavelengths the second harmonic, which is optically thick in both modes, enters the transition region; as a result the brightness increases further and the polarization decreases. At still longer wavelengths, where both harmonics are in the corona, the emission has a coronal brightness temperature and little overall polarization.

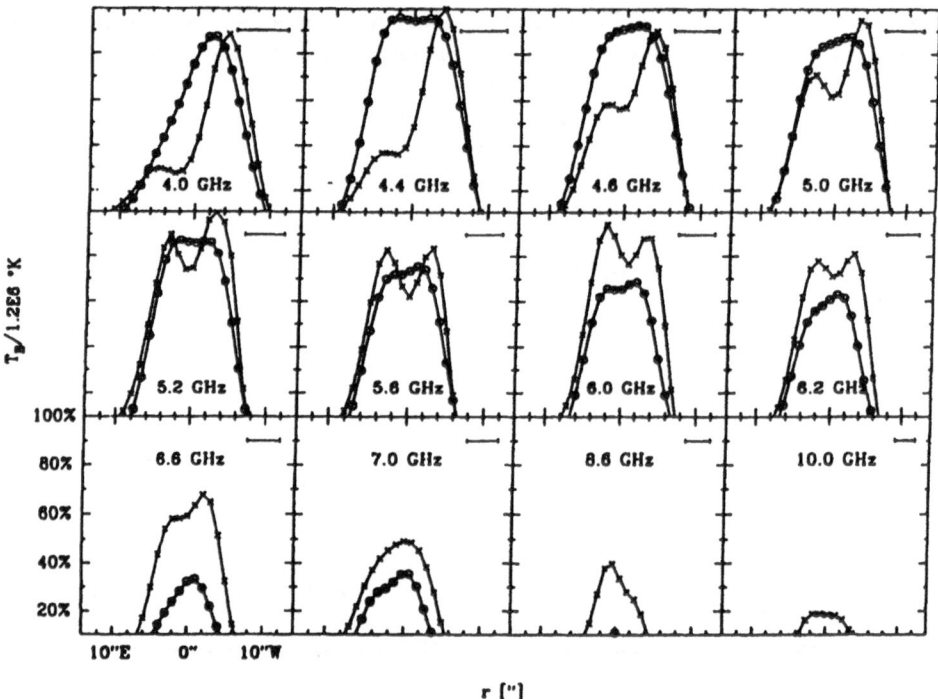

Fig. 2. Brightness distribution over a simple spot in the microwave range. Crosses show the left and circles the right circularly polarized emission (Lee *et al.*, 1993)

Thus in the microwave range sunspots appear brighter than their surroundings! This is just the consequence of the fact that the radiation above sunspots comes from the corona while around sunspots it comes from the chromosphere. High spatial resolution mapping of active regions (Kundu and Alissandrakis, 1975) has confirmed the standard model and has also shown a wealth of details due to the dependence of the opacity on the angle between the field and the line of sight, the local temperature and density of the plasma (Alissandrakis and Kundu, 1982, 1984; Lang and Willson, 1982; White et al., 1991). The mere presence of high brightness emission shows the presence of strong magnetic field in the corona, the strength of which can be measured from the wavelength of the observation and the harmonic number (Equation 3). In some cases it is also possible to measure the direction of the magnetic field (Alissandrakis and Kundu 1984). However a radio map is not a magnetogram; it gives the magnetic field as a function of temperature rather than as a function of position and this only to a first approximation. Detailed modeling is required (Alissandrakis et al., 1980; Krüger et al., 1986; Chiuderi-Drago et al., 1987; Bogod et al., 1992) to extract more information.

Extensive work on the spectrum of sunspot-associated emission has been done by the Pulkovo group and the RATAN-600 radio telescope (e.g. Akhmedov et al., 1982, 1990). Figure 2 shows a set of recent observations of a simple sunspot, obtained with the frequency-agile receiver of the Owens Valley Radio Observatory (Lee et al., 1993). The left and right circularly polarized emission corresponds to the e- and o-modes respectively, which in turn come from the third and second harmonics. At 10 GHz the third harmonic is already in the TR, which implies that the magnetic field at its base is at least 1200 G and no more than 1800 G, since there is no second harmonic emission. The second harmonic breaks up at 8.6 GHz, which puts the magnetic field at the base of the TR to 1550 G. The third harmonic emission stabilizes around 6 GHz; thus the magnetic field near the top of the TR and the low corona is ∼720 G, while the second harmonic stabilizes around 5 GHz, implying a value of ∼900 G for the same region. Similar studies have shown the presence of fields as high as 1800 G in the low corona (Alissandrakis et al., 1993).

Equally interesting are the diagnostic capabilities of gyroresonance emission in cases deviating from the standard model (Zheleznyakov and Zlotnik, 1980a, 1980b), which can lead to the formation of narrow band emission (cyclotron lines). One such case is when a hot region is embedded in the ambient coronal plasma (Willson, 1985; Zheleznyakov and Zlotnik, 1988); the identification of cyclotron lines gives a direct measurement of the magnetic field.

2.4 Non-thermal Gyrosynchrotron Emission

The characteristics of gyrosynchrotron emission from mildly relativistic electrons trapped in flaring loops depend strongly on the magnetic field. The emission coefficient has a quasi-continuous spectrum with maximum in the low harmonics of the gyrofrequency. The peak wavelength of the observed intensity spectrum is mainly determined by self absorption effects; these shift the peak to the 3rd-4th harmonic (Takakura, 1967). For a peak at 6 cm this corresponds to 450-600 G; the magnetic field is obviously higher in bursts with spectra that peak at shorter wavelengths, sometimes in the millimeter range.

It should be noted that the magnetic field in bursts is highly inhomogeneous, thus these values should be considered as gross estimates only. Detailed model computations by Preka-Papadema and Alissandrakis (1988) showed that the peak can occur between the second and sixth harmonic; the spectral maximum shifts to shorter wavelengths between the top of the loop and its footpoints as a result of the variation of the magnetic field strength and direction along the flaring loop.

2.5 Wave Propagation Effects

The observed polarization of the radio emission is largely determined by propagation effects. As the e- and o-mode waves propagate towards the observer their polarization characteristics change in such a way that, at each point along the ray path, they reflect the polarization properties imposed by the local magnetic field and the local density; inversion of the sense of circular polarization can occur if the radiation crosses a region of transverse magnetic field, where the longitudinal component of the magnetic field changes sign. This happens as long as the two waves are weekly coupled. Higher up in the corona, where the plasma density is low, the difference between the properties of e- and o-mode waves gets smaller and their mutual coupling increases. When the coupling becomes strong, the waves lose their separate identity and the polarization no longer changes along the path but rather attains a limiting value (Cohen, 1960; Bandiera, 1982).

As a result, circular polarization maps of active regions show two oppositely polarized areas, separated by a line of zero circular polarization. This line marks neither the neutral line of the photospheric magnetic field, nor the neutral line at the region of formation of the radiation; it is the locus of points where the magnetic field is perpendicular to the line of sight and the wave coupling is critical. Its position depends on the wavelength, the position of the active region on the solar disk and the characteristics of the large scale, dipole component of the magnetic field. Alissandrakis and Kundu (1984) measured the distance of the zero polarization line from the photospheric neutral line as a function of the heliocentric position of an active region and derived a magnetic field of 10-20 G at a height of 110-140 10^3 km; similar values were obtained by Gelfreikh *et al.* (1987), who measured the position of the zero polarization line as a function of wavelength.

2.6 Plasma Emission

Although detailed computations of the emission of metric radio bursts are rather uncertain, they have been used to provide information on the magnetic field in the middle/upper corona. One such diagnostic is the circular polarization of type III bursts which, for emission at the second harmonic of the plasma frequency, is proportional to the ratio of the gyrofrequency to the plasma frequency. Using the Nançay Radioheliograph, Mercier (1990) found that the degree of polarization increased from ∼5% at 164 MHz to ∼15% at 435 MHz. Assuming a model density distribution he deduced magnetic field strengths between 50 and 8 G at the 435 and 164 MHz levels.

Another diagnostic is the frequency drift of type II bursts; this is interpreted in terms of the radial velocity of the associated shock, which must be higher than the local Alfvén speed. Using this argument, Dulk and McLean (1978) obtained estimates of a few Gauss for the magnetic field between 0.2 and 1 $R_\odot$.

3 Qualitative Information

When direct measurements of the coronal magnetic field are impossible, the observer turns to the coronal structure in order to extract information about the topology of the magnetic field and its evolution with time. In principle all spectral ranges where the corona is visible are appropriate for the extraction of qualitative information about the magnetic field; however each spectral range has its own advantages and disadvantages, due to instrumental peculiarities and the different way that the physical conditions affect the radiation. It is thus desirable to combine information obtained in different ranges with different methods.

3.1 Magnetic Cartography

Any high resolution image of the corona is a map of magnetic field lines. This is beautifully illustrated by soft X-ray pictures, such as those of the Yohkoh satellite, which show a wealth of loops and other structures, large and small, revealing the topology of the magnetic field in all scales. Since there is a line of force of the magnetic field passing through each point in the corona, it is natural to pose the question: *Why do some flux tubes appear brighter than others?*

We should first note that the Lorentz force is perpendicular to the magnetic field; consequently, along the field, gravity is balanced by the pressure gradient and mass motions. Moreover, in low β plasmas, the gas pressure is much less than the magnetic pressure; consequently changes of pressure in one flux tube have little effect on its neighbours. Finally each flux tube is efficiently insulated from its surroundings by the magnetic field. In short, each field line has its own atmosphere and is little affected from whatever happens around it. You can thus heat a single flux tube and make it radiate. Then the real question is: *Why is there energy deposition in some flux tubes and not in others?* The answer to this question should be looked for in the lower layers, where the magnetic field is shaped and provided with energy.

3.2 The White Light Corona

Solar eclipse observations and spaceborne coronagraphs provide high spatial resolution and show the corona up to a few solar radii (Koutchmy, 1992). The corona presents a rich and complex structure with plumes, loops, arches and streamers which reveal the geometry of the magnetic field. Streamers are probably the most fascinating coronal structures, with their base made up of closed magnetic field lines and their tops stretched by the solar wind. Sharp changes in density are also observed, called *tangential discontinuities* by Koutchmy (1972) who, on the basis of pressure balance, estimated the associated jump of the magnetic field to be between $0.8\,\mathrm{G}$ at $1.5\,R_\odot$ and $0.12\,\mathrm{G}$ at $2.4\,R_\odot$.

Eclipse observations show beautifully the global configuration of the magnetic field, which changes dramatically during the solar activity cycle as illustrated in Fig. 3 (Loucif and Koutchmy, 1989): The nearly axisymmetric minimum corona with extended open polar regions and equatorial streamers is transformed to the highly irregular maximum corona with streamers almost everywhere. The low opacity of the corona, combined with the long integration along the line-of-sight inherent to

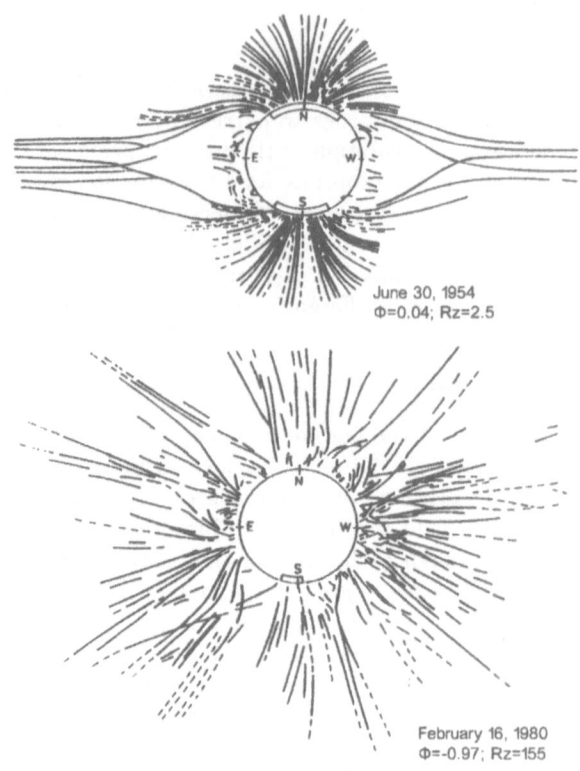

June 30, 1954
Φ=0.04; Rz=2.5

February 16, 1980
Φ=-0.97; Rz=155

Fig. 3. Coronal drawings near the minimum (top) and the maximum (bottom) of the solar cycle. Φ is the phase and R_z the Wolf number (Loucif and Koutchmy, 1989)

observations beyond the photospheric limb often make the reconstruction of the three-dimensional geometry of the structures difficult.

The scarcity of eclipse observations is complemented by ground based coronameters, which provide an almost uninterrupted data base but with much lower spatial resolution and at much lower altitudes. Synoptic maps of the K- corona reveal the large scale, long lived structures and in particular the base of streamers which form a belt around the Sun (Fig. 4). The streamer belt reveals the location of the neutral line of the large scale component of the coronal magnetic field. It follows closely the solar equator during the minimum but it is significantly displaced to the north and south near the maximum, in conformity with the eclipse results.

Both eclipse and coronameter observations are ideal for testing the theoretical extrapolation of the photospheric magnetic field into the corona. Potential field models were successful in predicting open and closed magnetic field configurations (e.g. Newkirk and Altschuler, 1970) and in associating the neutral line of the extrapolated field with the streamer belt as seen in the K-corona as well as with the base of the interplanetary current sheet (the heliosheet), responsible for the sector structure of the interplanetary magnetic field (Hoeksema, 1984). More elaborate computations,

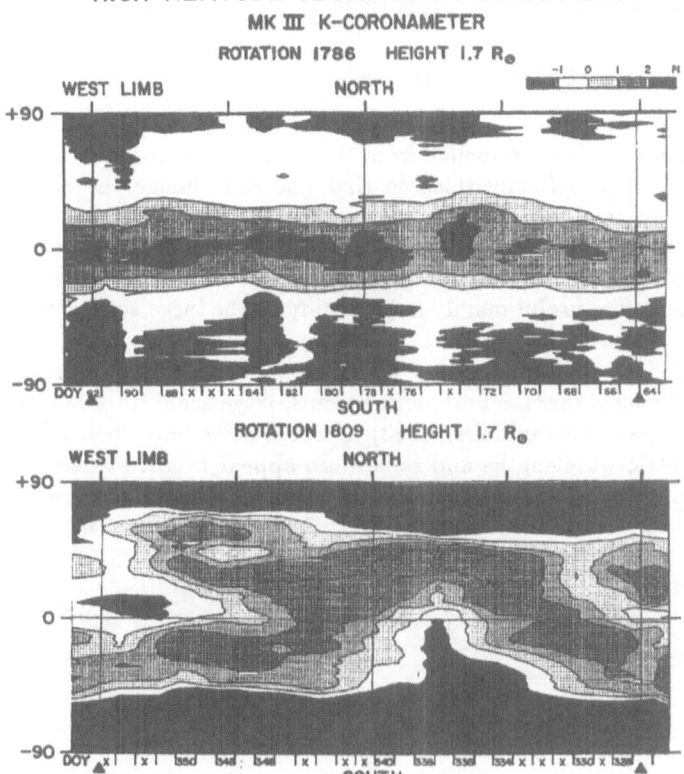

Fig. 4. Synoptic charts of the K-corona from the Mauna Loa station of the High Altitude Observatory near the minimum (top) and the maximum (bottom) of the solar cycle

based on the model of Bogdan and Low (1986) have been made recently (Bagenal and Gibson, 1991; Zhao and Hoeksema, 1992), which provide estimates of the magnetic field from K-coronameter data. It has also been pointed out that better agreement with the observations is obtained if the photospheric magnetic field is assumed to be radial (Wang and Sheeley, 1992).

3.3 Soft X-rays

Soft X-ray imaging observations were the first to show the corona in front of the solar disk. The extensive series of observations from Skylab, twenty years ago, were of fundamental importance for our understanding of the coronal structure, the clear identification of coronal holes and loops and their association with the magnetic field.

We are fortunate enough now to have the Yohkoh spacecraft in orbit (Bentley, 1993). It has already provided a wealth of coronal pictures with excellent spatial and temporal resolution, sufficient not only for mapping the structure of the magnetic field but also for studying its dynamics on short and long time scales.

3.4 Long Wavelength Radio Emission

The radio range at wavelengths longer than a few tens of cm is ideal for the study of the coronal structure: it is accessible from the ground, the corona can be observed on the disk and it is optically thick so that one can change the effective height of observation by changing the frequency. These advantages are partly offset by refraction effects at longer wavelengths and by the limited spatial resolution which, at present, is of the order of one minute of arc. As a consequence, long wavelength radio observations are useful mainly in the study of the large scale structure of the magnetic field.

Long wavelength radio maps show little resemblance to chromospheric images. The principal features observed are noise storms, large scale coronal loops, coronal holes and streamers (Alissandrakis, 1993). Coronal holes have their maximum contrast at decimetric wavelengths and sometimes appear bright at decametric wavelengths (Lantos *et al.*, 1987). Streamers are more prominent at decametric wavelengths, where they have been observed up to 2.5 $R_\odot$ (Kundu *et al.*, 1987).

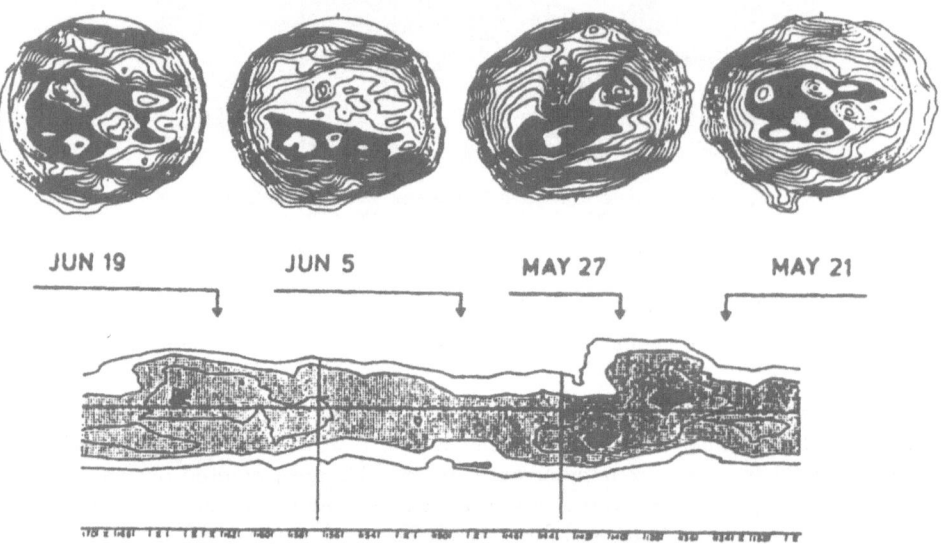

Fig. 5. Daily coronal maps at 408 MHz together with the K-corona synoptic chart. The black region is the *coronal plateau* (Lantos *et al.*, 1992)

Lantos *et al.* (1992) found that all quiet-Sun sources they observed, both at 164 and 408 MHz, were located inside the zone of enhanced K-corona emission. This shows that the quiet Sun at long wavelengths is associated with the large scale structure of the coronal magnetic field. They also identified a region of enhanced brightness around the local sources which they called *coronal plateau*. The plateau was found to be stable over at least a solar rotation at 164, 327 and 408 MHz and it followed closely the shape of the coronal neutral sheet as observed with the K-coronameter (Fig. 5). Thus long wavelength radio observations were the first to detect the base of the heliosheet on the disk.

3.5 Fast Drift Radio Bursts

Fast drift radio bursts are produced by mildly relativistic electrons travelling in the corona along the magnetic field. Without any spatial resolution, from dynamic spectra alone, one can identify open and closed magnetic structures associated with type III, J, U and N-bursts; together with model computations one can further deduce some physical parameters (e.g. Hillaris *et al.*, 1990). Using dynamic spectra and metric radioheliograph data, Klein and Aurass (1993) deduced a magnetic geometry consistent with that of a coronal streamer; a similar method was used by Aschwanden *et al.* (1992) to derive the path of electrons in a U burst observed at 21 cm. More information is expected to come from two dimensional mapping at several frequencies, which is now possible with the Nançay radioheliograph and can be used to measure the position of the electron beam as a function of time and to trace the magnetic field lines.

4 Conclusions

Our understanding of the coronal magnetic field is continuously improving, following the improvements of instrumentation and a better understanding of its interaction with the plasma. Yet it is far from being complete, due to the inherent difficulties in its measurement. No single method and no single spectral range can provide adequate information. In the near future we should expect exciting new results from continued high resolution microwave observations, from long wavelength observations with the multifequency Nançay Radioheliograph, from the Yohkoh soft X-ray telescope and from the forthcoming SOHO satellite.

References

Akhmedov Sh.B., Gelfreikh G.B., Bogod V.M., Korzhavin A.N. (1982): *Solar Phys.*, **79**, 41
Akhmedov Sh.B., Bogod V.M., Korzhavin A.N., Aurass H., Hildebrandt J., Krüger A. (1990): *Solar Phys.*, **129**, 351
Alissandrakis C.E. (1993): *Adv. Space Res.*, (*in press*)
Alissandrakis C.E., Kundu M.R., Lantos P. (1980): *Astron. Astrophys.*, **82**, 30
Alissandrakis C.E., Kundu M.R. (1982): *Ap. J. (Letters)*, **253**, L49
Alissandrakis C.E., Kundu M.R. (1984): *Astron. Astrophys.*, **139**, 271
Alissandrakis C.E., Gelfreikh G.B., Borovik V.N., Korzhavin A.N., Bogod V.M., Nindos A., Kundu M.R. (1993): *Astron. Astrophys.*, **270**, 509
Aschwanden M.J., Bastian T.S., Benz A.O., Brosius J.W. (1992): *Astrophys. J.*, **391**, 380
Bagenal F., Gibson S. (1991): *J. Geophys. Res.*, **96**, 17663
Bandiera R. (1982): *Astron. Astrophys.*, **112**, 52
Bentley R.D. (1993): These Proceedings

Bogdan T.J., Low B.C. (1986): *Astrophys. J.*, **306**, 271
Bogod V.M., Gelfreikh G.B., Willson R.F., Lang K.R., Opeikina L.V., Shatilov V., Tsvetkov S.V. (1992): *Solar Phys.*, **141**, 303
Bogod V.M., Gelfreikh G.B. (1980): *Solar Phys.*, **67**, 29
Chiuderi-Drago F., Alissandrakis C.E., Hagyard M. (1987): *Solar Phys.*, **112**, 89
Cohen M.H. (1960): *Astrophys. J.*, **131**, 664
Dulk G.A., McLean D.J. (1978): *Solar Phys.*, **57**, 279
Gary D.E., Zirin H. (1988): *Astrophys. J.*, **329**, 991
Gelfreikh G.B., Peterova N.G., Ryabov B.I. (1987): *Solar Phys.*, **108**, 89
Hagyard M.J., Teuber D., West E.A., Tandberg-Hansen E., Henze W., Beckers J.M., Bruner M., Hyder C.L., Woodgate B.E. (1983): *Solar Phys.*, **84**, 13
Henze W., Tandberg-Hansen E., Hagyard M.J., Woodgate B.E., Shine R.A., Beckers J.M., Bruner M., Gurman J.B., Hyder C.L., West E.A. (1983): *Solar Phys.*, **81**, 231
Hillaris A., Alissandrakis C.E., Caroubalos C., Bougeret J.-L. (1990): *Astron. Astrophys.*, **229**, 216
Hoeksema J.T. (1984): Thesis, Stanford University
Kakinuma T., Swarup G. (1962): *Astrophys. J.*, **136**, 975
Klein K.-L., Aurass H. (1993): *Adv. Space Res.*, (*in press*)
Koutchmy S. (1972): *Solar Phys.*, **24**, 374
Koutchmy S. (1992): *Proc. of First SOHO Workshop*, ESA SP-348, 73
Krüger A., Hildebrandt J., Bogod V.M., Korzhavin A.N., Akhmedov Sh.B., Gelfreikh G.B. (1986): *Solar Phys.*, **105**, 111
Kundu M.R., Alissandrakis C.E. (1975): *Nature*, **257**, 465
Kundu M.R., Rao A.P., Erskine F.T., Bregman J.D. (1979): *Astrophys. J.*, **232**, 1122
Kundu M.R., Gergely T.E., Schmahl E.J., Szabo A., Loiacono R., Wang Z., Howard R.A. (1987): *Solar Phys.*, **108**, 113
Lantos P., Alissandrakis C.E., Gergely T., Kundu M.R. (1987): *Solar Phys.*, **112**, 325
Lantos P., Alissandrakis C.E., Rigaud D. (1992): *Solar Phys.*, **137**, 225
Lang K.R., Willson R.F. (1982): *Ap. J. (Letters)*, **255**, L111
Lee J.W., Hurford G.J., Gary D.E. (1993): *Solar Phys.*, **144**, 45
Leroy J.L., Ratier G., Bommier V. (1977): *Astron. Astrophys.*, **54**, 811
Leroy J.L., Bommier V., Sahal-Bréchot, S. (1984): *Astron. Astrophys.*, **156**, 223
Loucif M.L., Koutchmy S. (1989): *Astron. Astrophys. Supp. Ser.*, **77**, 45
Mercier C. (1990): *Solar Phys.*, **130**, 119
Newkirk G.A., Altschuler M. (1970): *Solar Phys.*, **13**, 131
Preka-Papadema P., Alissandrakis C.E. (1988): *Astron. Astrophys.*, **191**, 365
Sahal-Bréchot S., Bommier V., Leroy J.L. (1977): *Astron. Astrophys.*, **59**, 223
Takakura T. (1967): *Solar Phys.*, **1**, 304
Wang Y.-M., Sheeley, N.R. (1992): *Astrophys. J.*, **329**, 310
White S.M., Kundu M.R., Gopalswamy N. (1991): *Ap. J. (Letters)*, **366**, L43.
Willson R.F. (1985): *Astrophys. J.*, **298**, 911
Zhao X., Hoeksema J.T. (1992): *Proc. of First SOHO Workshop*, ESA SP-348, 117
Zheleznyakov V.V. (1982): *Astron. Zh.*, **39**, 5 (*Soviet Astron.*, **6**, 3)
Zheleznyakov V.V., Zlotnik E.Ya. (1980a): *Solar Phys.*, **68**, 317
Zheleznyakov V.V., Zlotnik E.Ya. (1980b): *Astron. Zh.*, **57**, 778 (*Soviet Astron.*, **24**, 448)
Zheleznyakov V.V., Zlotnik, E.Ya. (1988): *Pis'ma Astron. Zh.*, **14**, 461 (*Sov. Astron. Lett.*, **14**, 195)
Zlotnik E.Ya. (1968a): *Astron. Zh.*, **45**, 310 (*Soviet Astron.*, **12**, 245)
Zlotnik E.Ya. (1968b): *Astron. Zh.*, **45**, 585 (*Soviet Astron.*, **12**, 464)

The Control of the Corona by the Convective Zone Magnetic Fields

P. Démoulin

Observatoire de Paris, URA 326 (CNRS), 92195 Meudon Cédex, FRANCE

Abstract: A review is presented on the physics of the convective zone and the implications at the coronal level. Solar magnetic fields are created in the convective layer from the kinetic energy of the dense plasma. At the coronal level, the magnetic field controls the plasma and is forced to evolve according to the time-dependent boundary conditions given at the photospheric level by the convective zone. The coronal field cannot find a smooth equilibrium when its topology is complex and current sheets are formed. These are the preferred regions where reconnection can occur. Present development of 3D reconnection is reviewed, and we show how observed flare kernels are related to the magnetic field topology. Then we describe how our present theoretical understanding of flares can help us to understand both large and small scale coronal events.

1 Introduction

The physics of the corona, which is magnetically dominated, contrasts strongly with the physics of the Convective Zone (CZ), where the plasma pressure dominates the magnetic field (for a review of the solar dynamo see Schmitt, this issue). The solar magnetic field is accurately measured mainly at the photospheric level. What is observed is both a remnant of the sub-photospheric processes and a time-dependent boundary condition for coronal physics. The convective motions are then thought to produce the magnetic field and also the heating, thus leading to the formation of the corona itself. The precise physical processes occurring within this overall framework are still being debated (see Einaudi, this issue). The convective motions are also responsible for the larger events, such as flares and Coronal Mass Ejections (CME), but again there is still some dispute regarding the physics involved.

To understand the above phenomena, we must know how the CZ evolves (Sect. 2). Subphotospheric motions are particularly efficient in building up coronal currents on surfaces, called separatrices, separating coronal fluxes coming from different photospheric regions (Sect. 3). Small scale lengths in the field are naturally formed, leading to reconnection. Recent work has evolved from 2D to 3D reconnection, giving a completely new view (Sect. 4). These theoretical developments can be better tested observationally for flares. Many well-resolved observations are available (radio, visible, UV and X-ray) giving strong constraints on the location of energy release. The importance of separatrices in flares is highlighted in Sect. 5. This scenario may also be valid on larger scales as on smaller scales, including the smallest ones involved in nanoflares and heating. In Sect. 6, speculations on the processes appearing in the quiet Sun are derived from what we have learned about flares.

2 Main Features of Magneto-Convection

2.1 Stratified Plasma Convection

In a region, 200 Mm thick, below the photosphere, radiation cannot efficiently transport the energy from the hot lower boundary layer to the cold upper one. Convective motions are set up which are so efficient that they maintain a near-adiabatic temperature gradient in the whole CZ (except in the thin overshoot upper and lower layers). The radiative exchange of energy is only important at the top, in a very thin layer (≈ 0.1 Mm) where the opacity is of the order of unity. The solar energy escapes mainly from this layer; this strong cooling is responsible for the motions in the entire CZ.

The density scale height in the CZ goes from ≈ 100 Mm at the bottom to 0.2 Mm at the top. Compared to the CZ height, this implies a density ratio between the bottom and top of $\approx 10^6$. This strong density stratification (coupled to mass conservation) creates a strong asymmetry between upward and downward motions: the ascending plasma must rapidly expand, then it must overturn within a density scale height while the descending plasma must rapidly contract and continue to fall down, becoming more concentrated and more dense. It implies that the CZ has gentle ascending motions and strong concentrated downflows (just the opposite to the hot ascending plumes in the earth's upper mantle !).

The downward motions start at the border of the photospheric granules. The density scale height increases with depth and motions are organised on larger horizontal scales: the downward motions meet and form stronger and more distant downdrafts at successively larger depths (Fig. 1). In the above scenario of Nordlund & Stein (1990) the larger convective scales are driven by the merging of smaller scale downward filaments, leading to an inverse cascade. This picture is, however, still under debate: Zahn (1988) suggests that the energy cascades from the large scales (supergranules) to the smaller ones (granules).

2.2 Magneto-Convection

The interaction of the magnetic field with the plasma flow is mainly described by the equations of motion and of induction:

$$\rho \frac{D\boldsymbol{v}}{Dt} = -\nabla P + \boldsymbol{j} \times \boldsymbol{B} + \rho \boldsymbol{g} + \rho \nu \boldsymbol{\Delta} \boldsymbol{v} \,, \tag{1}$$

$$\frac{\partial \boldsymbol{B}}{\partial t} = \nabla \times (\boldsymbol{v} \times \boldsymbol{B}) + \eta \nabla^2 \boldsymbol{B} \,. \tag{2}$$

In the kinematic approach, where the back reaction of the fluid, $\boldsymbol{j} \times \boldsymbol{B}$ in (1), is neglected the field is transported to the vertices of the convective cells in a time of the order of the turn over convective time (Galloway & Proctor 1983). The field is

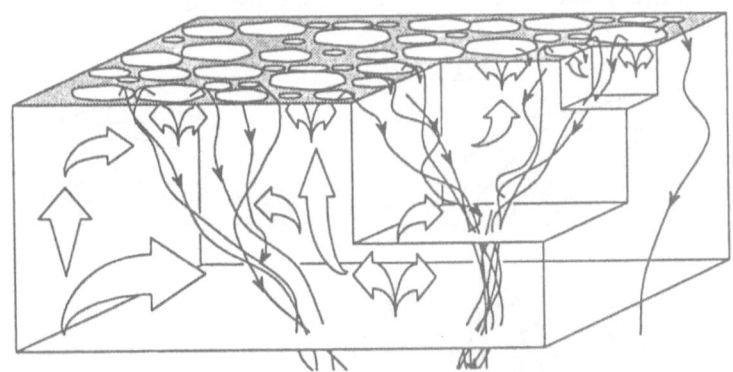

Fig. 1. Flow lines showing the merging of the downdrafts on successively larger scales (derived from numerical simulation results). The cool and concentrated downdrafts are immersed in a gentle ascending background. The boxes cut out illustrate how the same process occurs on three different scales. It should be noted that there is no clear cell structure in the vertical direction. From Spruit, Nordlund & Stein (1990).

also concentrated at the bottom center of the convective cells (bottom of the CZ on the sun). Strong magnetic fields are created and the Laplace force ($j \times B$) becomes important there. When included, it stops further concentration of the field when the equipartition of energy, between the magnetic field and the flow, is reached.

Nordlung & Stein (1990) have set up 3D numerical simulations of magneto-convection. The magnetic field is concentrated by the flows, as in the kinematic approach, but it reacts on the fluid and stabilizes the turbulent convection (it decreases the competition between adjacent convective cells, making the convective pattern more quiet, and the downward motions are extended to larger depths). The concentration with depth of the downflows implies the increase of the angular velocity of the flows (angular momentum conservation): the down-drafts are then natural places where the magnetic field is twisted. The field is also accumulated at the base of the CZ, and from time to time emergence of flux occurs. All these aspects (and others) are clearly visible in the new simulations of Brandenburg (this issue).

The evolution of the CZ plasma fixes the spatial and temporal evolution of the currents and magnetic field at the bottom of the corona. We will now study the reaction of the coronal field to coherent motions.

3 Current Sheet Formation

In ideal MHD, the formation of current sheets in a 2D potential magnetic configurations with an X point has been known for a long time (see the reviews of Amari 1991 and Van den Oord 1993). When new boundary conditions are imposed by converging photospheric motions, a smooth solution cannot be achieved without reconnecting field lines. Parker (1972, 1985) generalized this approach: he claimed that, under ideal MHD conditions, current sheets are necessarily formed as the con-

4   P. Démoulin

sequence of arbitrary smooth boundary motions. Currently there is no general proof of this hypothesis.

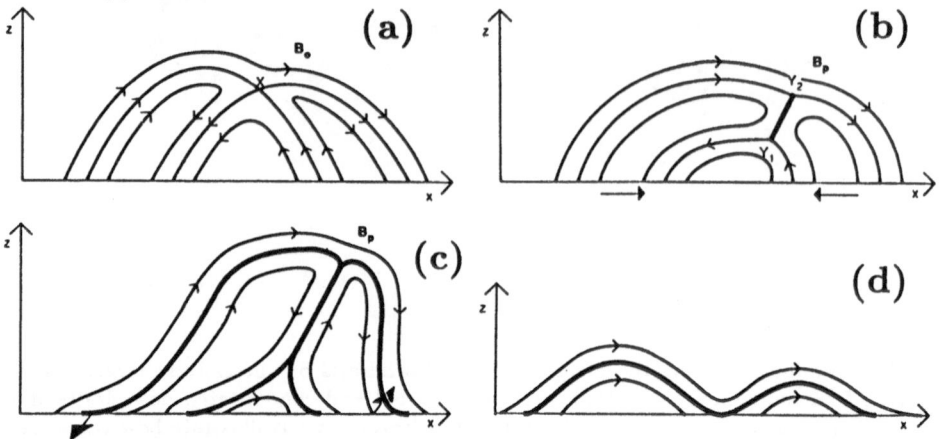

Fig. 2. Formation of current sheets. (a) In a 2D potential configuration with an X point, converging photospheric motions induce the formation of a current sheet around the X point ((b)) while shearing motions create a current sheet all along the separatrices ((c)). Replacing the initial potential field (a) by a force-free field, general photospheric motions create current sheets all along the separatrices as in (c). The presence of a separator is not needed provided some field lines are tangent to the photosphere as in (d). From Vekstein et al. (1991).

As it was shown (Van Ballegooijen 1985, Aly 1987, Antiochos 1987, Zweibel & Li 1987), the formation of current sheets is probably not as general as Parker first thought (Parker believed it could take place in an initial homogeneous field and for small footpoint displacements). But in 2D field configurations having separatrices, several analytical examples have been calculated in magnetic configurations invariant to translation where the field is force-free or in equilibrium with the plasma pressure (Low 1987, 1992, Low & Wolfson 1988, Wolfson 1989, Vekstein et al. 1991, 1992).

Indeed this mechanism is fairly general. A current sheet is in general formed all along the separatrices when continuous (but arbitrary) photospheric motions are applied to a magnetic configuration having a separator (the intersection of two separatrices) or field lines tangent to the photosphere (Fig. 2). From the force equilibrium and $div \boldsymbol{B} = 0$, a 2D field is expressed as:

$$\boldsymbol{B}(x, z) = \left(\frac{\partial A}{\partial z}, B_y(A), -\frac{\partial A}{\partial x} \right) . \tag{3}$$

The photospheric shearing displacement d_y, imposed at field-line foot-points, is related to the field by the equation governing the field lines:

$$d_y = B_y(A) \int_{\text{field line}} \frac{ds}{\sqrt{B_x^2 + B_z^2}}, . \tag{4}$$

The integral in (4) is discontinuous across a separatrice. So even with a continous displacement d_y imposed by coherent subphotospheric motions, $B_y(A)$ is necessarily discontinous on both sides of a separatrix in order to satisfy (4). Current sheets are then naturally formed along the separatrices.

A few counter examples exist which are structurally unstable (any slight departure in the boundary conditions implies the formation of a current sheet all along the separatrices). For an initial potential configuration and converging motion a current sheet is only formed at the separator (Priest & Raadu 1975). For configurations having field lines tangential to the photosphere (as in Fig. 2d) an asymmetric shear creates no current sheet (exept if the separator emerges from the photosphere). Some numerical simulations fail to detect current sheets (Karpen et al. 1990, 1991) because they are in this precise situation (see the answer of Low 1991, to Karpen et al.).

A rigourous extension of the mathematical proof to 3D configurations does not yet exist, but the above discontinuities are general (Zwingmann et al. 1985, Moffatt 1987, Van Ballegoijen 1988, Amari & Aly 1990 and ref. above). Because small scale lengths are created, the resistive term (last term of (2)) becomes important and separatrices are then natural places where reconnection can occur.

4 Reconnection Processes

Magnetic reconnection has been studied for thirty years. The initial very slow reconnection models have been replaced by models where the field can be reconnected at nearly the Alfvén speed (Jardine et al. 1988, 1992, Priest & Forbes 1992b). The rate of reconnection depends mainly on the externally-imposed boundary conditions. In the new models the magnetic field lines are strongly curved, there are jets along the separatrices and reversed current spikes at the ends of the diffusion region such as those obtained in numerical simulations. The advances in 2D reconnection processes are reviewed by Malherbe (1987) and Priest (1992), and these are linked to solar phenomena by Aly (1992) and Forbes (1992). But the most promising advance is the exciting new work of 3D reconnection which began few years ago.

Following Axford (1984), magnetic reconnection is usually defined in terms of the violation of the field line conservation property. It results in a localized diffusion region with an electrical field component parallel to the magnetic field. The physical conditions leading to reconnection are, however, controversial. At least four view points have been presented:

1. Hesse & Schindler (1988) set up the theoretical basis of 3D magnetic reconnection using a local description of the field given by the Euler potentials (α, β): $\boldsymbol{B} = \nabla\alpha \times \nabla\beta$. Because, at a given time, Euler potentials are constant along each field line, they allow us to determine whether two plasma elements stay on the same field line (no reconnection) or not (reconnection). With this formalism, the induction equation has a clear physical meaning. In particular, in order to have unbroken field lines and flux conservation, Hesse & Schindler derived the constraints satisfied by the resistive term.

2. Priest & Forbes (1989) generalised the 2D reconnection approach around an X point to 3D by looking for a field line which has an X-type topology in a plane orthogonal to it. They called such a line a "potential singular field line". They showed that, in general, there is a continuum of such lines and they proposed

that the imposed boundary flows select the particular line on which reconnection occurs.

3. Lau & Finn (1990) used a kinematic reconnection approach where the field is 3D and time dependent (however, since the field is imposed, the force balance equation is not solved). In the absence of nulls, singularities in the velocity field are present on the separatrices with the same dependence as in models invariant to translation with a non vanishing field at the separator. With nulls, the singularities in the velocity field are weaker on the separatrices, but stronger ones are present at the separator and at the nulls.

4. Priest & Forbes (1992a) showed how 3D reconnection without null points can occur. Reconnection takes place both at the singular line (separator) and on separatrices surfaces (where a rapid flipping of field lines occurs).

The above works are in fact complementary. Hesse & Schindler set up the equation to describe, in a simple manner, the local reconnection processes (if the magnetic field does not vanish). In their view, 3D reconnection occurs when the resistivity is locally enhanced but they do not explain the cause of this enhancement. However the formation of a current sheet along separatrices (Sect. 3) is a natural way to increase locally the resistive term in (2); then separatrices are natural locations for reconnection (Lau & Finn, Priest & Forbes). It is worth noting that a local analysis cannot be used to localize the reconnection region (a continuum of potential singular field lines is present locally). These theoretical developments can be tested not only in laboratory experiments (see for example Bogdanov et al, this issue) but also in flaring active regions, by comparing the topology of the extrapolated photospheric field with the locations of the energy release.

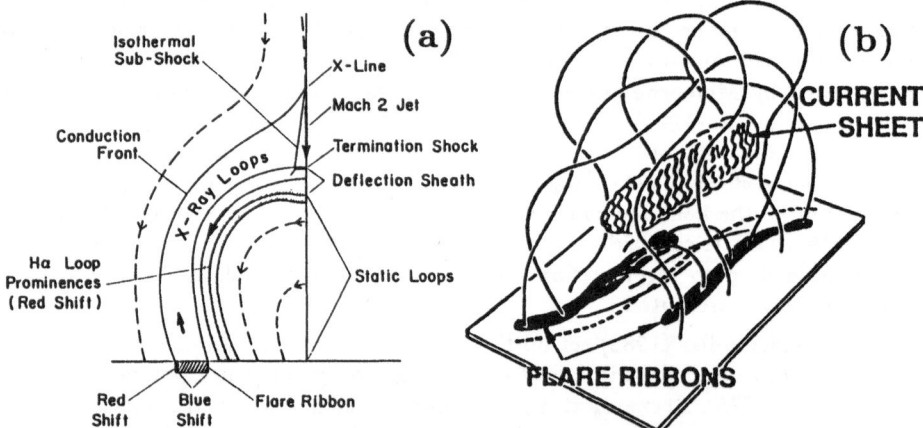

Fig. 3. (a) Schematic representation showing the MHD shocks, the H_α loops and ribbons generated by reconnection. Dashed lines denote magnetic field lines and dashed-dot lines the conduction front generated by field annihilation. From Forbes & Malherbe (1986) and Schmieder et al. (1987). (b) Magnetic field configuration of a typical two-ribbon eruptive flare during the eruption. From Moore & Roumeliotis (1992).

5 How Solar Flares Happen ?

Flare models have to explain how and where magnetic energy is released in solar flares and in particuliar the existing link between chromospheric flare ribbons sometimes separated by more than 100 Mm. "Post" flare loops are seen to form between the flare ribbons during the development of most flares (see Schmieder 1992 for a review) suggesting a model with reconnection (Fig. 3). In order to model an observed region we need to extrapolate the magnetic field from the photosphere to the corona. Important difficulties (linked to the presence of concentrated currents and of separatrices) are present in 3D force-free field computations using observed magnetograms. Currently they are still not solved (Amari & Démoulin 1992), and magnetic extrapolation is limited to linear force-free fields. This prevents us from actually testing quantitatively models for eruptive flares (Fig. 3) with observed data. However, much can be done for moderately sheared configurations because the magnetic topology, being a global property, is weakly sensitive to the exact distribution of coronal currents.

Following the theoretical suggestions of Sweet (1969), Baum & Brathenal (1980) and Hénoux & Somov (1987), Gorbachev & Somov (1988, 1989) were the first to investigate the link between observed flare ribbons and the magnetic topology of the corresponding active region. Since then an increasing number of investigations have clearly related the separatrices with the location of Hα flare ribbons (Fig. 4). They are magnetically connected by field lines passing close to the separator. Since reconnection is expected to take place on the separatrices (Sect. 4), these studies strongly suggest that this is the process driving energy release in solar flares. The above is true not only for 3D quadrupolar regions, Fig. 4a,c (similar to theoretical 2D1/2 configurations of Fig. 2), but also for bipolar regions with an "S" shaped inversion line (Fig. 4b) and even for bipolar regions with a near-potential field and a near-straight inversion line (Fig. 4d) where a simple arcade topology seems, at first, more appropriate !

Lin & Gaizauskas (1987), Ding et al. (1987), Hagyard (1988), Romanov & Tsap (1990) and Canfield et al. (1991) have found photospheric current concentrations at the border of flare ribbons. Mandrini et al. (1993), Démoulin et al. (1993a,b), Van Driel et al. (1993) confirmed this for other flares and showed that some currents are close to the separatrices. Moreover they found two currents of opposite sign, linked in the corona by field lines. This indicates that the energy is presumably stored in these field-aligned currents. Two scenarios are then compatible with the observations. First, the energy can be accumulated at the separatrices in current sheets. At some point in the evolution the current sheet become unstable, as in the emerging flux model of Heyvaerts et al. (1977), and the flare starts. In the second scenario, the energy is accumulated in a twisted flux tube. In this case the flare starts when the evolution brings the twisted flux tube to the separatrix locations.

The above observational results show that magnetic energy is released at the separatrices. This is exactly the theoretical picture which has emerged recently (Sect. 3 and 4). Note that the current sheets, described in Sect. 3, are thin and smooth only when viewed at the active region scale. At lower scales, tearing instabilities coupled to turbulence destroy this ideal view. Reconnection certainly has a very different rate depending on the position on the separatrix, and the MHD shocks generated have a

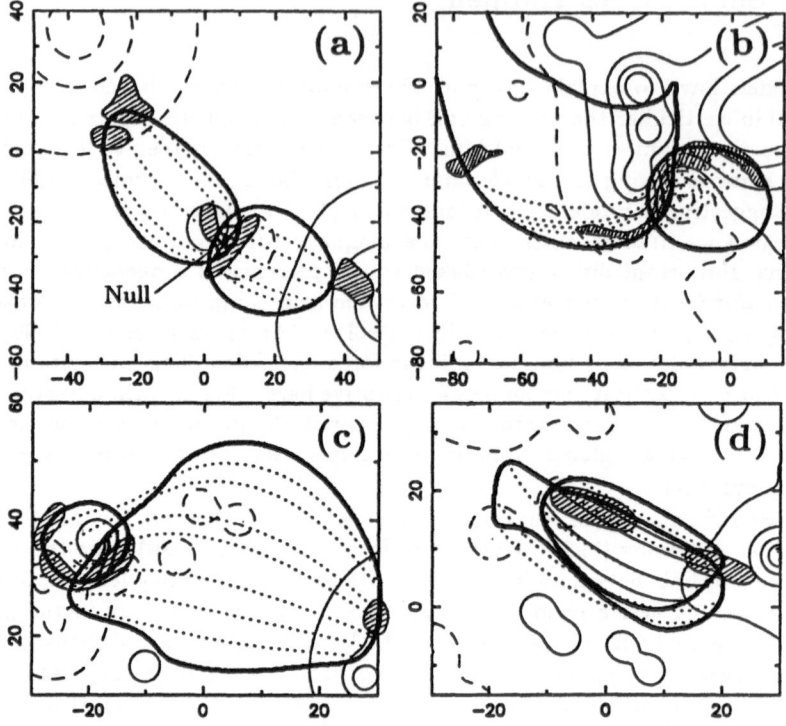

Fig. 4. Intersection of the computed separatrices with the chromosphere (thick lines) for flaring regions: (a) AR 2372 (Mandrini et al. 1993), (b) AR 2776 (Démoulin et al. 1993a), (c) and (d) AR 2511 (Démoulin et al. 1993b). The H_α ribbons are reported by hatched regions. Some field lines close to the separatrices have been drawn in the internal (external) connectivity cell with continuous (dotted) lines. The isocontours of the vertical field are $\pm 100, 500, 1000, 1500$ G.

disturbing effect on the medium. Moreover, like ribbons, separatrices change place during the eruption, and then a large volume of the active region field is involved in the whole flare. The present large-scale scenario is consistent with the constraints imposed by particle acceleration (Vlahos 1992). Particles can be accelerated at many reconnecting points and MHD shocks in the vicinity of separatrices.

Several improvements are needed to constrain even more the physics involved. How are coronal currents formed ? Is it by spot motions (Gesztelyi et al. 1986), or by photospheric twisting motions (Martres et al. 1970, Brandt et al. 1988, Hénoux & Somov, 1993). Do flux tubes emerge twisted, or do current sheets accumulate energy on separatrices ? Do we have neutralized or un-neutralized currents (Melrose 1991, Wilkinson et al. 1992) ? The answers to these questions are severely limited by the accuracy of the transverse field measurements and the 180° ambiguity. Moreover extrapolation of the field configuration with concentrated currents is still out of reach; this prevents us from modelling the time evolution of flare ribbons.

6 Does Reconnection Occur at all Scales in the Corona ?

While less demonstrative than the flare case, some evidence of reconnection processes also exists at other spatial scales on the Sun.

Poletto et al. (1993) used topological considerations to explain the large-scale spread of X-ray activity seen in a complex of two active regions. They showed that the "sympathetic activity" observed could be explained by the topology of linkages between the two regions.

An instability in a large twisted flux tube is thought to be at the origin of eruptive flares and CME (see for example Moore & Roumeliotis 1992, Forbes 1990, Steele & Priest 1990). In 2D models the two ribbons of eruptive flares are at the intersection of separatrices with the chromosphere (Fig. 3). As the twisted flux tube erupts, field lines are transported to the separatrices and reconnected. At the chromospheric level the energy is deposited in two expanding ribbons. Due to the difficulties outlined in Sect. 5, there have been as yet no 3D quantitative studies.

The presence of cancelling fluxes at photospheric level below prominences is well known (Martin 1990). The presence of sharp and bright lower boundaries in some well resolved quiescent prominences may be due to stronger condensation induced by reconnection (the lower boundary of the prominence being the separator of the magnetic configuration) as in the simulation of Malherbe et al. (1984). The energy released may explain the bright chromospheric corridor seen below some quiescent filaments (Martres et al. 1966; Zirin 1988 p. 274; Martin 1990). Finally during a prominence eruption, prominence feet are progressively detached from the chromosphere. Faint H_α brightenings are sometimes seen in eruptive prominence events; but it is in the He 10,830 line that Harvey & Recely (1984) have observed them best (for up to 60 hours !).

At lower scales, photospheric magnetic fields are concentrated in small flux tubes (Stenflo 1990). From time to time there is emergence of new flux, or cancellation between opposite polarities. Muller & Roudier (1992) have followed the rapid and complex motions of photospheric network bright points (usually interpreted as magnetic flux tubes). Aly (1990) and Strauss (1991) showed that such complex velocity patterns lead to the formation of separatrices (field lines, even initially as close as we want, will be separated if they lie on both sides of the surface separating two velocity cells). We can recover the situation observed in flares (Sect. 5) but, apart from integration along the line-of-sight, the time and spatial resolution of present instrumentation is not sufficient to observe the locations of energy release.

In the attempt to analyse the observed small-length scale features, it is worth noticing that the spatial resolution of magnetograms is crucial since it implies not only a flux spreading but also a net loss of flux when opposite polarities are present. A particular example can be found in Mandrini et al. (1991, 1993), where a fifth ribbon cannot be explained by the topology deduced from the observed magnetogram closest in time to the flare. In fact an emerging bipole was present and, 2 days later, each polarity size was well above the spatial resolution of the instrument. Then the deduced magnetic topology can explain the presence of a similar ribbon (for an homologuous flare). Without the time sequence observations, this kind of flare ribbon would not have been related to the magnetic topology !

There is indirect evidence that the same processes occur over a large range of spatial length scales. Dennis (1985)and Lin et al. (1984) show that the distribution of flare occurrence rate is a power law of the peak count rate of hard X-rays with logarithm slope 1.8. The slope is independent of the solar cycle. Lu & Hamilton (1991) explain these characteristics by invoking a self-organized critical state as found in avalanches. They defined a magnetic field vector on a 3D grid. Randomly a grid vertex is chosen and the field is increased. When the difference between the field, at some vertice V with its neighbour average, is greater than a critical value then reconnection is supposed to occur locally. The magnetic stress is then decreased at V, but increased at neighbouring points, where the critical value can be exceeded. It results in an avalanche of reconnections. Statistically, they can be any size. Their energy spectrum has no characteristic length, and it does not depend on the critical value assumed. However in such a model, the energy is liberated at random places in successive events, and consequently cannot explain homologous flares. The observed position of flare ribbons shows that flares are not purely statistical events. They reflect the coherence of the magnetic field, current and proper-motion pattern observed at the photospheric level over several days. From Sect. 3, coronal magnetic stresses are preferentially located near the separatices. With this coherence restriction, flares may be initiated according to the Lu & Hamilton model. The coronal field is "ready to go" (see Einaudi, this issue), but only near precise locations !

7 Conclusion

By imposing the time-dependent flux concentration pattern, the convective zone imposes a magnetic topology on the corona. Because many concentrated fluxes of mixed polarity are usually present, the topology of the coronal field is necessarily complex. The imposed photospheric motions of field line foot points overdetermine the magnetic field equations in the ideal MHD limit, and then current sheets are formed on separatrices. There, the resistive term becomes important and reconnection occurs. Such evolution has been tested successfully on flares, and it is thought to be general, going from micro-flares to eruptives flares and CME. It may also be the origin of the coronal heating. From nano-flares to eruptive flares the physics of the corona is a competition between the convective zone stress, which accumulates energy, and reconnection processes, which decrease this stress and liberate energy. The physics involved is still far from being understood, particularly in the 3D context.

Acknowledgements

I thank Tahar Amari, Jean Claude Hénoux and George Simnett for hepful comments on this paper.

References

Aly, J.J. (1987): Interstellar Magnetic Fields, eds. R. Beck, R. Grave, Springer, Springer Verlag, New York, p. 240

Aly, J.J. (1990): in "The Dynamic Sun", ed. L. Dezsõ, Public. Debrecen Obs., p. 176

Aly, J.J. (1992): in Proc. "International Conference on Plasma Physics", Innsbruck, Austria

Amari, T. (1991): in Advances in Solar System Magnetohydrodynamics, Eds. E.R Priest, A. W. Hood., Cambridge University Press, p. 173

Amari, T. & Aly, J.J. (1990): A&A, 227, 628

Amari, T: & Démoulin, P. (1992): in Proc. of the Workshop "Méthodes de détermination des champs magnétiques solaires et stellaires", Observatoire de Paris, p. 187

Antiochos, S. (1987): ApJ, 312, 886

Axford, W.I. (1984): Geopys. Monogr. Ser., vol 30, ed. E.W. Hones, AGU, Washington D.C., p. 1

Baum, P.J. & Brathenal, A. (1980): Solar Physics, 67, 245

Brandt, P.N., Scharmer, G.B., Ferguson, S., Shine, R.A., Tarbell, T.D. & Title, A.M. (1988): Nature, 335, 238

Canfield, R.C., de La Beaujardière, J.F. & Leka, K.D. (1991): Phil. Trans. R. Soc. Lond. A, 336, 381

Démoulin, P., Mandrini, C.H., Rovira, M.G., Hénoux, J.C. & Machado, M.E. (1993a): submitted to Solar Phys.

Démoulin, P., van Driel-Gesztelyi, L., Schmieder, B., Hénoux, J.C., Csepura, G. & Hagyard, M.J. (1993b): A&A, 271, 292

Dennis, B.R. (1985): Solar Phys, 100, 465

Ding, Y.J., Hagyard, M.J., de Loach, A.C., Hong., Q.F. & Liu, X.P. (1987): Solar Phys., 109, 307

Forbes, T. (1990): JGR, 95, A8, 11, 919

Forbes, T. (1992): in Proceedings of the Iguazú Meeting, eds. Z. Svestka, B.V. Jackson & M.E. Machado, Lecture Notes in Physics, Springer-Verlag, p. 79

Forbes, T. & Malherbe, J.M. (1986): ApJ, 302, L67

Galloway, D.J. & Proctor, M.R.E. (1983): Geophys. Astrophys. Fluid Dynamics, 24, 109

Gesztelyi, L., Karlicky, M., Farnik, F., Gerlei, O. & Valnicek, B. (1986): in "The Lower Atmosphere of Solar Flares", ed. D.F. Neidig, Sacramento Peak, p. 163

Gorbachev, V.S. & Somov, B.V. (1988): Solar Phys., 117, 77

Gorbachev, V.S. & Somov, B.V. (1989): Sov. Astron, 33, 1, 57

Hagyard, M.J. (1988): Solar Phys., 115, 107

Harvey, K.L. & Recely, F. (1984): Solar Phys., 91, 127

Hénoux, J.C. & Somov, B.V. (1987): A&A, 185, 306

Hénoux, J.C. & Somov, B.V. (1993): Adv. in Space Res., in press

Hesse, M. & Schindler, K. (1988): JGR, 93, A6, 5559

Heyvaerts, J., Priest, E.R. & Rust, D.M. (1977): ApJ, 216, 123

Jardine, M. (1991): in "Mechanisms of Chromospheric and Coronal Heating", Springer-Verlag, p. 588

Jardine, M., Allen, H.R., Grundy, R.E. & Priest, E.R. (1992): JGR, 97, A4, 4199

Jardine, M. & Priest, E.R. (1988): J. Plasma Phys., 40, 143

Karpen, J.T., Antiochos, S.K. & DeVore, C.R. (1990): ApJ, 356, L67

Karpen, J.T., Antiochos, S.K. & DeVore, C.R. (1991): ApJ, 382, 327

Lau, Y.T. & Finn, J.M. (1990): ApJ, 350, 672

Lin, Y. & Gaizauskas, V. (1987): Solar Phys., 109, 81

Lin, R.P., Schwartz, R.A., Kane, S.R., Pelling, R.M. & Hurley, K.C. (1984): ApJ, 283, 421

Low, B.C. (1987): ApJ, 323, 358

Low, B.C. (1991): ApJ, 381, 295

Low, B.C. (1992): A&A, 253, 311

Low, B.C. & Wolfson R. (1988): ApJ, 324, 574

Lu, E.T. & Hamilton, R.J. (1991): ApJ, 380, L89

Malherbe, J.M. (1987): Thèse de doctorat d'état, Université Paris VII

Malherbe, J.M., Forbes, T. & Priest E.R. (1984): ESA SP-220, p. 119

Mandrini, C.H., Démoulin, P., Hénoux, J.C., Machado, M.E. (1991): A&A, 250, 541

Mandrini, C.H., Rovira, M.G., Démoulin, P., Hénoux, J.C., Machado, M.E. & Wilkinson, L.K. (1993): A&A, 272, 609

Martin, S.F. (1990): in Proc. IAU Colloq. No. 117, eds. V. Ruždjak & E. Tandberg-Hanssen, Springer-Verlag, p. 1

Martres, M.J., Michard, R. & Soru-Escaut, I. (1966): Ann. Astrophys., 29, 249

Martres, M.J., Soru-Escaut, I. & Rayrole, J. (1970): in Solar Magnetic Fields, IAU 43, ed. R. Howard, p. 435

Melrose, D.B. (1991): ApJ, 381, 306

Moffatt, H.K. (1987): in Advances in Turbulence, eds. G. Comte-Bellot & J. Mathieu, Springer Verlag, New York, p. 228

Moore, R.L. & Roumeliotis, G. (1992): in Proceedings of the Iguazú Meeting, eds. Z. Svestka, B.V. Jackson & M.E. Machado, Lecture Notes in Physics, Springer-Verlag, p. 69

Muller, R. & Roudier, T. (1992): Solar Phys. 141, 27

Nordlund, A. & Stein, R.F. (1990): in "Solar Photosphere: Structure, Convection and Magnetic Fields", Proc. IAU Colloq No 138, ed. J.O. Stenflo, Dordrecht, Kluwer

Parker, E.N. (1972): ApJ, 174, 499

Parker, E.N. (1985): Geophys. Astrophys. Fluid Dyn., 34, 243

Poletto, G., Gary, G.A. & Machado, M.E. (1993): Solar Phys., 144, 113

Priest, E.R. (1992): Proceedings of the Iguazú Meeting, eds. Z. Svestka, B.V. Jackson & M.E. Machado, Lecture Notes in Physics, Springer-Verlag, p. 15

Priest, E.R. & Forbes, T. (1989): Solar Phys., 119, 211

Priest, E.R. & Forbes, T. (1992a): JGR, 97, A2, 1521

Priest, E.R. & Forbes, T. (1992b): JGR, 97, A11, 16757

Priest, E.R. & Raadu, M.A. (1975): Solar Phys., 43, 177

Romanov, V.A. & Tsap, T.T. (1990): Sov. Astron., 34, 656

Schmieder, B. (1992): Proceedings of the Iguazú Meeting, eds. Z. Svestka, B.V. Jackson & M.E. Machado, Lecture Notes in Physics, Springer-Verlag, p. 124

Schmieder, B., Forbes, T.G., Malherbe, J.M., Machado, M.E. (1987): ApJ, 317, 956

Spruit, H.C., Nordlund, A. & Stein, A.M. (1990): Ann. Rev. Astron. Astrophys., 28, 263

Steele, C.D.C. & Priest, E.R. (1990): Solar Phys., 125, 295

Stenflo, J.O. (1989): A&AR 1,1

Strauss, H.R. (1991): ApJ, 381, 508

Sweet, P.A. (1969): Ann. Rev. Astron. Astrophys., 7, 149

Van Ballegooijen, A.A. (1985): ApJ, 298, 421

Van Ballegoijen, A.A. (1988): Geophys. Astrophys. Fluid Dyn., 41, 181

Van den Oord, G.H.J. (1993): Proc. of COSPAR meeting, in Advances in Space Research

Van Driel-Gesztelyi, L., Hofmann, A., Démoulin, P., Schmieder, B., Csepura, G. (1993): submitted to Solar Phys.

Vekstein, G.E. & Priest E.R. (1992): ApJ, 384, 333

Vekstein, G.E., Priest E.R. & Amari, T. (1991): A&A, 243, 492

Vlahos, L. (1993): Proc. of COSPAR meeting, in Advances in Space Research

Wilkinson, L.K., Emslie, A.G. & Gary, G.A. (1992): ApJ, 392, L39

Wolfson R. (1989): ApJ, 344, 471

Zahn, J.P. (1988): in "Solar and Stellar Physics", eds. E.H. Schröter, M. Schüssler, Lecture
    Notes in Physics, 292, Springer-Verlag, p. 55
Zirin, H. (1988): Astrophysics of the Sun, Cambridge University Press
Zweibel, E.G. & Li, H.S. (1987): ApJ, 312, 423
Zwingmann, W., Schindler, K. & Birn, J. (1985): Solar Phys., 99, 133

Line-Tying in a Gravitationally Stratified Atmosphere

R.A.M. Van der Linden and A.W. Hood

University of St. Andrews, Dept. of Mathematical and Computational
Sciences, St Andrews KY16 8PP, Scotland

Abstract: We study the influence of the gravitational stratification of the photosphere
on magnetic line-tying. It has been demonstrated (Goedbloed et al, 1991) that in ideal
MHD with zero gravity, the anchoring of magnetic field lines in the dense photosphere
(modelled as a jump in density) cannot completely remove *any* coronal MHD instability.
Consequently, the commonly used rigid wall line-tying conditions have to break down
close to marginal stability. We demonstrate, using the (localized) ballooning ordering, that
when the effects of gravity are included, coronal modes can be completely stabilized by the
density stratification. Therefore, the marginal stability points predicted using rigid wall
conditions are still relevant.

1. Introduction

The anchoring of the footpoints of the coronal magnetic field lines in the photosphere
is often used to explain the long-term stability of magnetic structures in the solar
corona (Raadu, 1972). Usually, line-tying is simulated by using boundary conditions
on the coronal perturbations at the interface with the photosphere. Two sets of *ideal*
MHD boundary conditions have been commonly used: the *rigid wall* conditions,
where the plasma displacement is set equal to zero, and the *flow-through* conditions
where a flow across the boundary is allowed along the magnetic field lines, but
motion perpendicular to the field is still required to be zero. The rigid wall conditions
normally represent the most stabilizing constraint.

Hood (1986a) demonstrated that for modes far from marginal stability, applying
rigid wall conditions is equivalent to solving the equations for corona and photo
sphere with a large density jump. This suggests that the rigid wall boundary conditions constitute the best representation of the line-tying effect. Still, the transition
from marginally unstable to stable equilibria (as ρ_p is increased) remained to be
clarified. Hood et al (1989) and Hardie et al (1991) showed that when an unstable
mode approaches marginal stability in the high photospheric density limit, rigid wall
boundary conditions are no longer satisfied. The violation of line-tying conditions
close to marginal stability may account for the results of Goedbloed et al (1991).
These authors showed, using the energy principle, that in ideal MHD, and *in the*

absence of gravity, density cannot have any influence on the pure stability of the plasma. Whatever the density value and stratification be, a stable plasma remains stable and an unstable plasma remains unstable (though the *size* of the growth rate can be strongly reduced).

In view of the above, a common statement has been that line-tying conditions may be violated if the motions are slow enough to allow the photosphere to respond (Hood et al, 1989, Hassam, 1991). However, if the photosphere is indeed allowed to respond to coronal motions, it is imperative that the calculated response takes proper account of gravity, which is an important effect in the photosphere. For instance, gravity enters in the energy principle and introduces an explicit density dependence. Imposing perturbations with zero vertical components removes gravity again, but also implies that the most debated rigid wall line-tying condition is satisfied.

2. Model and Method of Solution

In this paper we use a simple model for a magnetic arcade to study the effect of the gravitational stratification of the solar atmosphere on perturbations of coronal origin. In particular we aim to demonstrate that using rigid wall conditions can give reasonable estimates of the marginal stability points and growth rates. We follow the analysis used by Hood (1986a) and Hardie et al (1991), but include a constant, vertical gravitational acceleration and a more realistic stratification in density and temperature. Only the thin temperature transition region is idealized as a jump in (temperature and) density across a single interface. In cylindrical coordinates (r, θ, z) we take for the dimensionless magnetic field and pressure

$$\boldsymbol{B} \equiv (0, B_\theta(r), 0) = \left(0, \frac{r}{1 + r^2}, 0\right), \tag{1}$$

$$p = \frac{1}{2(1 + r^2)^2} + p_0 \exp\left(-\frac{y}{H_i}\right). \tag{2}$$

Density and gravitational scale height are given by

$$\rho = \rho_i \exp\left(-\frac{y}{H_i}\right) \qquad H_i = \frac{p_0}{\rho_i g}. \tag{3}$$

where $y \equiv r \sin\theta$ is a vertical (Cartesian) coordinate, and $i = c$ or p respectively to indicate coronal or photospheric values. The linear stability equations are obtained by linearizing the equations of ideal magnetohydrodynamics (MHD). We consider normal mode solutions with 'growth rate' s. Despite the simplicity of the above equilibrium, it is still really two-dimensional, and solving the full linear stability equations would require a substantial effort. The relevant information can however be obtained from the analysis of localized modes using the ballooning ordering (Connor et al, 1979, Hood, 1986a,b). In this method all linear quantities are assumed to have the form

$$f(r, \theta, z) = f(\theta) \exp\left[in(S(r) + z)\right], \tag{4}$$

where $n \gg 1$ (producing a rapid variation across field lines). The stability equations are then expanded in powers of n to determine the leading order solutions for the amplitude function and $S(r)$. If the interest is not in the radial structure of the modes, we put $S'(r) = 0$ and solve the highest order equations for the growth rate s as a function of radius. This yields ordinary differential equations (along the field lines) for localized modes (on individual magnetic surfaces). How these local modes relate to global modes of the system is clarified in e.g. Connor et al (1979) and Van der Linden et al (1992). The relevant equations for the present study are obtained by neglecting non-ideal effects in the general derivation of Hood et al (1989) and are given in Van der Linden et al (1993).

3. Some Numerical Results

We solve these equations for stable and unstable configurations, and increase the photospheric density from coronal to realistic values. The local growth rate curves are labelled with mode numbers m, generally reflecting the number of nodes in the solution, and relating to the (integer) wave numbers for the uniform case. We only consider modes for which ξ_r is even about the summit ($\theta = \pi/2$) of the arcade.

When $\gamma = 1$ (i.e. only isothermal perturbations are allowed), the equilibrium considered is unstable even when rigid wall boundary conditions are applied. Taking the typical values $g = 0.001$ and $p_0 = 0.01$ for the dimensionless gravity and ambient pressure constants, we find (as in Hood, 1986a) that the instability growth rate in the periodic field tends towards the rigid wall value as the ratio of photospheric to coronal density ρ_p/ρ_c is varied from 1 to its realistic value, (see Fig. 1). Also, both parallel and perpendicular components of the plasma displacement (or velocity) at the boundary tend to zero. In this case therefore the rigid wall conditions are clearly recovered. For more details of this calculation see Van der Linden et al (1993).

For $\gamma = 5/3$, the equilibrium is stable when rigid wall conditions are used, but unstable (to $m = 0$ and $m = 1$) in the periodic field with uniform density. By the energy principle argument of Goedbloed et al (1991), if $g = 0$, increasing the photospheric density in the periodic field reduces the growth rate but cannot completely remove the instability (this result can also be quite easily recovered in the present analysis). Thence, the remaining unstable mode cannot satisfy the rigid wall boundary conditions, and it must have non-zero plasma displacement at the boundary and in the photosphere. To properly describe this mode then, gravity has to be included, because it has a strong effect in the photosphere. Also, because gravity introduces a stable stratification and the energy principle argument no longer holds in general (density appears in the energy integral), we may expect that the marginal stability points predicted by rigid wall conditions do correspond to a change in sign of s^2.

In Fig. 2 the local growth rate of the $m = 0$ mode is plotted versus the radial coordinate r for increasing values of the photospheric density parameter ρ_p. It is obvious that the region of instability is getting smaller, and even for the relatively low value $\rho_p \approx 30$ the equilibrium is stable to this mode of perturbation.

In Fig. 3 the same curves are shown for the $m = 1$ mode. Again the marginal stability point is shifting, and magnetic flux surfaces are changing from unstable to stable because of the inertial line-tying effect. For this mode, full stabilisation over the entire arcade could not be obtained for reasonable values of the density jump, but this just reflects the weakness of the model, in that the field lines in the low

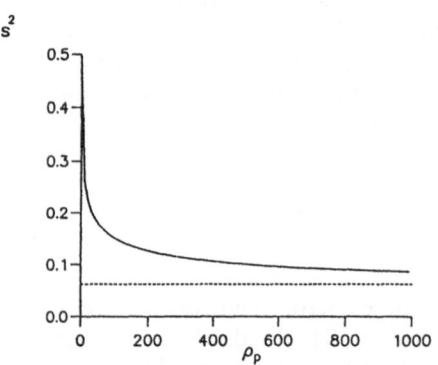

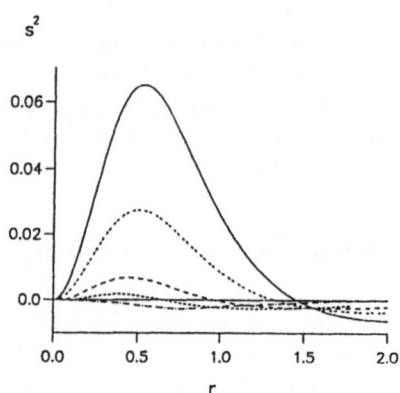

Fig. 1. Local growth rate at $r = 0.5$ plotted versus the photospheric density ρ_p for $\gamma = 1$, $g = 0.001$ and $p_0 = 0.01$. The horizontal dashed line represents the growth rate obtained for the same parameters but using rigid wall conditions on the coronal disturbances.

Fig. 2. Local growth rate curves as a function of radius for the even $m = 0$ mode with $g = 0.001, p_0 = 0.01, \gamma = 5/3$, for $\rho_p \approx 1.0$ (full line), 3.16 (-- --), 10.0 (– –), 17.0 (---), 31.6 (–·–·–).

r region do not penetrate more than a few scale heights into the photosphere. In reality, of course, the field lines should not be imagined to be periodical, but do extend deep down to the base of the convection zone, and consequently also the low r region should become stable (as it would if we further increased ρ_p). But even as it is now, it is clear that with gravity included, the inertial line-tying effect has fully stabilized the outer region of the magnetic arcade.

Although the results presented above indicate that when gravity is included, the rigid wall line-tying conditions correctly predict the stability properties, the modes in Figs. 2 and 3 do not as such correspond to the eigenmodes obtained when rigid wall conditions are applied. Indeed, in Fig. 4 it is shown that the local eigenmode at $r = 0.5$ does not satisfy rigid wall conditions at all. On the contrary, close to marginal stability, the displacement amplitude starts to increase instead.

4. Conclusions

We have determined in this contribution the influence of (purely external) gravity on the photospheric line-tying effect. We have shown that the rigid wall line-tying conditions provide reasonable estimates of the stability of the system, although individual eigenmodes may not satisfy the rigid wall conditions close to marginal stability.

When rigid wall boundary conditions predict instability, then the full model is indeed also unstable, and the instability growth rate is approximately equal to the value obtained using the rigid wall boundary conditions. The unstable eigenmodes

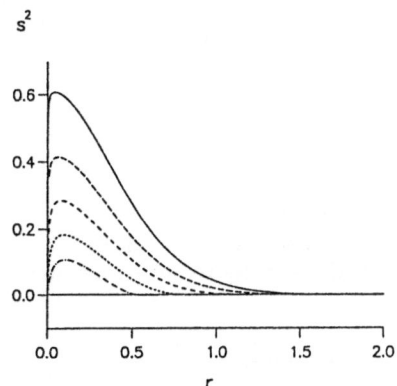

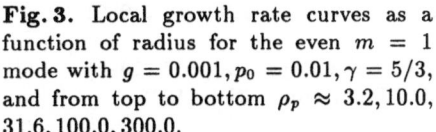

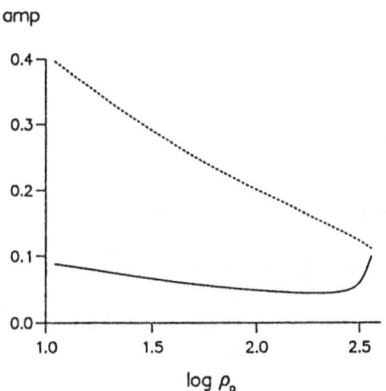

Fig. 3. Local growth rate curves as a function of radius for the even $m = 1$ mode with $g = 0.001, p_0 = 0.01, \gamma = 5/3$, and from top to bottom $\rho_p \approx 3.2, 10.0, 31.6, 100.0, 300.0$.

Fig. 4. Amplitude of ξ_r (full line) and $\xi_{\parallel}$ (dashed line) at the photospheric boundary $\theta = 0$ (normalised to $\xi_r(\theta = \pi/2)$) for the $m = 1$ mode with $\gamma = 5/3$ and $r = 0.5$ (stable for rigid wall conditions).

have negligible amplitude at the interface between corona and photosphere, and correspond one-to-one with the unstable modes for the reduced model with rigid wall conditions. Because the unstable modes are evanescent in the photosphere, and the gravitational scale height in the corona is typically much larger than the typical magnetic length scales, the influence of gravity is very small. The analysis for unstable modes therefore reduces essentially to that performed by Hood (1986a).

When the rigid wall line-tying conditions predict stability however, it is important to include gravity. If the coronal magnetic equilibrium is unstable in the absence of the density increase towards the corona, for $g = 0$ it remains unstable (albeit with reduced growth rates) whatever value is taken for the photospheric density. This feature has led to the common statement that some types of slowly growing instabilities cannot be stabilized by the anchoring of magnetic field lines in the dense photosphere, 'because the photosphere has time to respond' (e.g. Hood et al, 1989, Hassam, 1990). It should be kept in mind however, that a 'response of the photosphere' already implies that the modes are not purely coronal in nature any more, but have a non-negligable photospheric amplitude. In practice, since the photospheric density has been taken much higher than the coronal value, this implies that the instabilities are effectively driven in the photosphere, at least partly. It is then required to answer the question whether the photosphere *can* in reality drive the instabilities, which implies that the non-negligible effect of gravity and the gravitational stratification has to be included. This is what has been done in this paper, and it appears that indeed gravity removes the ability of the photosphere to drive instabilities, and therefore full stability is found when the photospheric density is sufficiently high and the field lines penetrate sufficiently deep into the photosphere.

It may not be obvious that the influence of gravity on other types of instabilities will be similar to what we found here, especially when non-ideal terms are included or even constitute the driving force for the instabilities. Is is interesting to point out, however, that in effect very similar conclusions have been reached for the thermal instability in coronal loops with rigid magnetic fields by Craig and McClymont (1987).

References

Craig, I.J.D., McClymont, A.N.: Astrophys. J. **318**, 421 (1987)

Connor, J.W., Hastie, R.J., Taylor, J.B.: Proc. Roy. Soc. London **A365**, 1 (1979)

Goedbloed, J.P., Halberstadt, G., Van der Linden, R.A.M.: Annales Geophysicae, Suppl. **9** (EGS), C561 (1991)

Hassam, A.B.: Astrophys. J. **348**, 778 (1990)

Hardie, I.S., Hood, A.W., Allen, H.R.: Solar Phys. **133**, 313 (1991)

Hood, A.W.: Solar Phys. **105**, 307 (1986a)

Hood, A.W.: Solar Phys. **103**, 329 (1986b)

Hood, A.W., Van der Linden, R.A.M., Goossens, M.: Solar Phys. **120**, 261 (1989)

Raadu, M.: Solar Phys. **22**, 425 (1972)

Van der Linden, R.A.M., Goossens, M., Hood, A.W.: Solar Phys. **140**, 317 (1992)

Van der Linden, R.A.M., Hood, A.W., Goedbloed, J.P.: in preparation (1993)

Magnetic Fields Surrounding Coronal Holes

Václav Bumba [1], Miroslav Klvaňa [1], Július Sýkora [2]

[1] Astronomical Institute of the Academy of Sciences of the Czech Republic,
251 65 Ondřejov, Czech Republic

[2] Astronomical Institute of the Slovak Academy of Sciences, 059 60
Tatranská Lomnica, Slovak Republic

Abstract: During July and August 1992 we succeeded in measuring photoelectrically longitudinal magnetic fields in areas surrounding four coronal holes, estimated from λ 10830 Å spectroheliograms. We were able to estimate values of the longitudinal magnetic flux in these areas and in their active regions, as well as in parts of coronal holes, covered by our measurements.

1 Introduction

Our knowledge of the concrete relations between coronal holes and solar magnetic fields, measured inside and around individual coronal holes, is not very comprehensive. Papers like the one published recently by Wang et al. (1992), which reported the amount of magnetic flux estimated in a coronal hole, are not very frequent.

Recently, McIntosh (1992) directly correlated formation of coronal holes not only with the global distribution of magnetic fields on the solar sphere, but also with the occurrence of large and very active (super) regions, claiming that the connection between such active centers and coronal holes can be described as "intimate". There are also many papers published earlier, during the seventies, which deal with the mutual relations of the large-scale distribution of solar magnetic fields and of solar activity, including that of the emission of the solar corona (see for example, Svalgaard and Wilcox, 1976). But all these studies were only very general.

This is the reason why we undertook the following, although still very preliminary study.

2 Observational Data

During July and August 1992 we succeeded in measuring, relatively systematically, longitudinal magnetic and Doppler velocity fields, as well as the continuum and spectral line intensity distribution in areas surrounding four low-latitude coronal holes photoelectrically with our reconstructed Magnetograph II (Klvaňa and Bumba, 1993) . We also studied the evolution of magnetic and velocity fields of active regions related somehow to coronal holes. We took the outlines of coronal holes from the Preliminary Report and Forecast of Solar Geophysical Data, and from the Solar-Geophysical Data prompt reports.

As we have already mentioned, we measure the longitudinal component of magnetic field only. Thus, the same applies to the values of magnetic flux obtained. But as regards the areas from which the flux is calculated, our program corrects these areas for their geometrical shortening.

The sizes of the areas in which we made the measurements vary, but the size of most of the measured areas oscillate around 300" × 200" (5' × 3.3'). This means that we usually cover a relatively large neighborhood of the active region. We have estimated the flux values for the whole area measured, as well as for the rectangle bordering the active region.

3 Coronal Holes under Consideration

During the summer months of July and August 1992 we used practically all clear days, suitable for magnetographic observations, which were relatively frequent that year. We measured in the Fe I 5253.47 Å line, usually in all active regions on the solar disk. In most interesting active regions or areas on the solar disk our measurements were more frequent. But during this time we knew nothing of the existence of a coronal hole on the disk.

Comparing the measured areas with the distribution of active regions and other events on Hα Synoptic Charts, we found that some of our measurements were displaced very favorably around four coronal holes, covering parts of their areas several times, and in one case most of the hole itself. All this concerns the following four coronal holes:

No 43 C. M. P. around July 9, 1992, extending between latitudes $-10°$ and $+20°$, of positive polarity;

No 44 C. M. P. around July 19, 1992; extending between latitudes $0°$ and $+30°$, again of positive polarity;

No 47 C. M. P. around August 5, 1992; latitudes: $-10°$ and $+10°$ and

No 48 C. M. P. around August 15, 1992; latitudes: $+10°$ and $+30°$, both of positive polarity too.

4 Long–Term Development of the Studied Corona Holes

If we study the global distribution of the background magnetic fields and the related distribution of coronal holes during the recent two years (Carrington's rotations Nos 1837 – 1859), we see not only the close relation of coronal holes to the regular patterns in the longitudinal magnetic field distribution, but above all, the fact that their existence and development is a long–lasting process, we can follow during the whole studied time interval. We also see that a coronal hole requires a specific magnetic situation to be able to form. It develops in a certain configuration of the large-scale distribution of solar magnetic fields.

In this way we see that coronal holes Nos 44 and 48 are not only the same subjects, recurring in Carrington's rotations No 1858 and 1859, but also that this hole had developed over many rotations before. During the period of our studies it was only going through the last phases of its existence.

The other two coronal holes Nos 43 and 47, representing also the same recurrent coronal area, seem to be on the opposite side of their development process. The coronal hole they represent, is just at the beginning of its existence. During Carrington's rotation No 1958 it even slowed down its formation, to be in full strength again in the next rotation No 1859.

5 Results of our Magnetic Flux Measurements

Wang et al. (1992) estimated the value of the total flux in a field of view with the dimensions of $5' \times 4'$ in a coronal hole as high as 3.1×10^{21} Mx, claiming that about 8 % of that value came from the flux contained in ephemeral active regions (EARs).

As regards our estimates of the longitudinal component of magnetic flux (corrected for the area geometrical shortening), we first have to say that we did not observe EAR magnetic fields in our measurements, with the exception of the largest ones, and we thus studied active regions with clearly visible magnetic fields only.

We present our estimates of the longitudinal magnetic flux around coronal holes following the concrete situation on the solar disk during the day of our measurements. We took the day with sufficient number of magnetic field measurements and with the coronal hole close enough to its central meridian passage (C. M. P.). We had at our disposal the values of areas of all measured regions in 10^{-6} of the visible solar hemisphere, the total positive and negative fluxes and their differences in 10^{21} Mx (10 TWb), as well as the same quantities for rectangular regions bounding only the active region closely in each measured area. As concerns the flux, we took the positive and negative flux difference as the resulting flux emerging out from the measured or bounded area. We also give the total area of sunspots in the studied regions.

In this way we found that during the day of the C. M. P. of coronal hole No 43 (July 9, 1992), the total area of all measured regions , but one (NOAA 7225, being too close to the east limb), was 0.0787 of the visible hemisphere only. Its total positive flux $F\Sigma^+$ was $\approx 33.61 \times 10^{21}$ Mx, total negative flux $F\Sigma^- \approx -51.12 \times 10^{21}$ Mx, the difference of both fluxes being $F\Sigma \approx -17.51 \times 10^{21}$ Mx. Flux F^+ of all active regions (but the mentioned one) is $\approx 30.68 \times 10^{21}$ Mx, their negative flux

$F^- \approx -41.17 \times 10^{21}$ Mx and the difference of both fluxes $F \approx -10.54 \times 10^{21}$ Mx. The area containing NOAA 7219 covered by ≈ 11 % of its size also a very small portion of the coronal hole. There we found $F^+ \approx 0.59 \times 10^{21}$ Mx, $F^- \approx -0.30 \times 10^{21}$ Mx and $F \approx +0.29 \times 10^{21}$ Mx. The area of all sunspots A was $\approx 3284 \times 10^{-6}$ of the visible hemisphere.

During the next passage of this coronal hole under number 47, about one day after its C. M. P. (on August 7, 1992), we again measured the surroundings of all active regions visible on the disk, including a part of the coronal hole itself. The total measured area represents 0.053828 of the visible hemisphere, with $F\Sigma^+ \approx 54.59 \times 10^{21}$ Mx, $F\Sigma^- \approx -51.04 \times 10^{21}$ Mx, $F\Sigma \approx +3.54 \times 10^{21}$ Mx. The fluxes in all active regions were as follows: $F^+ \approx 30.72 \times 10^{21}$ Mx, $F^- \approx -17.91 \times 10^{21}$ Mx, $F \approx +12.81 \times 10^{21}$ Mx, A $\approx 1776 \times 10^{-6}$ of the visible hemisphere. This time we covered more than 30 % of the coronal hole with the measured area around NOAA 7245, the magnetic field of which strongly influenced the magnetic field of the coronal hole. The characteristics of the magnetic field of this portion of the coronal hole were: $F^+ \approx 8.82 \times 10^{21}$ Mx, $F^- \approx -5.33 \times 10^{21}$ Mx and $F \approx +3.49 \times 10^{21}$ Mx.

We studied the second coronal hole No 44 for the first time on July 20, 1992, about 1–2 days after its C. M. P.; it occupied a very large part of the visible solar hemisphere. Maybe, it was so large because of the very weak activity around it, with relatively small active regions. The total measured area was ≈ 0.088596 of the visible hemisphere, its $F\Sigma^+ \approx 26.95 \times 10^{21}$ Mx, $F\Sigma^- \approx -46.99 \times 10^{21}$ Mx, $F\Sigma \approx -22.78 \times 10^{21}$ Mx. The values related to the active regions only were as follows: $F^+ \approx 15.82 \times 10^{21}$ Mx, $F^- \approx -25.01 \times 10^{21}$ Mx, $F \approx -9.18 \times 10^{21}$ Mx, and the area of all spots A $\approx 609 \times 10^{-6}$ of the visible hemisphere. This time three of the measured regions also partially covered the coronal hole, i. e. about one quarter of its whole area. The fluxes of this part of the coronal hole were: $F^+ \approx 7.06 \times 10^{21}$ Mx, $F^- \approx -4.51 \times 10^{21}$ Mx, $F \approx +2.55 \times 10^{21}$ Mx.

This coronal hole recurred under number 48 considerably diminished in size. We observed it on August 16, 1992 when its center of mass was about two days past its C. M. P.. Again we measured all active regions on the disk. Two of them (NOAA 7262 and 7263) occupied a considerable part of this coronal hole. The whole measured area covered ≈ 0.058793 of the visible hemisphere with $F\Sigma^+ \approx 39.25 \times 10^{21}$ Mx, $F\Sigma^- \approx -63.94 \times 10^{21}$ Mx and $F\Sigma \approx -24.70 \times 10^{21}$ Mx. In the active regions we obtained $F^+ \approx 26.48 \times 10^{21}$ Mx, $F^- \approx -38.30 \times 10^{21}$ Mx, $F \approx -11.82 \times 10^{21}$ Mx and A $\approx 2011 \times 10^{-6}$ of the visible hemisphere.

We monitored the fluxes in both active regions covering this part of the coronal hole for three subsequent days (August 16, 17, 18, when they were too close to the west limb to be corrected for geometrical shortening). On August 16 we obtained $F^+ \approx 4.37 \times 10^{21}$ Mx, $F^- \approx -3.63 \times 10^{21}$ Mx and $F \approx +0.74 \times 10^{21}$ Mx, on August 17 $F^+ \approx 7.42 \times 10^{21}$ Mx, $F^- \approx -6.28 \times 10^{21}$ Mx and $F \approx +1.13 \times 10^{21}$ Mx.

As regards the flux in the measured areas, active regions and parts of coronal holes, we see that although the maximal values of positive and negative longitudinal fluxes in large active regions may be as high as a few tens of 10^{21} Mx, the resulting flux (the difference between the positive and negative flux for the studied area) relatively rarely exceeds the value of 10×10^{21} Mx. Usually, it amount to a few units of 10^{21} Mx, or even less.

As concerns the coronal holes, all four studied were estimated as positive coronal holes. This is in agreement with our measurements for all cases, but one, in which we covered a part of a coronal hole with a part the area we scanned. The obtained values of fluxes vary from a few tenths to a few units of 10^{21} Mx.

6 Relations of Coronal Holes with Active Regions

What seems to be meaningful is the fact that coronal holes No 47, 44 and 48 are connected with magnetic fields of older active regions, or rather with their parts of mainly positive polarity, with relatively high flux values (about 10×10^{21} Mx). In the case of coronal hole No 47, the active region (NOAA 7245) in its base changed its magnetic flux relatively rapidly from one day to the next. The same is true as in the case of coronal hole No 44 and region NOAA 7229 which developed on the disk several days before July 20), as well as in the case of coronal hole No 48. As we will demonstrate elsewhere, it seems that every coronal hole has its attached active region.

With the exception of the largest coronal hole of the ones studied (No 44), we can also clearly see the close relation of coronal holes with largest active regions preceding and following each coronal hole at a distance of about $40° - 50°$, as mentioned by McIntosh (1992), and as can clearly be seen on the large-scale distibution of solar magnetic fields. Again it seems that this behaviour of coronal holes to alternate regularly with strong magnetic fields of the large-scale background field network is very important for their existence.

The fact that we do not see coronal hole No 44 accompanied by large active regions may be related to the process of redistribution of coronal holes on the solar surface taking place on a global scale.

Acknowledgements. This work was supported by Grant No 30302 of the Academy of Sciences of the Czech Republik.

References

Klvaňa, M., Bumba, V. (1993): Proceedings of the XIV Consultation on Solar Phys. Karpacz, in press
McIntosh, P. S. (1992): in "The Solar Cycle", ASP Conf. Series 27, Harvey K. L. ed., p.14
Svalgaard, L., Wilcox, J. M. (1976): *Solar Phys.* **49**, 177
Wang J., Wang H., Shi, Z. (1992): in "The Solar Cycle", ASP Conf. Ser. 27, Harvey K. L. ed., p. 108

IV

Large-Scale Structure of the Corona

Coronal Heating Mechanisms

G. Einaudi[1], M. Velli[1,2]

[1]Dip. di Astron. e Sc. dello Spazio, Università di Firenze, 50125 Firenze, ,
Italy
[2]Dep. de Rech. Spat., Observatoire de Paris-Meudon, 92195 Meudon, France

Abstract: Thermal energy must be continually supplied to the solar corona to maintain its 10^6 °K temperature. In the first part of this paper we review the efforts which have been made in the past twenty years to find a viable mechanism to explain coronal heating, with special emphasis on the conditions of applicability of the existing theories and on the possibility that coronal heating may be intimately linked to other manifestations of solar activity such as solar flares. The interplay of the different aspects of solar activity is discussed within the unifying framework of MHD turbulence.

1 Introduction

The fact that the solar corona is hotter, by several million degrees, than the underlying solar surface, where all the energy originates, has attracted the curiosity of many solar physicists since its discovery. When local thermodynamic equilibrium (LTE) holds, it is not possible to tranfer heat from one location to another at a higher temperature without doing work.

X-ray images of the corona from Skylab and more recently from Yohkoh and rocket flights with NIXT (Golub et al. 1990), showing the extreme structuring due to the coronal magnetic field, suggest that solar activity in general and coronal heating in particular are manifestations of the complex interaction between the magnetic field and the photospheric plasma motions. In this framework, however, the mechanism which non-radiatively tranfers the abundant mechanical energy in the turbulent photospheric velocity field up to hundreds of thousands of kilometers above, and then dissipates this energy within 1-2 solar radii, is still poorly understood.

What is clear is that there is more than enough energy present in the convection to supply total coronal losses. Energy is essentially injected from the photosphere as a Poynting flux $S = c/4\pi \, (E \times B)$ where the electric field is induced by the photospheric motions perpendicular to the magnetic field (if we consider the photosphere to be a perfect conductor), so that $E = -\delta v \times B/c$. The Poynting flux is typically 5×10^7 erg/cm^2/sec for a flow speed of, say, 0.5 km/sec and a field of 100G. The boundary motions of magnetic footpoints are due essentially to the solar granulation, with characteristic speeds $\delta v \simeq 0.25 - 2$ km/sec, sizes l_c of order $l_c \simeq 10^3$ km, and lifetimes τ_c of order $\tau_c \simeq 300$ secs, as well as the supergranulation, with characteristic speeds $\delta v \simeq 0.3$ km/sec, sizes $l_c \simeq 3 \times 10^4$ km and lifetimes $\tau_c \simeq 10^5$ seconds. On the other hand one may recognize different types of regions with different energy fluxes required to balance radiative and conductive losses (Withbroe and Noyes 1977, Withbroe 1988): for active regions and X-ray bright points the estimated energy flux required is $\epsilon \simeq 10^7$ erg/cm^2/sec, for the quiet corona $\epsilon \simeq 8 \times 10^5 - 10^6$ erg/cm^2/sec and for coronal holes $\epsilon = 5 \times 10^5 - 8 \times 10^5$ erg/cm^2/sec, including solar wind losses.

From the above arguments it follows that two major questions must be answered: how is magnetic energy supplied to the corona, and how is it dissipated?

It is highly likely that different mechanisms are acting in different parts of the corona, an important distinction being between the open corona, where the magnetic lines of force emerging from the photosphere connect to the interplanetary field, and the closed corona, where the field lines take the shape of loops with two endpoints in the photosphere.

In the next Section we review some of the ways that have been proposed to heat the average corona, and will show how in some sense such mechanisms must be considered as large-scale descriptions, applying to the global space and time averaged corona. In Section 3 we discuss a framework which has been recently gaining ground and which puts the large scale solar flare and coronal heating on the same footing, respectively at the high energy and low energy ends of the spectrum of solar activity. Finally, in the last Section we outline a research program which might validate or disprove such a scenario.

2 Heating the Average Corona

The lack of LTE is a possible explanation of the high coronal temperature. The corresponding mechanism has been recently explored in detail by Scudder(1992), who shows that if the velocity distribution below the transition region is non-maxwellian with extended high-energy tails, a transition region with scale heights as small as 200 km may be formed because the core of the ditribution function is removed in climbing the gravitational potential well. But in the presence of LTE, because of the enormous ($\sim 10^{12}$) kinetic and magnetic Reynolds numbers, and the coincidence of magnetic and thermal structures, one is led to the conclusion that MHD waves and/or turbulent current dissipation must play a fundamental role.

How the corona responds to perturbations on the time-scale τ depends on the ratio τ/τ_a, with $\tau_a = l/V_a$ the time-scale for typical (Alfvén) wave propagation along a coronal loop. For $B = 100$ Gauss, n $= 10^9$ cm^{-3} and a loop length L $= 10^9$ cm this amounts to 1.5-2 seconds. Roughly speaking, if $\tau/\tau_a \leq 1$ perturbations will propagate as waves, while if the inequality is reversed, the corona will respond quasistatically, developing force-free magnetic fields with a consequent storage of magnetic energy.

Most heating theories can be classified as either wave theories, where waves (essentially Alfvén waves) are generated in the photosphere and above, carrying and dissipating energy into the corona, or current dissipation theories, which are based on the idea that the coronal magnetic field should spontaneously relax to lower energy once some kind of threshold has been exceeded. We also group some recent large scale turbulence (LST) models, which aim at describing coronal fields using turbulent eddy viscosities in this category.

2.1 Wave Dissipation Mechanisms

Three types of MHD waves exist: a) slow waves, which are essentially sound waves constrained to propagate along the magnetic field; these may be ruled out as a significant source of heating because of the small observed flux. b) Fast magneto-acoustic waves, which are however totally reflected at transition region heights, unless their wave-vectors are strictly aligned with the magnetic field, in which case they are indistiguishable from c) Alfvén waves. These appear to be the most promising candidates for coronal heating: because of their anisotropic dispersion relation, they do not suffer from reflection in the geometrical optics approximation. Also, Alfvén waves propagating away from the Sun are observed to be a dominant component of the turbulence observed in the solar wind, which could well be the remnant of a wave flux with sufficient energy for coronal heating (Belcher and Davis 1971, Mangeney et al. 1991).

It has been argued that Alfvén waves arising from photospheric motions could not contain a high enough flux because of the poor coupling between the photospheric Alfvén speed and the characteristic phase velocity of the dominant Fourier components of granular convection (Collins 1989). Also, Alfvén waves of periods greater than a few minutes no longer propagate according to geometrical optics and are strongly reflected in the chromosphere and transition region (Leroy 1981, Velli et al 1991). However, as pointed out in the next section, energy stored by slow motions in the photosphere might be released in bursts in the corona, where as a consequence large amplitude Alfvén waves would be generated (Longcope and Sudan 1992).

These problems notwithstanding, one must face the question of how to dissipate such waves within the corona. Because of the extremely small dissipation coefficients large spatial gradients must develop for damping to become significant. Since there is very little energy directly available at the dissipative scales, the generation of such scales must rely on the linear or nonlinear evolution of the waves. Hence, we must invoke either a nonlinear cascade (leading to shock formation and/or turbulence) or the interaction of waves with the non-uniformities of the background magnetic field.

For waves propagating along magnetic fields which are inhomogeneous in a transverse direction, two different effects have been shown to be important. The first one, known as phase mixing (Heyvaerts and Priest 1983), is due to the frequency (or wavelength) detuning between neighbouring oscillating magnetic field lines due to the Alfvén velocity gradients. As a result the oscillations become rapidly out of phase as they propagate (or evolve in time) and the wave fronts are rapidly deformed and corrugated transversely as the perturbation moves upwards. The second process, known as resonant absorption (Mok and Einaudi 1985; Lee and Roberts 1986; Hollweg 1987; Einaudi and Mok 1987 and references therein), is due to the pressure gradients associated with the wave, which have the tendency to concentrate the energy in vicinity of the point where the frequency of the wave is equal to the local Alfvén frequency, i.e., where $\omega = k_\parallel V_a$.

Corresponding to such processes there exist two different types of normal modes (Califano et al. 1992), with a dissipation rate which is very much enhanced with respect to that of the corresponding uniform case: in fact, it is independent of the magnetic Reynolds number S, defined here as $S = lV_a/\eta$, $\eta = c^2/4\pi\sigma$, σ being the electrical conductivity. We conclude that a random initial perturbation is efficiently dissipated when it reaches its asymptotic normal mode state. The question is how long it takes to generate the normal mode structure, i.e. the transient time (Malara et al. 1992). For open magnetic structures this is of great importance, since the transient time translates directly into an equivalent height for the formation of small scales which may be as great as a few solar radii.

The transient time can be exceedingly long if only linear effects are taken into account and no external drivers are present (Einaudi et al. 1993). Since nonlinear couplings for Alfvén waves derive from the interactions of waves travelling in opposite directions (Dobrowolny et al. 1980), the reflection back by the stratified atmosphere of the outgoing waves produced by photospheric effects seems important for heating open field regions by MHD waves. The possibility of the reflection of waves in radially stratified atmospheres has been recently discussed by several authors, see Velli (1993) and references therein.

The results just described, obtained in 1D configurations, indicate that wave dissipation might be a relevant heating mechanism for coronal holes and probably also for quiet region loops. For active regions, where more complex fields are likely to be present, not many results are presently available. We mention just two of them: propagation of Alfvén waves in a magnetic configuration containing X-points (Craig and Watson 1992: Bulanov et al. 1992) leads to normal modes with a dissipation rate (normalized to the real frequency) $\gamma\tau_a \sim -lnS$, and a similar damping rate is obtained if the coronal magnetic field is chaotic (Similon and Sudan 1989). In this case it is the exponential separation of neighbouring field lines which causes the rapid deformation of the wave fronts. For a detailed discussion of the propagation of MHD waves under solar conditions, see also Ulmschneider et al. 1991 and Einaudi et al. 1993.

2.2 Quasi-Static Current Dissipation Mechanism

Gold (1964) first put forward the idea that coronal heating is due to the dissipation of field-aligned electric currents. Convective flows below the solar surface cause a random footpoint shuffling of magnetic field lines, which in regions of closed magnetic topology cause a secular increase in the stresses within the coronal magnetic field.

Parker (1972) conjectured that such motions must lead to singularities (current sheets or filaments) appearing in the coronal field configuration. The relaxation of such currents would then result in coronal heating. Although Parker's conjecture has not been proved for continuous footpoint motions and fields lacking separatrices or X-points, it is certainly verified for the random motion of finite-sized flux tubes (discontinuities appear when different flux tubes come into contact).

Parker (1983,1988), Sturrock & Uchida (1981), Van Ballegooijen (1986) and Berger (1991a,1991b) among others, have further developed such 'stochastic' models of coronal heating. Numerical simulations by Mikic et al. (1989) confirm that the transverse magnetic field cascades to small scales, leading to an exponential growth of the coronal current. Neglecting the details, all of the above referenced papers obtain a heating function due to the processes considered which has the form $F_H = q\,B^2/4\pi\,\delta v$, where B is the normal (vertical) component of the magnetic field, δv

the rms photospheric velocity field and q an efficiency factor, which, depending on the model, may vary, as we shall now see, over a few of orders of magnitude.

Let us consider for example the model of Sturrock & Uchida (1981), who studied the random twisting of an isolated flux tube. The mean square winding angle $\theta^2(t)$ will depend on the characteristics of the turbulent photospheric velocity field. Considering for simplicity the flow to lie within the plane of the photosphere, the vorticity ω will contribute to the rotation in the form $d\theta^2/dt = \tau_c < \omega^2 >$ with τ_c the correlation time of the flow. A comparable contribution to the twist comes from shearing the fluxtube footpoints (as long as the tube is not perfectly circular). In terms of the correlation length λ_c and the rms velocity δv, $< \omega^2 >$ may be written as $4\delta v^2/\lambda_c^2$. For motions on the granular scale $\tau_c \simeq 800$ secs, $\lambda_c \simeq 8 \times 10^7$ cm, $\delta v \simeq 10^5$ cm/sec, the root mean square twist reaches about 4 turns in 24 hours.

How much energy is injected in this way? Consider a flux tube of length L, radius R and axial field strength B, the energy of the non-potential magnetic field in such a tube, which is associated with its azimuthal component created by the twist, is $\delta E = R^4 B^2 < \theta^2 > /16L$ and the input power per unit area becomes

$$F_H = \frac{3R^2 B^2 \tau_c \delta v^2}{16\pi\lambda_c^2 L}, \tag{1}$$

so $q = 3/4R^2/(\lambda_c L)$. For $B = 100$ Gauss, $R = \lambda_c$ and $L=10^{10}$ cm we obtain a heating rate $F_H = 5 \ 10^5$ ergs/cm^2/sec, an order of magnitude lower than the 10^7 required for active region heating.

Parker (1983) considered the braiding of several flux-tubes. For simplicity, consider a cartesian geometry in which the photosphere is represented by a pair of conducting planes separated by a distance L. The x, y axes will be taken in the planes parallel to the photosphere, while the z axis is orthogonal to the photosphere. The photospheric motions will move the fluxtube footpoints around in a random walk, generating a transverse magnetic field $B_\perp$, given by $B_\perp/l = B/L$, where $l(t)$ is the total transverse length of the field lines connecting two footpoints on the photosphere projected on the $x - y$ plane. When the transverse distance between the top and the bottom end points of the field lines, $s(t)$, is much less than the initial distance d_0 between footpoints of separate flux tubes, the field lines will be practically straight, and $l(t) = s(t)$. The mean square transverse distance separating two end points in random motions is $< s(t)^2 >= 8Kt$, where K is the diffusion coefficient (Berger 1991). The average input power then is given by:

$$F_H = \frac{KB^2}{2\pi L}, \tag{2}$$

yielding $3 \ 10^5$ergs/cm^2/sec. If however $s(t) \geq d_0$, the footpoints are close enough so that a significant amount of tangling between different flux tubes occurs and the field lines cannot be straight, then $l(t) > s(t)$. Taking the extreme case where $l(t)$ equals the total distance travelled by the two end points, $l(t) = 2\delta vt$, we get:

$$F_H = \frac{B_\perp^2 L}{4\pi t} = \frac{\delta v^2 B^2 t}{2\pi L}. \tag{3}$$

Parker (1983) considers supergranular motions with a correlation time $\tau_c = 5.0 \ 10^4$ secs and an axial field strength of 100 Gauss ($\delta v = 0.5$ km/sec). In this way one finally obtains the required energy flux of 10^7 ergs/cm^2/sec. A consequence of this choice is that reconnection should set in once the angle between neighbouring field lines exceeds $\theta = 14^0$, beyond which current sheet dissipation limits any additional current build up. The magnetic field variation across the current sheet is $\delta B_\perp = 25$ Gauss. We will come back to this point in the next Section.

Let us now describe some examples of large-scale turbulence models for coronal heating. The physics is essentially the same: what changes is the way in which the fields are described.

Heyvaerts and Priest (1984), consider solar coronal loops to be made up of force-free MHD turbulence undergoing a cyclic process of energization and relaxation subject to the boundary straining motions. They point out that not all of the energy above potential would be dissipated in such a turbulent relaxation, because of the approximate conservation of the total helicity during the relaxation process (Taylor 1974). They envisage a series of discrete steps during which a non-linear force-free field is built up separated by a set of relaxation events in which the field decays to a linear force-free field with the same total helicity. The power obtained in such a way depends on the ratio of dissipation time to energy build-up time. Applied to coherent photospheric motions, such as the shearing of an arcade, the model can produce a heating rate consistent with observations. It is unclear however whether the magnetic field should relax all the way down to a linear force-free field, because constraints besides total helicity conservation might conceivably play a role during relaxation.

Van Ballegooijen (1986) also considered the nonlinear cascade driven from boundary motions. Starting from an initially uniform field between two plates, representing the photosphere, he investigates the properties of the force-free field which is created by the boundary motions as a first order correction to the uniform field. In order to calculate the spectrum of the generated transverse magnetic field, van Ballegoojin assumes isotropy and homogeneity in planes parallel to the photosphere and also that the velocity and its derivative normal to such planes are independent statistical variables, with Gaussian statistics. Moreover the photospheric random motions are considered statistically stationary and the time-behaviour of the correlation length of the velocity gradients is specified in a way consistent with Parker's idea of a quadratic increase in the stored magnetic energy. Calculation of the magnetic power spectrum as a function of time then shows that an efficient cascade occurs; dissipation sets in when the diffusion time at the peak of the current density spectrum becomes of the same order as the braiding time. The resulting heating rate is a factor of ten lower than required for the active region corona, a discrepancy which might arise from an underestimate of the time at which the energy input saturates (which he takes to be the cascade time described above).

More recently Heyvaerts and Priest (1992) and Gomez and Ferro-Fontan (1992) have presented models for MHD turbulence in the solar corona. The approach of Heyvaerts and Priest attempts to achieve self-consistency: the photosphere drives a coronal turbulence, which is treated as an eddy viscosity term in the large-scale description of a viscous and resistive coronal equilibrium. The amount of energy driven into the corona by the photospheric motion depends on the coronal 'effective' Reynolds number. On the other hand the effective Reynolds number must have just the value required to dissipate the input energy. They consider the simple case of a

1-D incompressible photospheric shear flow and the EDQNM simulations of MHD turbulence in the Kraichnan (or mean magnetic field dominated) regime by Pouquet et. al. (1976): the velocity and magnetic field spectra are taken to depend on the wavevector k as $E_v(k) \sim E_b(k) \sim k^{-3/2}$. For a photospheric shear flow with a Kolmogorov spectrum, and characteristic size of 10^8 cm (the loop length being 10^9 cm), typical heating rates of order 10^6 ergs/cm^2/sec. are obtained.

Another fully turbulent model for coronal heating has been presented by Gomez and Ferro-Fontan (1992), who adapted the large-scale turbulence model of Canuto et al. (1987) to two-dimensional MHD. In this computation the goal is not to obtain the correct heating rate, which is fixed by the photospheric turbulent forcing (a broad band Kolmogorov-type spectrum), but rather to derive the coronal magnetic and velocity spectrum on the basis of a given photospheric velocity spectrum.

3 Heating and Flares : Different Aspects of the Same Physics?

3.1 Time and Space Scales in an Active Region

One problem with all the theories we have discussed so far is that their main purpose is to reproduce the required energy flux for heating, which is derived assuming a stationary corona. Therefore we must consider them as "average" theories which adopt a phenomenological treatment of turbulence.This approach can probably be appropriate for regions such as coronal holes and quiet coronal loops, where no other activity than heating takes place, but it must be taken with much care elsewhere. When dealing with active regions it seems unreasonable to build up a theory for coronal heating which neglects the existence of the energetically fundamental flares. To sustain a large active region modelled as a cube of side L $= 10^{10}$cm we need on average a flux equivalent to a power $W \sim 10^{27}$ erg/sec. A large flare occurring in the same region gives off $\sim 10^{32}$ ergs on a typical time scale of a few minutes, profoundly affecting the energetics. The measured distribution of flare numbers as a function of energy and duration might be considered as evidence of the highly intermittent nature of coronal MHD turbulence, both in time and in space, which a realistic active region heating theory must take into account. Otherwise (when the coronal turbulence is assumed stationary and homogeneous) the corresponding model actually considers averages over a time scale longer than the time T necessary to store 10^{32} ergs with the above quoted power W, that is $T \simeq 1$ day: but the observed timescales in an active region range from the few seconds typical of the impulsive phase of a flare down to the microseconds of structured microwave bursts.

The magnetic Reynolds number for a macroscale of 10^{10} cm, a field of 50 G, density $\sim 10^9$ cm^{-3} and temperature $\sim 2 \times 10^6$ ^oK is S $\sim 10^{13}$. Assuming a homogeneous and stationary coronal turbulence, one finds that the Taylor microscale, defined as the energetically weighted average length over the entire inertial range, is $\lambda \sim S^{-1/2}L \sim 3 \times 10^3$ cm and is spectrum independent provided the energy decreases with scale. The dissipative scale depends on the spectrum, $l \sim S^{-2/3}L \sim 20$ cm adopting the Kraichnan description of MHD turbulence. These numbers are of course indicative and represent just orders of magnitude, but there is no doubt that in order to explain solar activity in terms of dissipation of magnetic energy on

required strength of such fields can be obtained by examining the energetics of a
big flare. In fact, with the volume described above, to obtain a 10^{32} erg flare we
need to be able to dissipate a 50 Gauss magnetic field throughout the whole volume,
$\delta b^2/8\pi \times 10^{30} \simeq 10^{32}$ ergs.

The structure of the currents giving rise to such a field has been examined by
Berger (1991b), who argues that the field with the available energy should consist
of relatively few large scale tubes separated by narrow current sheets rather than
uniformly distributed small-scale flux tubes. Consider the latter case: the field lines
in each tube spiral about the central axis with a pitch angle θ. For a spiral a small
distance r away from the axis $\delta B_\perp/r\theta = B/L$, where L is the length of the tube
extending from one footpoint to the other and B is the vertical component of the
magnetic field. Adopting a particular model for the perpendicular field in a cross
section of the tube, Berger shows that $\theta = \delta B_\perp \pi L/4Bl_\perp$, where $l_\perp$ is the typical
island radius. Balancing now the energy input given by eq. 2 with the dissipative
term $\eta J_z^2 L$, it is easy to obtain an estimate for $B_\perp/l_\perp$: $\eta\pi B_\perp^2 L/32l_\perp^2 = KB^2/2\pi L$,
or $\theta = \sqrt{K/\eta}$, which, for $K = 200$ Km/sec and $\eta = 10^{-6}$ km/sec, gives a twist
$\theta = 1.4\ 10^4$ radians. Since from the random photospheric motions an rms twist
of a few turns is expected, as previously seen, Berger concludes that the current
must flow essentially in thin current sheets in a small fraction of the volume. The
essential point here is that it is not sufficient to generate through photospheric
motions the correct energy flux for coronal heating, but it is also necessary to worry
how to dissipate such flux: it is paramount to generate sufficiently small scales,
or sufficiently high twist in this example. In order to produce significant heat, or
magnetic energy dissipation, the current distribution must be highly intermittent,
and this is the only configuration compatible with observational constraints. In the
Van Ballagooijen cascade model (1986) for example, this occurs if the statistics
governing the cascade are non-Gaussian. Of course, this argument becomes even
stronger when applied to a big flare, since the power needed in this case is about
300 times larger than the power needed for heating.

What should be the intensity of typical current sheets? Numerical simulations
(in 2-D) appear to demonstrate that current sheets typically form with a thickness
of the order of the dissipative scale l, and width of the order of the Taylor microscale
λ (Biskamp and Welter 1989; Diamond and Biskamp 1990). Although we really have
no direct information about the appropriate scale to choose in the third dimension,
we may, on the basis of the arguments presented above, assume a length of order
of the macroscopic scale. Then the volume occupied by a single sheet is $\Delta V = l\lambda L$.
Numerical evidence would seem to imply a typical current sheet separation also
of the order of λ. The effective volume occupied by the individual current sheet
is therefore more like $dV = \lambda^2 L$. So the total current sheet number is at most
$N = V_{tot}/dV$ and the volume physically occupied by such sheets $V^* = N\Delta V = V_{tot}\Delta V/dV \simeq V_{tot}\lambda/L$. With the values of λ and l given above, $V^* \simeq 6 \times 10^{27}$ cm^3
, $\Delta V \simeq 6 \times 10^{14}$ and $N \simeq 10^{13}$.

It is easy to estimate the current densities flowing in the sheets by requiring that
the total energy dissipated therein match the energy released by a big flare, namely:

$$\frac{4\pi\eta}{c^2} J^2 V^* = \frac{E_{flare}}{\tau_{flare}}. \tag{4}$$

$$\frac{4\pi\eta}{c^2} J^2 V^* = \frac{E_{flare}}{\tau_{flare}}. \tag{4}$$

For the typical values considered above, we get $J \simeq 5 \times 10^8$ statamp/cm^2. The power released in each current sheet is then $W_{cs} \sim 3 \times 10^{16}$ erg/sec.

3.2 From a Thermodynamic to a Statistical Theory

The idea that solar coronal activity consists essentially of a superposition of elementary events or substructures was put forward by a number of authors (Gold 1964; Parker 1983,1988,1991; Haerendel 1987; Lu and Hamilton 1991; Vlahos 1993) with different details and on different bases.

As discussed in the previous Section, it is reasonable to believe that the corona is built up by a large number of small scale current sheets where dissipative processes take place. We shall call this local phenomenon an "elementary event" (Chiuderi 1993). Such a scenario is qualitatively similar to the one proposed by Parker, with the big difference that its elementary events leads to the observed nanoflares, whereas the one we are envisaging has a much less significant energy content. This difference is due to the recognition of the very low dissipative efficiency of the corona. It must be investigated whether, at some stage of the local build up of the current, some explosive phenomenon, such as a reconnective instability, can occur. If this is the case, the cascade could stop at a somewhat larger scale than the dissipative one defined above.

The fact that the corona is 'on average' heated to some million degrees is the signature that it must at all times contain a large number of dissipating current sheets at large number of locations. In fact, sustaining the average corona would require that 10^{13} elementary events take place incoherently in one day. On the other hand, explaining the occurrence of flares of all sizes, with a power law peak brightness distribution over 4 decades (Lin and Schwartz 1987), would need some form of coherent activation over time scales of few seconds, i.e. the duration of the explosive phase.

The complexity of the corona in such a highly dynamical scenario prompts us to move from a context of thermodynamical theories toward one of statistical theories. By analogy with the fact that a detailed knowledge of microscopic processes is not needed to describe the state of a complex dynamical system (such as an ideal gas), we may hope to be able to say something about solar activity without having to describe the basic processes causing the elementary event in detail.

It must be kept in mind, however, that some basic properties of the "molecules" of our "gas" (the elementary events) must be understood, and even if such events take place over regions of space very small compared to the overall size of the system, thermodynamical theories still apply perfectly to them. The statistical approach is necessary when trying to understand the global behaviour of an active region. To be more specific on this point, let us consider the problem of the coherent and incoherent activation of elementary events. Depending on the physics, still not understood, of single current sheets and their mutual interaction, the coherent activation producing flares may or may not be due to an external agent. In fact, recent studies concerning the notion of "self-organized criticality" have shown the possible decoupling of the importance of the cause from that of the effect.

numerical sandpile models where grains of sand are added at random sites (typically on a 2-d or 3-d Cartesian lattice). It is found that such extended, dynamical,
randomly-driven dissipative systems with many metastable states can evolve into a
statistically stationary situation where no intrinsic length or time scales are present.
In such a critical state, the system has a distribution of minimally stable regions of
all sizes: small perturbations give rise to avalanches or disruptions which can attain
regions of all dimensions, from the smallest (a single unstable site), to the largest
possible, i.e., the whole system. Because there are no preferred time or length scales,
the distribution of avalanche sizes and duration display a power-law behaviour. The
most interesting property of the self-organized critical state which distinguishes it
from other critical phenomena is its robustness. By definition, it is an attractor of
the dynamics of the system in response to a random external forcing.

An application to solar flares has been worked out by Lu and Hamilton (1991)
and Lu et. al. (1993), based on the ideas of Parker for current sheet dissipation. In
this model, the corona is discretized as a 3-D cubic lattice, on each point of which a
magnetic field vector is assigned. The sites are 'energized' (throughout the volume)
on a time-scale much greater than the nearest neighbour propagation time τ_a by
randomly perturbing the magnetic field, subject to a preferred direction of increase
(if magnetic field were added and subtracted symmetrically, there would be the
possibility of energy removal as well as addition, and a critical state would not be
attained). When the difference between the magnetic field on a site i and the average
value over nearest neighbours $dB_i = B_i - 1/6\sum_{nn}B_{nn}$ is greater than a critical
value dB_c, a 'burst' of reconnection occurs and the magnetic field is redistributed
according to the rule $B_i \rightarrow B_i - 6/7dB_i$, $B_{nn} \rightarrow B_{nn} + 1/7dB_i$, i.e. the gradient is
redistributed to its nearest neighbours, causing additional reconnection bursts and
so on (a flare is any collection of such events). At the end of each flare, the process
of random energization is repeated. A statistically stationary state is reached after
an initial transient where all the energy additionally injected is dissipated. The
configuration of the field is a non-trivial minimally stabilized state: small noise
leads to flares of all sizes distributed according to a power law.

In such a scenario no large amplitude external drivers are needed. However the
rules of the game have been conjectured and must be confirmed by a understanding
the physics of energy release in the elementary current sheets. The possibility of a
driven flare, in which the coherent activation of all current sheets is due, for example,
to a large amplitude photospheric perturbation (of which a typical case which comes
to mind is the triggering by emerging flux), can not be ruled out.

In any case there is strong observational support for the correctness of a picture
of the corona built up by a large number of "elementary events". For example,
"turbulent events" in EUV and by CIV line brightenings in the magnetic network
over bipoles of dimensions of $2 - 4 \times 10^3$ km, with durations around 100 sec (Dere et
al. 1991). Another example comes from the rapid fluctuations in brightness observed
in the transition region and from the short bursts of X-rays observed by Lin et al.
(1984). Finally, the idea that nanoflares are due to a coherent activation of a fraction
of the existing current sheets is supported by the observed morphology of nanoflares,
microflares and flares (Machado et al. 1988a,1988b).

4 Conclusions

a) A lack of local thermodynamical equilibrium at photospheric levels can account for the high coronal temperatures, but the degree of realism of the consequent model must still be carefully checked.

b) Open regions and quiet coronal loops can be heated by waves if their amplitudes are such that, at least locally, nonlinear (local) effects become important, which along with linear (non-local) interactions of the travelling waves with the background nonuniformities produce small scales and consequent dissipation. In magnetically open regions the need of nonlinear mode mode coupling requires reflection of the outgoing waves due to stratification. This aspect of the problem must be further investigated.

c) In active regions, where the magnetic field topology is much more complex, we believe that the existing theories can explain the heating of the "average" corona, but do not describe the detailed physics of the phenomenon. We have discussed what we mean by the "average" corona and why the explanation of coronal heating may not be considered independently from other manifestations of solar activity such as solar flares. If the elementary building block of solar activity is in fact the dissipating current sheet with scales of the order of meters or less, it will remain hidden from direct observation for the foreseeable future. In such a scenario, progress is required in understanding the physics of such current sheets, both as far as their origin (here waves can play an important role) and their energy release properties are concerned. The existence of a critical shear for effective dissipation is at present still just a conjecture, and the differing estimates of such a critical shear are no more than free parameters allowing to fit different models of coronal heating.

A fundamental concept is that of the coherent activation of all "elementary events" and a key point would be the comprehension of the cause of this coherent activation, whether a statistical fluctuation or a driven phenomenon. In this respect a search for stronger observational correlations between photospheric and coronal events could be a good strategy to improve our understanding of coronal activity.

Acknowledgements

We would like to thank C. Chiuderi and L. Vlahos for stimulating discussions.

References

Bak, P., Tang, C. and Wiesenfeld, K.: 1987, *Phys. Rev. Letts.*, **59**, 381

Bak, P., Tang, C. and Wiesenfeld, K.: 1988, *Phys. Rev. A*, **38**, 364

Belcher, J.W. and Davis, L.: 1971, *J. Geophys. Res.*, **76**, 3534

Berger, M. A.: 1991, *Astron. Astrophys.*, **252**, 369

Berger, M.: 1991, *Advances in Solar System MHD*, eds. E.R. Priest and A.W. Hood, (Cambridge), p.241

Biskamp, D. and Welter, H.: 1989, *Phys. Fluids*, **B1**, 1964

Bulanov, S.V., Shasharina, S.G. and Pegoraro, F.: 1992, *Plasma Phys. and Contr. Fusion*, **34**, 33

Califano, F., Chiuderi, C. and Einaudi, G.: 1992, *Astrophys. J.*, **390**, 560

Canuto, V.M., Goldman, I. and Chasnov, J.: 1987, *Phys. Fluids*, **30**, 3391

Chiuderi, C.: 1993, *Scientific Requirements for Future Solar- Physics Space Missions*, eds. P. Maltby and B. Battrick, (ESA SP-1157), p.25

Collins, W.: 1989, *Astrophys. J.*, **343**, 499

Craig, I.J.D. and Watson, P.W.: 1992, *Astrophys. J.*, **393**, 385

Dere, K.P. et al.: 1991, *J. Geophys. Res.*, **96**, 9399

Dobrowolny, M., Mangeney, A. and Veltri, P.: 1980, *Phys. Rev. Letts.*, **45**, 144

Diamond, P.H. and Biskamp, D.: 1990, *Phys. Fluids*, **B2**, 681

Einaudi, G. and Mok, Y.: 1987, *Astrophys. J.*, **319**, 520

Einaudi, G., Chiuderi, C. and Califano, F.: 1993, *Adv. Space Res.*, **13**, 85

Golub, L. et al.: 1990, *Nature*, **344**, 842

Gold,T.: 1964, *The Physics of Solar Flares*, ed. W.Hess, (NASA SP-50), p.389

Gomez, D.O. and Ferro-Fontan, C.F.: 1992, *Astrophys. J.*, **394**, 662

Haerendel, G.: 1987, *Proc. 21st ESLAB Syposium*, (ESA SP-273)

Heyvaerts, J. and Priest, E.R.: 1983, *Astron. Astrophys.*, **117**, 220

Heyvaerts, J. and Priest, E.R.: 1984, *Astron. Astrophys.*, **137**, 63

Heyvaerts, J. and Priest, E.R.: 1992, *Astrophys. J.*, **390**, 297

Hollweg, J.: 1987, *Astrophys. J.*, **312**, 880

Lee, M.A. and Roberts, B.: 1986, *Astrophys. J.*, **301**, 430

Leroy, B.: 1981, *Astron. Astrophys.*, **91**, 136

Lin, R.P. et al.: 1984, *Astrophys. J.*, **283**, 421

Lin, R.P., Schwartz, R.A.: 1987, *Astrophys. J.*, **312**, 462

Longcope, D.W. and Sudan, R.N.: 1992, *Phys. Fluids*, **B4**, 2277

Lu, E.T. and Hamilton, R.J.: 1991, *Astrophys. J.*, **380**, L89

Lu, E.T. et al.: 1993, *Astrophys. J.*, "Solar flares and avalanches in driven dissipative systems", in press

Mangeney, A. M., Grappin, R. and Velli, M.: 1991, *in Advances in Solar System MHD, Cambridge*, MHD turbulence in the solar wind, pg. 327

Malara, F. et al.: 1992, *Astrophys. J.*, **396**, 297

Machado, M.E. et al.: 1988a, *Astrophys. J.*, **326**, 425

Machado, M.E. et al.: 1988b, *Astrophys. J.*, **326**, 451

Mikic, Z., Schnack, D.D. and Van Hoven, G.: 1989, *Astrophys. J.*, **338**, 1148

Mok, Y. and Einaudi , G.: 1985, *J. Plasma Phys.*, **33**, 199

Parker, E.N.: 1972, *Astrophys. J.*, **174**, 499

Parker, E.N.: 1983, *Astrophys. J.*, **264**, 642

Parker, E. N.: 1988, *Astrophys. J.*, **330**, 474

Parker, E.N.: 1991, *Astrophys. J.*, **372**, 719

Pouquet, A., Frisch, U., Léorat, J.: 1976, *J. Fluid Mech.*, **77**, 321

Scudder, J.D.: 1992, *Astrophys. J.*, **398**, 299

Similon, P.L. and Sudan, R.N.: 1989, *Astrophys. J.*, **336**, 442

Sturrock, P.A. and Uchida, Y.: 1981, *Astrophys. J.*, **246**, 331

Taylor, J.B.: 1974, *Phys. Rev. Letts.*, **33**, 1139

Ulmschneider P., Priest, E.R. and Rosner, R. eds.: 1991, *Mechanisms of Chromospheric and Coronal Heating*, (Springer Verlag, Heidelberg), Chapt. 3

van Ballegooijen, A.A.: 1986, *Astrophys. J.*, **311**, 1001

Velli, M., Grappin, R. and Mangeney, A.: 1991, *Geophys. Astrophys. Fluid Dyn.*, **62**, 101

Velli, M.: 1993, *Astron. Astrophys.*, **270**, 304

Vlahos, L.: 1993, *Adv. Space Res.*, **13**, 122

Withbroe, G. and Noyes, R.W.: 1977, *Ann. Rev. of Astron. & Astrophys.*, **15**, 363

Withbroe, G.: 1988, *Astrophys. J.*, **325**, 442

Coronal Heating via Nanoflares

Giannina Poletto [1], Roger Kopp [2]

[1] Osservatorio Astrofisico di Arcetri, Largo Enrico Fermi 5, 50125 Firenze, Italy
[2] Los Alamos National Laboratory, Los Alamos, NM 87545, USA

Abstract: It has been recently proposed that the coronae of single late-type main sequence stars represent the radiative output from a large number of tiny energy release events, the so-called nanoflares. Although this suggestion is attractive and order of magnitude estimates of the physical parameters involved in the process are consistent with available data, nanoflares have not yet been observed and theoretical descriptions of these phenomena are still very crude. In this paper we examine the temporal behavior of a magnetic flux tube subject to the repeated occurrence of energy release events, randomly distributed in time, and we show that an originally empty cool loop may, in fact, reach typical coronal density and temperature values via nanoflare heating. By choosing physical parameters appropriate to solar conditions we also explore the possibilities for observationally detecting nanoflares. Although the Sun is the only star where nanoflares might be observed, present instrumentation appears to be inadequate for this purpose.

1 The Coronal Heating Puzzle

The coronal heating mechanism has been a subject of debate over the last 40 years and still is not unambiguously identified. Following the acoustic wave heating hypothesis proposed around 1950, alternative processes, in which magnetic fields play a crucial role, were advanced as a result of observations acquired from space-borne experiments. Satellite data revealed that the solar corona consists of a hot gas whose radiative output occurs largely in the previously unaccessible X-ray region and is mainly concentrated in magnetically active areas. Subsequent observations of X-ray emission from other objects showed the presence of hot coronae in middle and lower main sequence stars. All this information was consistent with the hypothesis that the ultimate source of the energy responsible for the formation of stellar coronae resides in the convection zone. However, although the coronal energy requirement is easily met, neither acoustic, slow-mode, nor Alfvén waves seem capable of depositing the right amount of energy in the correct location. The situation had come nearly to a standstill, when new observations suggested a different approach to the problem.

Beginning in the late 70's Skylab, OSO-8, and subsequent missions reported the ubiquitous presence on the Sun of UV and X-ray small-scale, short-duration impulsive events that appeared to be the low-energy analogy of "classical" flares.

At present this family of mini-events includes a variety of members – UV jets, UV turbulent events, and hard X-ray events, among others – whose mutual interrelationship is not clear. These data also show that UV microflares, for instance, are much more frequent and widespread than are larger flares: Porter et al. (1984) reported almost continuous activity at the brightest sites of solar active regions over time intervals on the order of hours. More recently, Porter et al. (1993) have also shown that CIV microflares have impulsive counterparts in 3.5-5.5 keV X-ray emission. Hence, microflares are "real" flares, reaching temperatures in excess of that of the ambient corona.

These data raise a question: can this poorly explored population of small flares contribute significantly to the formation of coronae? Can an unobserved swarm of still smaller events, the nanoflares, – with energy releases typically on the order of 10^{-9} that of a large solar flare – be responsible for the existence of the solar corona itself? This question addresses not only the solar case: according to Butler et al. (1986), EXOSAT observations of flare stars show evidence for a microflare-heated corona in dMe stars and possibly in all late-type stars. Although neither the solar nor the stellar data are unambiguous, they seem to lead toward the same conclusion.

¿From a theoretical viewpoint, the formation of the corona as a result of energy release from a multitude of small-scale events has sporadically shown up in the literature, starting from a suggestion of Gold (1964) and being subsequently pursued by Levine (1974), Glencross (1975), Heyvaerts and Priest (1984), and Parker (1983, 1988, 1991). The latter author has pointed out that, whenever field-line footpoints wander randomly around each other (as a consequence of photospheric motions), a large number of localized tangential discontinuities arise. The nanoflares that purportedly build up the corona may originate via magnetic reconnection and dissipation processes localized at these discontinuities. An order of magnitude estimate of the nanoflare energy release, based on typical values of scale-length, footpoint velocity, and magnetic field strength in the solar atmosphere, supports the view of the solar corona as being maintained in a quasi-steady state by an ongoing succession of a great many nanoflares.

However, much remains to be done, both observationally and theoretically. In the solar atmosphere, neither individual nanoflares nor moving magnetic fibrils have been detected. From a theoretical point of view, there are a number of unexplored issues – not even the physics of the reconnection process giving rise to nanoflares has been adequately developed. In the present work, we consider the following problem. Assuming a nanoflare population – whose origin is not discussed here – we investigate whether these events are capable of building up a high-temperature, low-density, quasistatic corona. This question is not a trivial one, as it has still to be demonstrated that a cool empty loop, subject to repeated energy release events, may eventually reach a quasisteady state with the observed coronal properties, via the complex interplay of conduction, radiation, evaporation and gravitational settling. The next section offers a solution to this problem; our conclusions are drawn in Section 3.

2 Modeling a Nanoflare-Heated Corona

The temporal behavior of a magnetic loop subject to an energy release event has been dealt with by a number of authors, who described flaring loops both in the Sun and in stars (see, e.g., Vesecky et al. 1979; Pallavicini et al. 1983; Fisher et al. 1985; MacNeice 1986; Mariska 1987; Reale et al. 1988; Cheng and Pallavicini 1991). Usually detailed hydrodynamical codes are used to integrate numerically the full time-dependent partial differential equations for conservation of mass, momentum, and energy. These models predict the temporal profiles of temperature (T), density (ρ), and velocity (v) as functions of position along the flaring loop. The complexity of the method is justified when, as in the solar case, model predictions can be checked against high-spatial resolution observations. However, for stellar flares as well as for solar and stellar nanoflares, spatial resolution is entirely lacking and a less sophisticated, more flexible method can be adopted. Namely, analytically integrating the conservation equations for mass and energy over the loop semilength L, the temporal variations of the spatially averaged loop pressure and density [1] are given by the following set of total differential equations:

$$L\frac{d\overline{\rho}}{dt} = (\rho v)_b \approx \overline{\rho v} \ . \tag{1}$$

$$\frac{L}{\gamma - 1}\frac{d\overline{P}}{dt} = \overline{Q_F(t)}L + \frac{\gamma}{\gamma - 1}\overline{P}\overline{v}_d - (F_r - F_{r0}) \ , \tag{2}$$

where $\overline{P}$ is the loop pressure; γ is the specific heat ratio ($= 5/3$); F_r and F_{r0} are, respectively, the flare and pre-flare radiative loss rates [2]; $\overline{Q_F(t)}$ is the transient nanoflare energy input; v_b is the net plasma velocity at the base of the flux tube and the downflow velocity $\overline{v}_d$ is expressed in terms of the free-fall speed v_{ff} for a semicircular loop of height $h = 2L/\pi$:

$$\overline{v}_d = -(1 - \overline{\rho}_0/\overline{\rho})\,v_{ff} \ , \tag{3}$$

where $v_{ff} \equiv \sqrt{2gh}$. We refer the reader to Kopp and Poletto (1993) for a detailed description of the technique and assumptions which lead to Eqs. 1–3.

In order to predict the behavior of a particular nanoflare-heated loop over an extended time interval, we need to know how many nanoflares per unit time occur on the loop. This is given by the nanoflare energy distribution and by the time-averaged power input to the corona. The latter is known to be $\approx 10^6 \ ergs \ cm^{-2} \ s^{-1}$, for the quiet solar corona (Withbroe and Noyes 1977). The nanoflare distribution function, on the other hand, is poorly defined, even for the Sun (Hudson 1991). We know that the number N of flares with total radiated energy W follows approximately a power-law distribution

$$dN/dW = \mathcal{A}W^{-\alpha} \ (ergs \ s)^{-1} \ , \tag{4}$$

where $\mathcal{A}$ is a constant and $\alpha \approx 1.8$. However, the value of α is well-defined only for the population of "classical" flares; it becomes uncertain when microflares are

---

[1] From here on an overbar is used for all quantities averaged over the loop semilength
[2] These quantities are expressed in terms of an equivalent energy flux at the footpoint

considered and is totally unkown for nanoflares. A knowledge of its precise value is critical: only distributions with $\alpha > 2$ support the hypothesis of a nanoflare-heated corona. Distributions with $\alpha < 2$ are dominated by highly energetic flares, which are too few in number to make a significant contribution to the heating of the corona. Recent work by Porter et al. (1993) gives a tentative value $\alpha \approx 2.2$ in the microflare range. In the following we adopt this value also for the exponent of the power-law nanoflare energy distribution and we assume a value $10^{23}\,ergs$ for its low-energy cutoff.

If we make the simplifying hypothesis that the solar corona is entirely filled with loops with a semilength L=2000 km and aspect ratio (R/L) = 0.1, we can estimate the rate of nanoflare occurrence per loop corresponding to the coronal energy requirement. It turns out that, on the average, a given loop experiences one nanoflare every 239 seconds. Figure 1 shows the behavior predicted by our model for an initially cool ($T = 10^5\,K$) and nearly empty ($P \leq 0.001\,dynes\,cm^{-2}$) loop, for which, over the time interval of 2388 s covered by the Figure, 10 nanoflares occur randomly distributed in time and energy, although following the aforementioned power-law energy distribution function. Figure 1 shows that temperature and pressure quickly reach values typical of the coronal plasma and support our contention that nanoflares are a viable means to heat the solar corona.

3 Discussion and Conclusions

The simulations shown in Section 2 illustrate how a quasi-static corona may be created via the nanoflare heating mechanism. However, considering the capabilities of present-day solar instrumentation, one has been unable, so far, to image a single loop, if its size is that assumed in our modeling. An instrument with a spatial resolution of 3", for instance, will include $\approx$ 6 such loops within a single pixel. Obviously, the brightness fluctuation from such a loop ensemble is much smoother than would be predicted from the spiky behavior of T and P shown in Figure 1, and an observational test of the theory becomes difficult. Besides, Figure 1 gives an optimistic view of the real situation: longer loops or power-law distributions with higher values of α provide less favorable scenarios, as far as detectability is concerned; we refer the reader to Kopp and Poletto (1993) for a thorough discussion of this point. The situation is, of course, even more discouraging for stars other than the Sun.

Nevertheless, nanoflare heating offers an appealing alternative to more traditional approaches to the coronal heating problem, and it continues to be supported by theoretical and indirect observational evidence. Still, a number of steps should be taken to put this hypothesis on a firmer ground. We mentioned the need for higher spatial resolution observations both at photospheric levels – where fibril motions could be detected – and at coronal levels – where the energy release should be observable. Moreover, better knowledge of the flare energy distribution, of the relationship between energy release events in different wavelength bands, and of the loop length distribution and aspect ratio, is badly needed. Future ground based projects and space missions like LEST, THEMIS and SOHO, may lead to a substantial improvement in this respect.

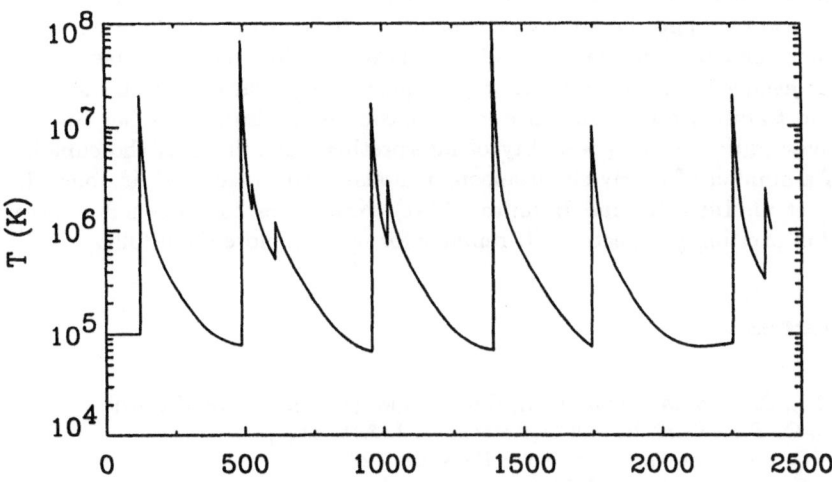

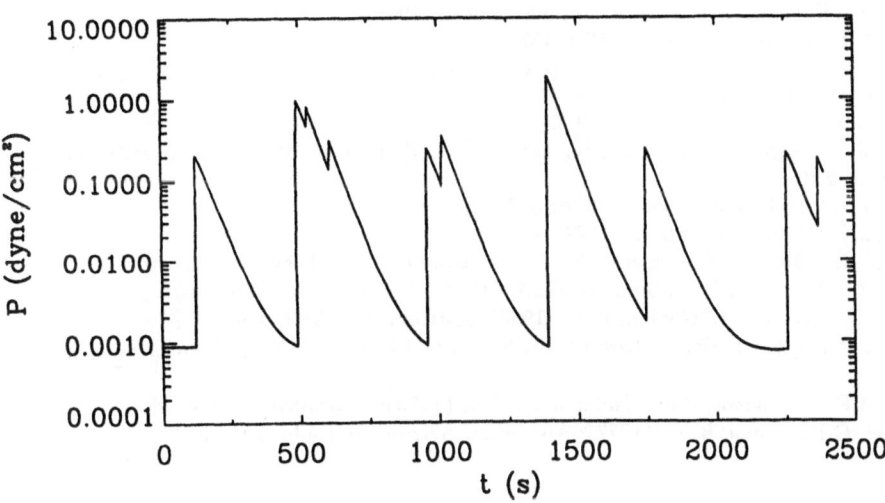

Fig. 1. Time profiles of temperature and pressure for an initially cool loop with L=2000 km. The simulation extends over a time interval of 2388 s, during which 10 nanoflares occur at random times and with a random (power-law) distribution of energies ($\alpha = 2.2$).

Among the problems to be addressed theoretically, we may mention also that the threshold behavior required in the nanoflare energy release process is still to be demonstrated and the value of the threshold has still to be established. The credibility of even a simple model like the one presented here will gain substantially from observational and theoretical developments. It is also worth pointing out that simulations analogous to these are still entirely lacking for stellar coronae: although subject to even greater uncertainties with respect to the solar case, such calculations would add to our overall confidence in the basic mechanism. A further topic that needs investigation is the possibility of interpreting large flares as the cumulative effect of a number of nearly simultaneous nanoflares (de Jager and de Jonge 1978; Sturrock et al. 1984; Lu and Hamilton 1991). Nanoflares have opened a research area full of exciting perspectives. It remains for us to explore them fully.

References

Butler, C.J., Rodonò, M., Foing, B.H., Haisch, B.M. (1986): Nature **321**, 679

Cheng, C.-C., Pallavicini, R. (1991): Astrophys. J. **381**, 234

de Jager, C., de Jonge, G. (1978): Solar Phys. **58**, 127

Fisher, G.H., Canfield, R.C., McClymont, A.N. (1985): Astrophys. J. **289**, 414

Glencross, W.M. (1975): Astrophys. J. **199**, L53

Gold, T. (1964): in "The Physics of Solar Flares", ed. by W. Hess, NASA-SP **50**, p. 389

Hudson, H.S. (1991): Solar Phys. **133**, 357

Kopp, R.A., Poletto, G. (1993): Astrophys. J. **418**, 496

Levine, R.H. (1974): Astropys. J. **190**, 457

Lu, E.T., Hamilton, R.J. (1991): Astrophys. J. **380**, L89

MacNeice, P. (1986): Solar Phys. **103**, 47

Mariska, J.T. (1987): Astrophys. J. **319**, 465

Pallavicini, R., Serio, S., Vaiana, G.S., Acton, L., Leibacher, J., Rosner, R. (1983): Astrophys. J. **270**, 270

Parker, E.N. (1983): Astrophys. J. **264**, 642

Parker, E.N. (1988): Astrophys. J. **330**, 474

Parker, E.N. (1991): in "Reviews in Modern Astronomy" **4**, ed. by G. Klare, p. 1

Porter, J.G., Fontenla, J.M., Simnett, G.M. (1993): Astrophys. J., submitted

Porter, J.G., Toomre, J., Gebbie, K.B. (1984): Astrophys. J. **283**, 879

Reale, F., Peres, G., Serio, S., Rosner, R., Schmitt, J.H.M.M. (1988): Astrophys. J. **328**, 256

Vesecky, J.F., Antiochos, S.K., Underwood, J.H. (1979): Astrophys. J. **233**, 987

Withbroe, G.L., Noyes, R.W. (1977): Ann. Rev. Astron. Astrophys. **15**, 363

Magnetic Structures
of the Intermediate Corona

S. Koutchmy [1], M. Molodensky [2]

[1]Institut d'Astrophysique de Paris, CNRS, 98bis Bd Arago, F-75014 Paris,
France
[2]IZMIRAN, Troitsk, 142092 Moscow, Russia

Abstract: The fine structures of the corona are analyzed using SXR and W–L eclipse
pictures. The measurement of the finest features observed at the CFHT on July 11, 1991
is discussed. We call the attention to the 3–D aspect of the magnetic structure which
can be apprehended and analysed assuming the plasma is concentrated in thin curved
sheets observed on eclipse W–L pictures. More geometrical parameters can be deduced
from the stereo–view based on the hypothesis of a rigid rotation. A preliminary 2.5–D
model calculation is put forward to explain the active region coronal rays. It is based on
the assumption of ad hoc chromospheric current distribution and analytic calculations.

1 Introduction

Recent soft X-ray (SXR) observations of the solar corona dramatically illustrated
the activity of the intermediate corona. They were performed during the years of
maximum solar sunspot activity, on the Yohkoh satellite and, to some extent, during
rocket flights (ex : NIXT observations by the L. Golub group). We want to turn our
attention to structures made of threads, rays, sharp edges, sheet–like enhancements
(curved or not), fan streams which are typical of what rises up above or in the vicinity
of active regions and which are well observed on white–light (W–L) eclipse pictures.
They are between the very inner corona which is almost exclusively made of loops
(we do not consider here flares nor small scale impulsive events), and the external
corona made of large streamers (see Koutchmy and Livshits, 1992). Moreover, above
the quiet Sun, outside large streamers, quite similar plumes and rays are seen and
their physics could also be similar.

The radial distances of these structures are typically 1.1 to 2 radii from the solar
center, i.e. it is an intermediate region situated far beyond the transition region but
before the structures stretched outward by the wind. In this part of the corona, not
made of "packed" closed loops, structures do not extend radially; sometimes a very
large departure from the radial direction is observed. One of the best example from
the Yohkoh data was analyzed by Shibata et al., 1992. Yet, these structures have
been known since a long time from eclipse observations and they were even analyzed

(e.g. see Koutchmy, 1969 and Burnichon et al, 1969). The resulting values of electron densities were above the background densities by at least 2 orders of magnitude, especially when the spatial resolution of W–L images was improved by the radial filter. Indeed these structures are made of tiny threads stretched out in one direction, almost parallel to each other or slightly divergent. The 3–D geometry is difficult to consider because of the projection effects and we will propose a new approach here. Theoretical problems connected with the appearance of these cannot be treated in the frame of a potential or force–free approximation with a "source"surface at 2 or more radii, as often done by modelers ; we should think about something new. We will eventually go through an original approach at the end of this short paper.

2 Eclipse Observations

As said before, there exists a lot of past eclipse observations reporting the description of fine structures above active regions. However, most of them do not have the adequate resolution. Authors often report the observation of an "enhancement", which is obviously made of the superposition of unresolved structures. Good correlations were reported with enhancements observed at metric radio–waves and, additionally, a slowly varying component and radio noise (e.g. see Lantos,1980).

2.1 Cross–Section of Threads

One intriguing question concerns the size of the smallest structure which is observable. This problem was recently dealt with at the time of the total solar eclipse of July 11, 1991, which was observed with the best possible spatial resolution, thanks to the use of the 3.6 m aperture CFHT telescope and video CCD imaging. Fast imaging permits to catch pictures free of atmospheric image motion and a great amount of pictures permits to improve the signal/noise ratio. By this way we picked up a typical elongated structure, seen in the vicinity of the E–limb, and we analyzed it to measure its cross–section (see Koutchmy et al, 1993). After having corrected for the seeing, we found 0.4 ± 0.1 arcsec, which means a 0.3 Mm diameter that is quite similar to the diameter of flux tubes or magnetic elements of the photosphere!

We noted that the thread rapidly evolved, "disapppearing" in the background in a few min. The proper motion seemed to occur along the local magnetic field, not far from the radial direction.

2.2 3–D Structures

The coronal plasma being optically thin, the integration along the l.o.s. corresponds to the superposition of different structures. However, when considering a curved sheet, there are special cases when the integration produces a useful effect, namely an edge–enhancement, see figure 1.

Sometimes, the edge is so sharp that a tangential discontinuity is suggested, especially when a void is seen on the other side. Sharp edges are easy to locate on a good radially filtered picture. Using 2 of these pictures taken at a 3.2 hour interval, we succeeded in determining their respective positions in space assuming that there was a rigid rotation at a large scale (see Koutchmy and Molodensky, 1992). A stereogram was produced, permitting to visualize in 3–D the main coronal structures of the intermediate corona. The method presents a great potential for the future, especially to compare a computed 3–D map of the magnetic field with the actual coronal structure. Transverse gradients of densities draw the structures quite well, when W–L images are processed with the Madmax algorithm (Maximum of Magnitude of the octo–directional secondary derivatives (see Koutchmy et al, 1993)). We note however that transient phenomena are also observed, including fast restructuring. Thus we need observations with a good enough temporal resolution to identify and analyze these restructuring.

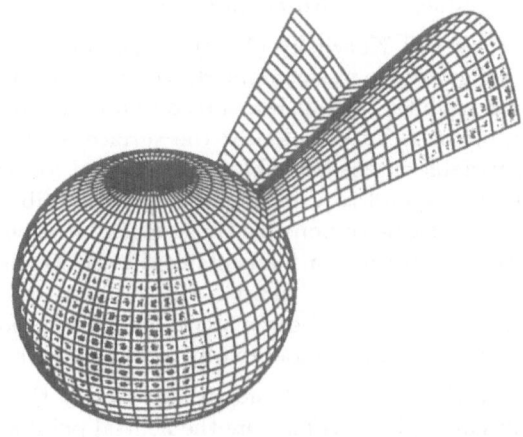

Fig. 1. Illustration using a 3-D model to show the occurrence of a sharp edge in a sheet-like stream. The plasma could be concentrated in a current sheet or at the place of a separatrice.

3 A Possible New Theoretical approach

The comparison of full disc coronal images made in SXR (Skylab, NIXT, Yohkoh mission) with Kitt–Peak magnetograms of the photospheric magnetic flux distribution over the whole disc shows some correspondance between enhancements of SXR emissions, including loop structures, and concentration of magnetic fluxes in active regions, including plages with different polarities. This is mainly due to the occurence of a stronger heating near the surface when the magnetic field is higher. Conversely, extended weak unipolar regions correspond to almost empty coronal regions, i.e. coronal holes. What is more difficult to understand is the coronal structure in the quiet Sun or out of active regions, where filaments are observed. In the place of a neutral sheet or current sheet above active regions, other structures were proposed. Reconnection of lines is often involved, including scenarii with flares, when the filament is inside an active region.

However, existing models of coronal neutral sheets do not reproduce the large density enhancement which is observed, and the streamer is essentially radial when a magnetic configuration rooted in the surface is stretched out by the wind. It is also interesting to note that models, after a large amount of iterations, could evolve toward a depleted or reduced density structure (see Suess, 1992). Because of these difficulties in predicting correctly what is observed, we want to come back to the beginning of the discussion about the distribution of currents at the surface and the corresponding magnetic field lines.

We note that on many occasions Alfvén (1975, 1981) stated that the picture of frozen–in plasma is misleading and, about the so–called magnetic reconnection he commented on : "... energy release by magnetic field line reconnection, this monstrous concept is a product of the frozen–in picture in absurdum...".

Following Solov 'ev and Kutvitsky (1992)'s suggestion and calculations, a set of magnetic field configurations was computed, based on the analytic solutions proposed in Landau and Lifchitz, 1959. It is assumed that there are 2 axially symmetric currents flowing in opposite directions near the surface, with rings of diameter in a ratio a/r and of variable intensities. (More details in a forthcoming paper). The inner current could correspond to what is partially responsible for a sunspot. The more external current could be responsible for faculae of reversed polarity. However, the case of currents in active region filaments could also be considered (see Molodensky and Filippov, 1992). Figure 2 shows a configuration obtained with a/r = 0.1 matching quite well with what we observe above an active region, with a system of loops of a coronal condensation surmounted by elongated rays.

It is conjecturated that density enhancements occur in the place of the separatrices, where the plasma rapidly drifts. Note the neutral point at the top of the loop system, on the axis; the plasma presumably avoids this point or alternatively, explosive phenomena would originate there in case of disconnection. Further analyses are needed to predict the behavior of some thermodynamical parameters, and first of all, the temperature.

4 Conclusion

Today, coronal structures of the intermediate corona present a challenge to theoreticians. Many observations are accumulated to give evidence of the importance of currents flowing in the low temperature inner corona and chromosphere. Very high resolution observations show large proper motions occurring in coronal threads and SXR observations brought evidence of magnetic restructuring occurring at these heights. In the future, the analysis of intermediate corona structures should be carried out at the light of possibly new idea on the role of surface currents.

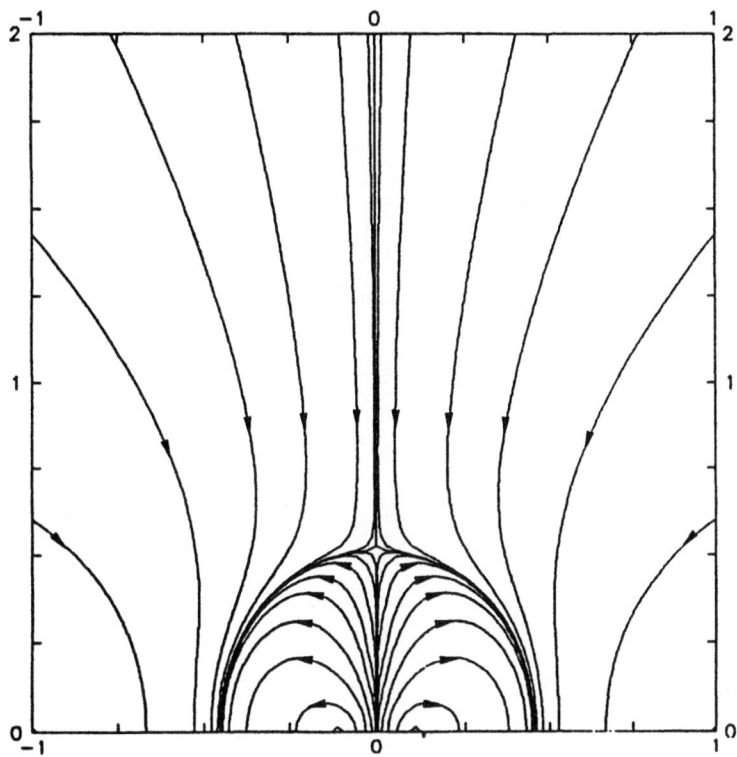

Fig. 2. Distribution of magnetic field lines in an axially symmetric model of currents flowing at the surface in rings with a/R = 0.1 and a neutral (singular) point at a/2 (Lambda =1).

References

Alfvén, H. in "Physics of the hot Plasma in the Magnetosphere : Electric Current Structure of the Magnetosphere", ed. by. B. Hultquist and L. Stenflo, Plenum, N.Y., 1975

Alfvén, H. in "Cosmic Plasma", Rendel, Dordrecht, 1981

Burnichon, M.L., Koutchmy, S. and Laffineur, M., 1969, C.R.A.S., **269 B**, 139

Koutchmy, S., 1969, Astrophys. Letters, **3**, 215

Koutchmy, S. and Livshits, M., 1992, Space Science Review, **61**, 393–417

Koutchmy, S. and Molodensky, M., 1992, Nature, **360**, 717

Koutchmy, S., Bouchard, O., Mouette, J. and Koutchmy, O., 1993, Solar Phys. Letter, in press

Landau, L.D. and Lifchitz, E.M., 1959, "Electrodynamica Sploshnykh Sred", ed. "Nauka", Moscow

Lantos, P., 1980, IAU Symp. **86**, 41

Molodensky, M.M. and Filippov, B.P., 1992, "Magnetic Fields in Active Regions", ed "Nauka", Moscow

Shibata, K. and the Yohkoh Team, 1992, PASJ **44**, L173
Solov ′ev, L.S. and Kutvitsky, V.A., 1992, private communication
Suess, S.T., 1992, in Proceedings of the first SOHO Workshop, ESA **SP–348**, 63

Oscillations in Quiescent Prominences

B. Roberts [1], P.S. Joarder [2,3]

[1]Department of Mathematical and Computational Sciences, University of St
Andrews, St Andrews, Fife KY16 9SS, Scotland
[2]School of Mathematics and Statistics, University of Birmingham,
Edgbaston, Birmingham B15 2TT, England

Abstract: Theoretical aspects of prominence oscillations are discussed in terms of the
basic modes of oscillation of a simple slab in the long wavelength limit. Oscillations with
a period of the order of one hour are produced by the *Alfvén string mode*. Intermediate
period oscillations, of ten to twenty minutes, arise from the *fast string mode* or possibly
an *internal Alfvén mode* or an *internal slow mode*. Short period oscillations, of the order
of a few minutes, are produced either by *internal fast modes* or by an *internal Alfvén
mode*. Sub-minute oscillations are generated by *guided fast magnetoacoustic waves* in fibril
structures embedded within the prominence.

1 Introduction

Prominences oscillate. That simple observational fact is potentially very important
for it affords us with the possibility of using data on the modes of oscillation of
a prominence to derive seismic information about the structure of the prominence
and its coronal environment. It is natural here to draw an analogy with the developments in helioseismology, which provide a spur to the quest for seismic information
about other solar objects such as sunspots or the corona. But it is clear that the
problem of obtaining such information about a prominence is more complicated
than the corresponding one of determining, say, the run of sound speed in the solar interior. Prominences are intrinsically complex three dimensional objects, with
magnetism and fine structure both playing important roles. So the seismic problem
is correspondingly more complicated. Nonetheless it is both interesting and potentially important for us to pursue this problem. Progress in this direction depends
critically on detailed observational evidence as to the character of the natural modes
of oscillation of a quiescent prominence. Such information is not yet available though
likely to become so in the near future.

To progress in this area it is evidently necessary to understand the intrinsic make-
up of a prominence. Theoretical models of prominence equilibria have been studied

---

[3] SERC research assistant

for many years but a detailed comparison of such models with the observations is difficult. The study of the modes of oscillation of a prominence offers the hope that progress in this direction may be possible. Such progress will require detailed observational studies of prominences, giving information as to their periods, character of oscillation, and how such features vary from prominence to prominence so that distinguishing aspects of the oscillations may be determined. For example, how does the period of oscillation of a prominence vary with its width or with its magnetic field strength? Complementary to such observational studies is the theoretical requirement of ascertaining how such features influence the periods of oscillation of the prominence.

Observations have indicated that there are broadly three distinct bands of modes of oscillation of a prominence (see the review by Tsubaki 1988). Long period vibrations have periods of about one hour but may in fact range from 40 to 90 minutes (Bashkirtsev et al. 1983; Bashkirtsev and Mashnich 1984; Wiehr et al. 1984; Balthasar et al. 1988; Mashnich and Bashkirtsev 1990). An intermediate period vibration occurs with a period of around 10 minutes, but possibly ranging from 8 to 20 minutes (Tsubaki et al. 1987; Balthasar et al. 1988, 1993; Yi Zhang and Engvold 1991; Yi Zhang et al. 1991). Short period vibrations have periods of 5 minutes or less (Wiehr et al. 1984; Balthasar et al. 1988, 1993; Thompson and Schmieder 1991; Yi Zhang et al. 1991). The short period modes may require sub-division into minute (2-5 min) and sub-minute modes, with the sub-minute modes being related to the fine-scale structure within the prominence.

How are we to understand this array of modes? It is clearly important to develop detailed theoretical models of the vibrations if we are to be able to utilize the seismic information that such oscillations contain. However, the inherent complexity of a prominence's equilibrium structure makes for a daunting problem. It seems worthwhile that we should first understand the fundamental modes of oscillation of an essentially simple structure before trying to understand the complexity of reality. This is the view-point adopted in Joarder and Roberts (1992a,b, 1993a,b) and further expounded upon here.

2 Long Wavelength Oscillations

The equilibrium state of our problem involves both stratification and structuring, two aspects that inevitably make for complexity. To describe the basic modes of oscillation of a prominence, however, we here ignore the role of stratification, setting the gravitational acceleration to zero. The strong structuring evident in a prominence, by which the temperature and density change by two orders of magnitude from within the prominence to its environment, argues that this feature is likely to be particularly important for defining the fundamental modes of oscillation. Studies of prominence oscillations in the presence of gravity offer support for this view (see Oliver et al. 1992, 1993; Joarder and Roberts 1993b).

Consider an equilibrium in which the gas pressure p and magnetic field $\mathbf{B}$ are both uniform, though the density $\rho(x)$ and temperature $T(x)$ are functions of a spatial scale x. The magnetic field is taken to lie purely in the xy-plane, so that $\mathbf{B} = (B_x, B_y, 0)$. We are thinking of a sheet-like structure for the equilibrium of the prominence, taken to be oriented with its long axis in the y-direction. We may view the xy-plane as being the horizontal and the z-axis as the vertical. Then the equilibrium magnetic field lies in the horizontal plane and makes an angle ϕ with the long axis of the prominence sheet, so that $B_x = B \sin \phi$ and $B_y = B \cos \phi$ for field strength $B (= (B_x^2 + B_y^2)^{1/2})$.

Linear isentropic disturbances about this equilibrium are readily investigated using ideal MHD. Combining the first time derivative of the linear momentum equation with the equations of mass conservation, adiabaticity and induction, we obtain the vector wave equation

$$\frac{\partial^2 \mathbf{v}}{\partial t^2} = [c_s^2(x) + c_A^2(x)]\mathrm{grad\ div\ } \mathbf{v} - \frac{1}{\mu\rho(x)}(\mathbf{B} \cdot \mathrm{grad})\{(\mathbf{B} \times \mathrm{curl\ } \mathbf{v}) + \mathbf{B}\ \mathrm{div\ } \mathbf{v}\}. \quad (1)$$

Here $c_s(x)(= (\gamma p/\rho(x))^{\frac{1}{2}})$ and $c_A(x)(= (B^2/\mu\rho(x))^{\frac{1}{2}})$ are the sound speed and Alfvén speed within the gas which has an adiabatic index $\gamma(= 5/3)$ and magnetic permeability μ.

Equation (1) possesses solutions with Fourier representations of the y and z dependence in terms of wavenumbers k_y and k_z. The scale of a prominence, with length $L (\approx 200,000 \mathrm{km})$ and height $H (\approx 50,000 \mathrm{km})$, is much larger than the observed width $2a (\approx 5000 \mathrm{km})$ and so we may expect that wavenumbers k_y and k_z (of order π/L and π/H, respectively) play only minor roles in determining the structure of solutions compared with the effective wavenumber for variations *across* the structure. Accordingly, we consider the form of (1) in the limit when $\partial/\partial y = 0$ and $\partial/\partial z = 0$, corresponding to $k_y = 0$ and $k_z = 0$. The general case of non-zero wavenumbers is discussed in Joarder and Roberts (1993a) and Joarder (1993).

With $\partial/\partial y = 0$ and $\partial/\partial z = 0$, the components of (1) yield

$$\frac{\partial^2 v_x}{\partial t^2} = [c_s^2(x) + c_A^2(x) \cos^2 \phi]\frac{\partial^2 v_x}{\partial x^2} - c_A^2(x) \cos \phi \sin \phi \frac{\partial^2 v_y}{\partial x^2}, \quad (2)$$

$$\frac{\partial^2 v_y}{\partial t^2} = c_A^2(x) \sin^2 \phi \frac{\partial^2 v_y}{\partial x^2} - c_A^2(x) \cos \phi \sin \phi \frac{\partial^2 v_x}{\partial x^2}, \quad (3)$$

$$\frac{\partial^2 v_z}{\partial t^2} = c_A^2(x) \sin^2 \phi \frac{\partial^2 v_z}{\partial x^2}. \quad (4)$$

We thus see that the system presents a simple wave equation for the vertical motions v_z, and coupled wave equations (2)-(3) for the horizontal motions. Equation (4) describes the Alfvén wave and the motions are incompressible. Equations (2)-(3) describe the magnetoacoustic modes (fast and slow) and are compressive.

The magnetoacoustic modes are decoupled in the special case of a purely transverse equilibrium magnetic field, corresponding to $\phi = \pi/2$. For this special case, all three magnetohydrodynamic waves in the long wavelength limit decouple to yield elementary wave equations:

$$\frac{\partial^2 v}{\partial t^2} = c^2(x)\frac{\partial^2 v}{\partial x^2}, \quad (5)$$

where v takes on its appropriate component and $c(x)$ is the appropriate speed: $v = v_z$ and $c = c_A(x) \sin \phi = c_A(x)$ in the Alfvén wave; $v = v_x$ and $c = c_s(x)$ in the slow wave; and $v = v_y$ and $c = c_A(x)$ in the fast wave. The fast wave is thus such that its motions are predominantly orthogonal to the magnetic field; the slow wave has motions that are predominantly aligned with the magnetic field. In classifying the magnetoacoustic string modes we have supposed that $c_A > c_s$; if $c_A < c_s$, then the names 'fast' and 'slow' should be interchanged.

3 String Modes

The occurrence of the wave equation (5) suggests that the basic modes of oscillation of a prominence sheet are closely analogous to the modes of oscillation of a stretched elastic string of non-uniform density (Roberts 1991; Joarder and Roberts 1992a,b, 1993a; Oliver et al. 1993). This analogy with waves on a non-uniform string led Joarder and Roberts (1993a) to refer to these waves as *string modes*, being either Alfvén or fast or slow in nature. Oliver et al. (1993) term these modes 'hybrid waves'.

Consider, then, the basic wave equation (5) subject to the boundary condition that $v = 0$ at $x = \pm l$. This is the appropriate boundary condition for an elastic string fixed at points $2l$ apart. For the prominence problem it represents the effect of inertia line-tying, whereby the dense photosphere effectively anchors the field lines of the corona.

The string analogy is most simply illustrated for the case of two uniform strings of differing densities joined together. Consider a section of string, extending from $x = -a$ to $x = +a$, of density ρ_p and propagation speed c_p joined to string of density ρ_c and speed c, with $c^2 \rho_c = c_p^2 \rho_p$; the ends are tied at $x = \pm l$. This represents a prominence sheet of width $2a$, density ρ_p and propagation speed c_p embedded in a corona of density ρ_c and propagation speed c. Across the joins at $x = \pm a$ we require that v and $\partial v/\partial x$ be continuous. It is evident that there are two types of modes, those that disturb the centre $x = 0$ and those that do not. Perhaps of most interest for prominence vibrations are those modes that disturb the centre of the prominence sheet. The solution of (5) for these modes leads to the dispersion relation

$$\tan\left(\frac{\omega a}{c_p}\right) = \left(\frac{\rho_c}{\rho_p}\right)^{\frac{1}{2}} \cot \frac{\omega}{c}(l-a). \tag{6}$$

Under prominence conditions, $a \ll l$ and $\rho_c \ll \rho_p$. Then (6) reduces to (Rayleigh 1877)

$$\frac{\omega l}{c} \tan\left(\frac{\omega l}{c}\right) = \frac{l\rho_c}{a\rho_p} = \frac{2\rho_c l}{M}, \tag{7}$$

where $M \equiv 2\rho_p a$ is the prominence mass per unit surface area in the slab. The fundamental solution of (7) for $a \ll l$ and $\rho_c \ll \rho_p$ is $\omega^2 \approx c_p^2/al$. Thus we obtain a mode with period $\tau (= 2\pi/\omega)$ given by (Roberts 1991)

$$\tau = \frac{2\pi}{c_p}(al)^{\frac{1}{2}} = \pi\left(\frac{2l\mathcal{M}}{c_p^2 LH\rho_p}\right)^{\frac{1}{2}}, \tag{8}$$

where $\mathcal{M} (\equiv MLH)$ is the total mass within the prominence sheet. Thus we obtain the fundamental period of a mass loaded string, the period being determined by the propagation speed c_p within the prominence and the geometric mean of the prominence width $2a$ and the line-tying distance $2l$.

4 Discussion

The analysis we have given points out the basic nature of a prominence's modes of oscillation in terms of the string analogy. Because there are in general three modes in MHD, there are three string modes. We consider each in turn. To illustrate the results we consider a sheet of width $2a = 5000$km with a temperature of 8000K and gas density $\rho_p = 2 \times 10^{-10}$kg m^{-3}. The slab is threaded by a horizontal magnetic field of strength $B = 12$G, making an angle ϕ with the long axis of the slab; the field is anchored at a distance $l = 8 \times 10^4$km either side of the prominence, measured perpendicular to the slab. Within the prominence slab the sound speed is $c_s = 15$ km s^{-1} and the Alfvén speed is $c_A = 75$ km s^{-1}.

Consider the *Alfvén string mode*. The Alfvén string mode produces vertical oscillations with a period τ given by

$$\tau = \frac{2\pi}{c_A \sin \phi}(al)^{\frac{1}{2}}. \tag{9}$$

The term $\sin \phi$ simply reflects the fact that, as measured between the two anchor points, a field line is correspondingly longer at smaller ϕ than at larger ϕ, and so the travel time (and hence the period) of an Alfvén wave–which is a field-aligned mode– is also correspondingly longer. For the prominence parameters chosen above, the Alfvén string mode gives a period of about 60 minutes for a skew angle of $\phi = 20°$, reducing to 47 min at $\phi = 25°$. At very small skew angles the period of the Alfvén string mode is increased considerably (e.g., to almost 4 hours at $\phi = 5°$). Also, the period of the Alfvén string mode scales inversely with B.

Similar estimates may be made for the *magnetoacoustic string modes*. The coupling between the fast and slow modes evident in (2) and (3) complicates the discussion somewhat, but estimates of the form (8) apply with $c_p \approx c_t \sin \phi$ in the *slow string* mode and $c_p \approx c_f (= (c_s^2 + c_A^2)^{1/2})$ in the *fast* mode (Joarder and Roberts 1993a; Joarder 1993). Here $c_t (= c_s c_A/c_f)$ is the cusp speed. The low value of the cusp speed (about 14.7 km s^{-1}) in a prominence makes for a very long period, of about 5 hours for $\phi = 20°$. Only a very long data set is thus likely to show the slow string mode. On the other hand, the high value of the fast speed, combined with the absence of any strong ϕ-dependence, makes for the shortest period of the three string modes. With $c_p = c_f = 76.5$ km s^{-1}, (8) produces a period of about 20 minutes for the *fast string mode*. The motions are predominantly perpendicular to the prominence long axis. Thus, the string modes are capable of producing both the long (one hour) and intermediate (20 min) periodicities observed in prominences.

The string modes do not exhaust all the periods of oscillation available to a prominence: there are also the *internal* modes given by the dispersion relation (6). These are modes that have appreciable kinetic energy because they disturb the dense prominence matter. The period of an internal mode is basically the transit time of that mode across the prominence slab and back again:

$$\tau^{\text{internal}} = 4a/c_p, \tag{10}$$

where c_p takes on the same values as for the string modes, namely $c_p = c_f$ in the *internal fast mode*, $c_p = c_A \sin \phi$ in the *internal Alfvén mode*, and $c_p = c_t \sin \phi$ in the *internal slow mode*. These modes produce periods of about 2, 6.5 and 33 minutes, respectively, for $\phi = 20°$. Thus periods of intermediate value (say 10 to 20 min) may be either the fast string mode or possibly the internal Alfvén or slow modes. Short period modes, in the 2-5 min range, are here associated with the internal fast wave or possibly an internal Alfvén wave.

Finally, we are left with the observations of sub-minute modes (Balthasar et al. 1993). These very low period oscillations are not given by the global modes discussed here. Instead, it seems plausible that these modes are the result of magnetoacoustic waves guided within fibrous structures embedded within the prominence (Joarder and Roberts 1992a; Joarder 1993; Balthasar et al. 1993). Local regions of low Alfvén speed within a prominence act as wave guides and produce periods determined by the transit times across the fibrous structure. These are the *magnetic Love* and *magnetic Pekeris waves* described in general terms by Edwin and Roberts (1983) and Roberts, Edwin and Benz (1984). For example, a fibrous tube of radius R and local Alfvén speed c_A^{fibre}, embedded within a prominence, supports quasi-periods of the order (Roberts et al. 1984)

$$\tau^{\text{Pekeris}} \approx 2.6R/c_A^{\text{fibre}}. \tag{11}$$

Thus a fibre of radius $200 - 400\,\text{km}$ with an Alfvén speed of $25\,\text{km s}^{-1}$, somewhat lower than the $75\,\text{km s}^{-1}$ speed of the bulk of the prominence and corresponding to an order of magnitude density enhancement, produces periods of $21-42\,\text{s}$, consistent with the $30\,\text{s}$ reported by Balthasar et al. (1993). Thus sub-minute periodicities in prominences may be the result of ducted fast waves in fibril structures.

References

Balthasar, H., Stellmacher, G., Wiehr, E. (1988): A&A **204**, 286
Balthasar, H., Wiehr, E., Schleicher, H., Wöhl, H. (1993): A&A, **277**, 635
Bashkirtsev, V.S., Kobanov, N.I., Mashnich, G.P. (1983): Solar Phys. **82**, 443
Bashkirtsev, V.S., Mashnich, G.P. (1984): Solar Phys. **91**, 93
Edwin, P.M., Roberts, B. (1983): Solar Phys. **88**, 179
Joarder, P.S. (1993): Ph. D. thesis, St Andrews University
Joarder, P.S., Roberts, B. (1992a): A&A **256**, 264 (Paper I)
Joarder, P.S., Roberts, B. (1992b): A&A **261**, 625 (Paper II)
Joarder, P.S., Roberts, B. (1993a): A&A **277**, 225 (Paper III)
Joarder, P.S., Roberts, B. (1993b): A&A **273**, 642
Mashnich, G.P., Bashkirtsev, V.S. (1990): A&A **235**, 428
Oliver, R., Ballester, J.L., Hood, A.W., Priest, E.R. (1992): ApJ **400**, 369
Oliver, R., Ballester, J.L., Hood, A.W., Priest, E.R. (1993): ApJ **409**, 809
Rayleigh, Lord (1877): The Theory of Sound, vol.1 (1945 edition), Dover, New York, p. 205
Roberts, B. (1991): Geophys. Astrophys. Fluid Dynamics **62**, 83
Roberts, B., Edwin, P.M., Benz, AA.O. (1984): ApJ **279**, 857
Thompson, W.T., Schmieder, B. (1991): A&A **243**, 501
Tsubaki, T. (1988): in Proc. 9th Sacramento Peak Summer Meeting, ed. by R.C. Altrock (Sacramento Peak, New Mexico), p. 140
Tsubaki, T., Ohnishi, Y., Suematsu, B. (1987): PASJ **39**, 179
Wiehr, E., Stellmacher, G., Balthasar, H. (1984): Solar Phys. **94**, 285
Yi, Zhang, Engvold, O. (1991): Solar Phys. **134**, 275

Pressure Diagnostics of Coronal Loops Observed by NIXT

G. Peres [1], F. Reale [2], L. Golub [3]

[1]Osservatorio Astrofisico di Catania, Città Universitaria, 95125 Catania,
Italy
[2]Istituto e Osservatorio Astronomico, Palazzo dei Normanni, 90134
Palermo, Italy
[3]Smithsonian Astrophysical Observatory, 60 Garden St., Cambridge Ma
02138

Abstract: The Normal Incidence X-ray Telescope (NIXT) sounding rocket payload - a set of multilayer telescopes of novel design - provides images of the corona at sub-arcsec angular resolution in narrow X-ray spectral bands centered at the wavelengths of particular coronal emission lines. The NIXT 63.5 *angstrom* coronal images show the well-known and ubiquitous coronal loops but also, mostly in active regions, shallow and bright areas of intense emission not resembling loops. We have explained such areas within the traditional physics of coronal loop models as intense emission in the NIXT band coming from a narrow region at the base of high pressure loops; the particular nature of the NIXT temperature sensitivity, with its bimodal temperature response, is the key to detecting such a feature. We discuss the implications of this finding and, in particular, we show the possibility of new diagnostics of plasma pressure independent of the traditional one based on the determination of emission measure. In addition, we show that there is very little dependence of the spatial distribution of the X-ray emissivity in the NIXT passband on the details of the spatial distribution of the heating function.

1 Introduction

Solar X-ray observations made with grazing incidence telescopes (Vaiana, Krieger and Timothy, 1973; Vaiana and Rosner, 1978) showed that the X-ray emitting corona apparently consists entirely of hot plasma confined inside magnetic loops; this view has been commonly accepted and extended also to the coronae of solar-like stars.

More recently the novel technology of *normal incidence* multi-layer X-ray mirrors has been applied to the construction of a new kind of X-ray solar telescope, in particular of the Normal Incidence X-ray Telescope (NIXT) (see Golub *et al.* 1990 and refernces theirein; also Golub 1992). The images collected with this telescope achieve *unprecedented* angular resolution in X-rays and show a high level of structuring of the whole corona. However, in addition to the detailed transverse fine structure of the corona observed by the new imaging method, the NIXT data have also shown the presence – in some locations – of low-lying bright patches of X-ray emission.

Detailed hydrostatic models of coronal loops help us to understand that these low-lying bright X-ray structures, which in NIXT pictures closely resemble corresponding Hα plages, correspond to regions of bright emission at the base of high pressure loops. The ability to identify these regions is mainly due to the bimodal nature of the NIXT spectral response: basically, what causes the effect is that a high temperature coronal loop will be seen by the NIXT in Fe XVI throughout the corona, but there will bee a thin, high-density region near the base where the loop temmperature passes through temperatures appropriate to Mg X, thereby causing a low-lying "spike" in the loop brightness. By measuring the distribution *along the loop* of emission in the NIXT band and taking into account the characteristics of the spectral response, we can determine the pressure of the plasma confined within individual coronal loops. Combining models of stationary loops with NIXT observations promises, therefore, to provide new diagnostics of pressure in X-ray emitting solar loops.

Some aspects of this work, such as examples of morphological similarity of some Hα and NIXT bright regions imaged simultaneously, are discussed in Peres, Reale and Golub (1993) to which we refer for many details. Here we stress other aspects of our work such as the relative unimportance of heating spatial distribution in determining our results.

2 The Observations and the Loop Model

The NIXT mirror utilizes a multilayer coating with 140 layer-pairs of Co and C, alternating in precise fashion; the layers act interferentially to produce high reflectivity in a narrow spectral band, determined by properly tuning the layer thickness (Barbee, 1981; Spiller, 1990). The spectral response of the mirror system analyzed in the present study is centered at 63.5 A and the bandwidth is 1.4 A; this region includes intense Mg X and Fe XVI soft X-ray lines within the band and therefore the instrument is most sensitive to emission of plasma at temperatures of 10^6 K and 3×10^6 K; however the temperature sensitivity extends above 10^7 K. The mirror diameter is 27.5 cm and the optical scheme is an f/8 prime focus.

This telescope is characterized by a much lower scatter than obtained with grazing incidence telescopes (Golub and Spiller 1992) and achieves better angular resolution, namely below 1 arcsec. So far the resolution has been limited to slightly less than 1 arcsec by the grain of the photographic film used to record the images.

The telescope has to date been flown five times and exclusively on sounding rockets. The instrumental apparatus is presently being upgraded as shown in Table 1.

NIXT pictures show loops everywhere on Sun but also broad and apparently shallow regions of bright emission similar to underlying Hα plages and certainly not to loops. Because of the mirror bandpass, the filter response and film spectral response, we can safely exclude that this morphology is due to UV contamination. Moreover, many bright Hα patches do *not* correspond to bright areas in the NIXT

Instrument	NIXT	Multilayer EUV Telescope
λ (A)	63.5	193
Prominent Lines	Mg X + Fe XVI	Fe XII
Mirror Diameter (cm)	27.5	6
Focal Length (m)	2	1.25
Detector Type	CCD	CCD
nr. of pixels	2000 $\times$ 2000	800 $\times$ 490
pixel linear size (arcsec)	0.9	2.5

The payload will carry also an Hα TV camera for pointing verification and control during flight.

images, therefore it is not just a leak of chromospheric emission into the X-ray images.

We have studied whether this morphology, apparently different from loops, can be explained with the physics of loops. Our studies included the question of whether we might need to consider a new class of structures to explain the new morphology, or whether loops, so far considered as the only building blocks of the X-ray emitting corona, are enough.

Since the observations by NIXT show that most of the emitting structures, and in particular the aforementioned low-lying structures, do not change significantly over 5 minutes - i.e. for longer than typical cooling or dynamical times - we assume a stationary configuration for them. We have computed a set of coronal loop models covering the typical physical conditions in coronal loops, and synthesized from them the emission observed in the NIXT passband, so as to realize whether we can reproduce the observed morphology.

We have used the Serio et al. (1981) hydrostatic model of a coronal loop, semi-circular, of constant cross section, symmetric with respect to its apex; the model is based on the equations of plasma hydrostatic equilibrium and of energy balance among radiative losses, a phenomenological term of local heat input in the plasma and thermal conduction according to Spitzer's formulation. The base of the model is at $2 \times 10^4 K$ and there we assign the base pressure and the heat flux which is set at a very low value, practically to zero.

Within any loop model the plasma temperature increases, and the density decreases, monotonically from the base to the apex of the loop, as shown in the example shown in Fig. 1.

Such models are fully characterized by the plasma pressure at the base of the arch and by the arch semi-length, L; these, in turn, are related to the maximum plasma temperature T_{max}, located at the loop apex, and to the plasma volumetric heating Q by the scaling laws of Serio et al. (1981). The heating has the functional form $Q(x) = Q_0 \, exp\,(x/x_Q)$ where Q_0 is the heating at the base of the loop, x is the field line coordinate along the loop reckoned from the base, and x_Q is the e-folding heating scale length. We refer to Serio et al. for further details on the model.

We have computed the emission B *vs field line coordinate in the NIXT band* from the density and temperature distributions, respectively $n(x)$ and $T(x)$ obtained

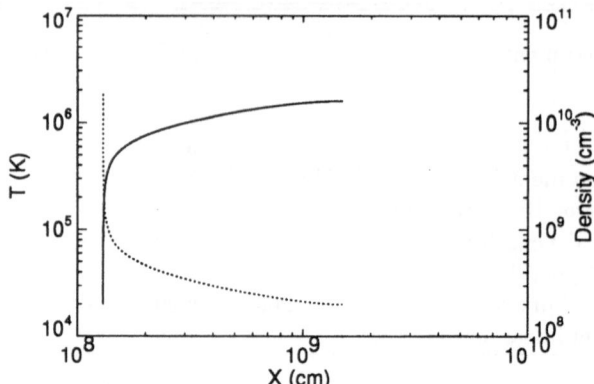

Fig. 1. Profiles of temperature (solid) and density (dashed) vs. field line coordinate, from the base to the apex of the loop in a representative model.

with the model as:

$$B(x) = \quad n(x)^2 \; G(T(x)), \tag{1}$$

where G(T) is the solar plasma emissivity, folded through the NIXT spectral response. G(T) has a peak around 10^6 K and another, lower by half a decade, at $3 \; 10^6$ K with a long tail extending to high temperatures; the emissivity in the million degree range is dominated by the Fe XVI and Mg X lines (Golub and Herant, 1989). Such features are reflected in the emission distribution along the loops which shows enhancements in correspondence of such temperatures.

3 Results and Discussion

Fig. 2 shows how the loop emission in NIXT observations depends on the loop base pressure. There we show, for loops of semilength $5 \; 10^8$ cm, $1.5 \; 10^9$ cm, and $5 \; 10^9$ cm, of spatially uniform heating and of pressure ranging from 0.3 to 30 dyne cm^{-2}, the distribution of emission along the loop, according to Eq. (1).

As the loop plasma pressure increases, so does the density and, therefore, the emission measure. As a consequence the loop emission increases at any place in the loop. The higher the pressure values, the closer to the base of the loop is the place of maximum emission, mostly due to the Mg X line. In particular the emission of the loops with higher base pressure has a sharp peak located immediately above the loop base. While the Mg X hump is squeezed at the base of the loop, a second smaller hump - the one due to Fe XVI - appears higher up in the loop (cf. the case with p= 30 dyne cm^{-2} and half length 5×10^9 cm) and behaves more and more like the Mg X one as the loop plasma pressure increases. The emission distribution changes also as a consequence of the change in the maximum temperature with pressure, according to the Serio et al. (1981) scaling law: for sufficiently high plasma pressure, the loop plasma will reach such a high temperature at the apex that the range of Mg X temperature formation is squeezed at the loop's lowest section, thus giving a very bright base, in the NIXT band.

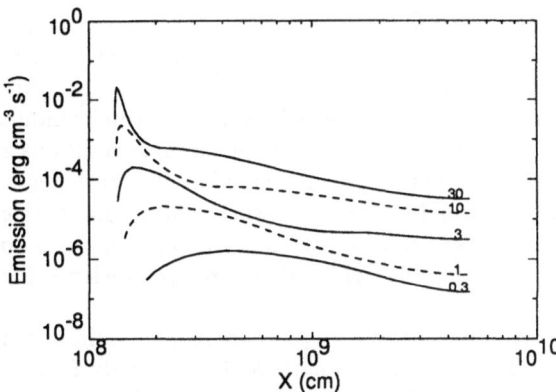

Fig. 2. Distribution of the emission in the NIXT band along the field line coordinate, from loop base to apex, in hydrostatic models of loops with semi-length $L = 5 \times 10^8$ cm, $L = 1.5 \times 10^9$ cm, $L = 5 \times 10^9$ cm and heating uniform along the loop. The labels of each set of curves yield the loop base pressure (dyne cm^{-2}).

We have found a similar dependence of emission profile on plasma pressure for all loop lengths. The lower sections of the emission distribution vs. x overlap very closely in loops of different length but with the same pressure. Fig. 2 provides an illustration of this effect. The sub-arcsec resolution of NIXT allows imaging of lower section of the loops at a high level of detail, yielding an accurate spatial distribution of emission.

We have also computed the emission distribution for loops of the same length and base pressure values as above, but for different values of x_Q, ranging from infinity (uniform heating) to x_Q as short as one third of the loop length.

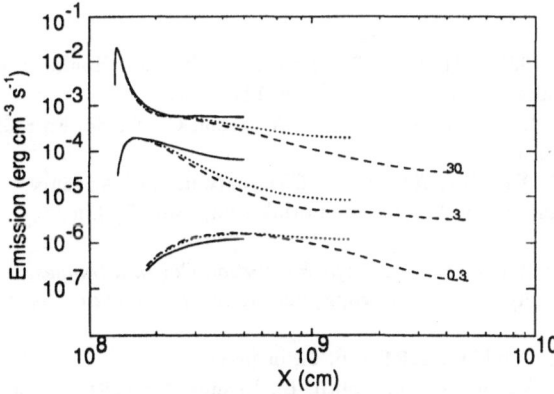

Fig. 3. Emission in NIXT band of loops with semi-length 5×10^9 cm, different base pressure (the labels of each set of curves yield the loop base pressure in units of dyne cm^{-2}), for uniform heating (solid) and e-folding heating scale length 1/3 of the loop length (dashed).

Fig. 3 illustrates the difference between the two sets of models with infinite heating scale height and those with heating scale height as short as one third of the loop. Differences are smaller than (although comparable to) those found among loops of same pressure and different length. In fact thermal conduction is an efficient mechanism of thermal energy redistribution in coronal conditions and, therefore, the emission distribution derived from the loop models cannot depend drastically on the spatial details of the heating function. It is worth noting that for significantly shorter x_Q no solution with energy balance can exist for some of the models here reported (cf Serio et al. for further details) which therefore provide lower limits of x_Q for loop models.

In conclusion footpoints of the high pressure loops should be very bright in NIXT pictures. A thick bundle of high pressure loops, typically found in active regions, should have a bright base whose shape should nearly match the underlying Hα plage. This characteristics should explain the low-lying bright regions detected by NIXT in addition to the usual filamentary loop structures. Thanks to this effect, it is possible to identify high pressure coronal regions in NIXT pictures.

More generally, the emission distribution along the loop (and, in particular, the emission contrast between the loop leg and its base) depends almost completely on the plasma pressure, being almost the same for loops of very different lengths, even for quite different heating distributions. This means that measurements of the brightness distribution along loops in NIXT band should *directly* provide the value of plasma pressure inside individual loops, in a way complementary to the traditional one of inferring the emission measure from X-ray data, and from it the plasma density. We plan to refine this diagnostic and apply it to data in future work.

Acknowledgments: We acknowledge partial support from the Agenzia Spaziale Italiana and from the Italian Ministero dell'Università e della Ricerca Scientifica e Tecnologica. At SAO, the work was supported by a Grant from NASA to the Smithsonian Institution.

References

Barbee, T. W. Jr.: 1981, AIP Proc. **75**, Atwood D. T., and Henke B. eds., 131

Golub, L. and Herant, M.: 1989, Proc. SPIE **1160**, 629.

Golub, L., Herant, M., Kalata, K., Lovas, I., Nystrom, G., Pardo, F., Spiller, E., Wilczynski, J.: 1990, Nature **344**, 842

Golub, L.: 1993, *NIXT High Resolution Observations*, in "Advances of Stellar and Solar Coronal Physics - G. S. Vaiana Memorial Symposium", Linsky, J. and Serio, S. eds., Kluwer, p. 71

Golub, L. and E. Spiller: *Obtaining High Resolution Coronal Images*, in "Proc. ESA Workshop on Solar Physics and Astrophysics at interferometric Resolution" (Paris, Feb. 1992), p. 221.

Peres G., Reale F., Golub L.: 1994, Ap. J., in press

Serio, S., Peres, G., Vaiana, G. S., Golub, L., Rosner, R.: 1981, Ap. J. **243**, 288

Spiller, E.: 1990, Opt Eng. **29**, 609

Spitzer, L.: 1962, *Physics of Fully Ionized Gases*, Interscience Publ.

Vaiana, G. S., Krieger, A. S., and Timothy, A. F.: 1973, Sol. Ph. **32**, 81

Vaiana, G. S., Rosner, R.: 1978, Ann. Rev. Astr. Ap. **16**, 393

Quiet Sun from Multifrequency Radio Observations on RATAN-600

V.N. Borovik

Main Astronomical Observatory, Pulkovo, St.Petersburg, 196140, Russia

Abstract: We present several characteristics of the quiet Sun radio emission in the range 2-32 cm: the center-to-limb variation of brightness temperature, the radio radius for various positional angles, the brightness temperature spectrum at the center of the disk. These were obtained with the RATAN-600 radio telescope at wavelengths of 2.0, 2.3, 2.7, 3.2, 4.0, 8.2, 11.7, 20.7 and 31.6 cm simultaneously during the recent sunspot minimum (1984-87). The new data are in good agreement with those obtained during previous solar activity cycles both on RATAN-600 radio telescope and on other large instruments.

1 Introduction

Regular multi-wavelength radio observations of the Sun during low solar activity using the same large instrument provide important information concerning the quiet Sun characteristics. Together with data in other spectral bands, e.g. UV, these are critical for the development of realistic solar atmospheric models. Similar observations were made by Zirin et al. (1990) with the 27 m diameter dish at Owens Valley simultaneously at 20 wavelengths over the range of 1.7-21 cm during two epochs in 1986-1987. The brightness temperature spectrum of the quiet Sun at the center of the disk is consistent with previous data but exhibits much less frequency-to-frequency scatter. However, the angular resolution of the radio telescope (2.6-33 arc.min) used by Zirin et al could not provide information on the brightness centre-to-limb variations and radio radius of the quiet Sun.

At Pulkovo Observatory quiet Sun studies were first made using the Large Pulkovo radiotelescope at wavelengths 3.2, 4.5, 6.6 and 9.0 cm with one-dimensional 1-3 arc.min resolution in 1964-1978. Later the RATAN-600 radiotelescope was operated at numerous wavelengths over the range 2-6 cm with spatial resolution of 18 arc.sec - 1 arc min in 1976-78 (Borovik 1979, 1981). The brightness temperatures at the center of the disk and radio radii obtained from these observations are presented in Fig.1 (a,b). A nearly circular-symmetry in brightness distribution at wavelengths of 2-4 cm and an ellipticity at longer wavelengths have been found, as well as an increase with wavelength in limb brightening less than ~10% at 2-3 cm, but approaching 40-50% at 9.0 cm.

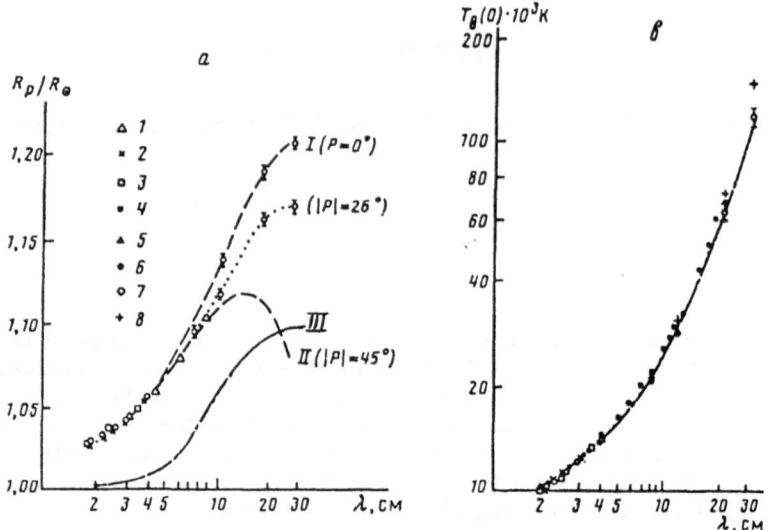

Fig. 1. The radio radius (a) at three positional angles P and brightness temperature (b) of the quiet Sun versus wavelength. The instruments or observatories and periods of observations are shown by numbers 1-8: 1-Large Pulkovo radio telescope, 1964-1978, 2-RATAN-600, 1976-1978 (Borovik 1981), 3-RT-22, Crimea,1972-1975, (Bachurin et al. 1976 and Bachurin 1982) 4-Toyokawa,1975 (Ishiguro et al. 1980), 5-Fleurs, 1977 (Dulk et al. (1977), 6-Owens Valley, 1986-1987 (Zirin et al. 1990), 7-RATAN-600, 1984-1987 (T_b at the center of the disk), 8-Ratan-600, 1984-1987 ($\overline{T}_b$ – averaged). The dotted lines I and II are experimental data summarized and separated for two positional angles by Fürst (1979). The solid lines are calculated data for the homogeneous model of Borovik et al. 1992.

In this work we present the radio characteristics of the quiet Sun obtained with the RATAN-600 radio telescope during the recent sunspot minimum (1984-87) at numerous wavelengths in the range of 2-32 cm.

2 Observations and results

For solar observations the southern sector of RATAN-600 was used in combination with the flat periscopic mirror. Two Stokes parameters (I and V) at 2.0, 2.3, 2.7, 3.2, 4.0, 8.2, 11.7, 20.7 and 31.6 cm were recorded simultaneously on most days as the Sun crossed the stationary diagram pattern at the local meridian. The dimensions of antenna pattern were $18'' \times 13'$ and $285'' \times 205'$ at extreme wavelengths of the range (2.0 cm and 31.6 cm). Examples of solar scans during the quietest periods are shown in Fig 2 (a-d).

The main goal of the data reduction was to separate the profiles of the quiet Sun radio emission from more than 500 solar scans at each wavelength. The data sets which were unreliable due to receiver instability, deviation of the electrical axis

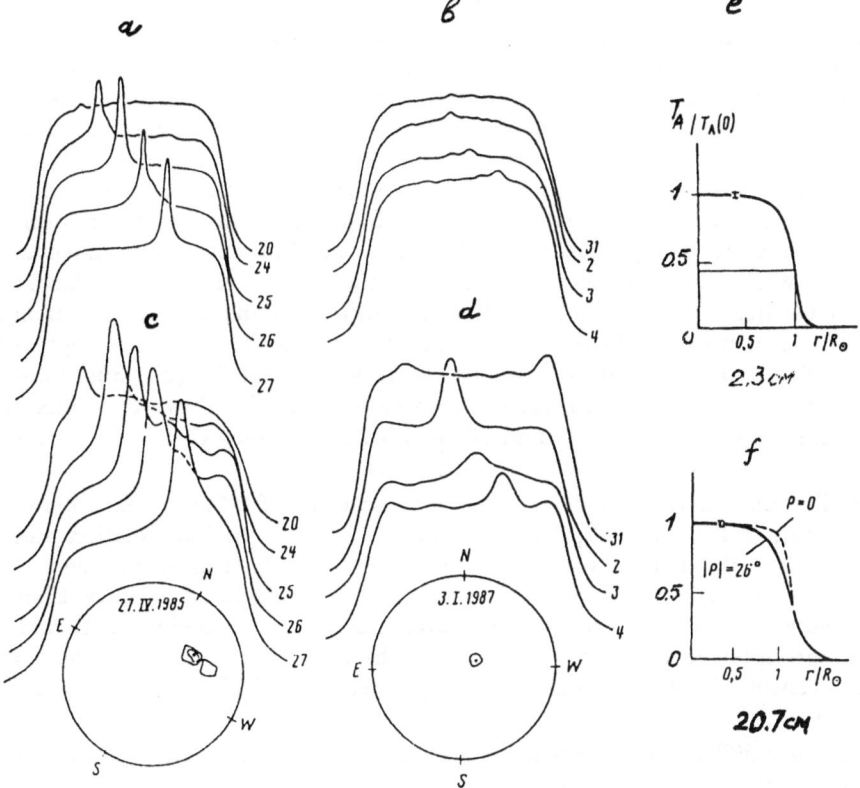

Fig. 2. One-dimensional solar scans obtained on RATAN-600 at 2.3 cm (a,b) and at 20.7 cm (c,d) during the periods of low solar activity (April 20-27, 1985 and December 31,1986-January 4, 1987); e,f - the quiet Sun profiles, separated from the solar scans obtained at two extreme solar positional angles P.

of the diagram pattern with respect to the center of the solar disk, or the high level of scattering due to reflection errors in the antenna have been excluded (the last problem was considered by Borovik, 1981).

"Solar Geophysical Data" and "Solnechnye Dannye" Bulletin were used to select the following periods, which had extreme values of the solar positional angle "P", for reduction:

I - September 12-15, 1984; II - October 11-12, 1984; III - April, 20-28, 1985; IV - March 17-19, 1987; V - April 7-10, 1987; VI - December 31, 1986-January,5,1987. During these periods large portions of the solar surface covering 50 — 90° of heliospheric longitude had no significant manifestations of activity. The associated parts of the solar scans have been used for the construction the quiet Sun profiles.

No equatorial coronal holes (CH) were found on the chosen solar sections according to the Catalogue of CH (1992). The setting and rising of the polar CHs 100, 133 and 149 (Catalogue of CH, 1992) in the periods I, II, IV and V were not observed because of the small fraction of the hole in the N-S strip due to its orientation relative to the beam pattern. In period VI the poles of the Sun were covered by the CH completely.

Quiet Sun profiles at wavelengths of 2.0, 2.3, 2.7, 3.2 and 4.0 cm, averaged over the supergranulation and small-scale features typically observed with interferometers, were constructed for each period from consecutive solar scans by the iteration method (the standard deviation of the ordinates of the normalized profiles was 1.5%). The widths of the profiles at extreme solar positional angles turned out to be the same (within the accuracy of measurements $\pm 0.005 R_o$) This confirms the absence of considerable ellipticity in the quiet Sun brightness distribution over this wavelength range. The profiles at 8.2, 11.7, 20.7 and 31.6 cm have been separated in the records as the lower envelopes . As shown above, the coronal holes have been excluded in this procedure while the filaments, which in some cases were seen on the drift curves at wavelengths of 8.2 ,11.7 and 20.7 cm as the brightness depressions (1-2 % of antenna temperatures) against the background of the quiet Sun, have not been excluded in constructing the lower envelopes. The Toyokawa total flux measurements made at 3.2, 8.0, 15 and 30 cm with an accuracy of 1% (Tanaka et al. 1973) have been used for scaling the solar scans.

Radio fluxes have been determined with an error of 2-3% for each period at wavelengths of 4.0, 8.2, 11.7, 20.7 and 31,6 cm through the interpolation and extrapolation of the daily Toyokawa spectra. The fluxes determined for the periods I - V, when the solar positional angle was $\pm 24 - 26°$, differed no more than 3% from the average value. The averaged profiles separated in periods I, II and IV, which corresponded to the lowest fluxes of the quiet Sun given in Table I, have been deconvolved using a maximum entropy method by M.S. Kurbanov and V.V. Makarov.

The most difficult problem was to determine the flux of the quiet Sun radio emission at wavelengths shorter than 3.2 cm because the solar patrol services at Sagamore Hill Observatory, at the Havana (Cuba) and the Kislovodsk stations showed the significant scattering in the total solar flux during sunspot minimum of 500 $\pm$ 100 s.f.u. at wavelengths of 1.95-2 cm. We used the results of measurements carried out at Pulkovo during 1966-1981. A flux of quiet Sun radio emission of 520 $\pm$ 20 s.f.u. at 2.0 cm has been obtained from comparison of lunar and solar observations made with the same 2-3 m diameter dishes during periods of solar eclipse (Gelfreich et al. 1972 and Korzhavin 1979) and in other periods. The absolute flux of the Moon was determined using the calibration method of " artificial Moon" developed in Nizhnij Novgorod (Krotikov 1961). The flux of 520 s.f.u at 2 cm is close to the result of extrapolation of quiet Sun flux spectrum at 3.2 - 31.6 cm given in Table 1. The fluxes at 2.3 and 2.7 cm were obtained through the interpolation the same spectrum.

The radial brightness distributions obtained from deconvolution of the separated one-dimensional quiet Sun profiles at wavelengths of 2-31.6 cm have been tested by using the lunar observations carried out with the RATAN-600 radiotelescope in the same wavelengths range and by model calculations: the same method was used by Borovik (1981).

The following parameters of the deconvolved radial brightness distributions of the quiet Sun are given in Table I:

$\Delta T_b/T_b$ – the peak brightening as a percentage of the central brightness temperature T_b; R_b/R_o – the distance of the peak brightening from the center of the disk(with the accuracy ± 0.01); R_o – the optical radius of the Sun; R_p/R_o

– the radio radius of the quiet Sun (the distance from the center of the disk where $T_b(r) = 0.5T_b(0)$); the accuracy of radio radius measurements is ± 0.005 at 2-4 cm and ± 0.015 at 8.2 - 31.6 cm. F_q – the average quiet Sun flux in the periods I,II and IV; $T_b(0)$ – the brightness temperature at the center of the quiet Sun corresponding to the deconvolved brightness distribution and flux F_q; $\overline{T}_b$ – the averaged brightness temperature of the quiet Sun corresponding to total flux and radio radius

The second values of $\Delta T_b/T_b$ and R_p/R_o at wavelengths of 11.7, 20.7 and 31.6 cm corresponds to positional angle P=0°.

Table 1. The parameters of the deconvolved brightness temperatures distributions

λ, cm	$\Delta T_b/T_b$ (%)	R_b/R_o	R_p/R_o	F_q	$T_b(0)$	$\overline{T}_b$
2.0	< 2	1.00	1.026	520 ± 20	10.4 ± 0.4	10.5
2.3	3 ± 2	1.00	1.030	417 ± 10	10.9 ± 0.3	11.0
2.7	5 ± 2	1.00	1.035	324 ± 8	11.5 ± 0.3	11.7
3.2	10 ± 2	1.00	1.041	253 ± 3	12.4 ± 0.4	12.7
4.0	15 ± 2	1.00	1.052	180 ± 5	13.2 ± 0.6	13.8
8.2	34 ± 5	0.98	1.095	74 ± 2	21.1 ± 0.6	24.1
11.7	65 ± 5	0.98	1.106	56 ± 2	29.6 ± 1.2	33.2
	95		1.135			
20.7	113 ± 10	0.95	1.155	46 ± 1	59.6 ± 1.5	78.3
	200		1.188			
31.6	92 ± 10	0.93	1.166	44 ± 1	131.7 ± 3.5	171.3
	148		1.205			

3 Discussion

The spectra of the central radio brightness of the Sun obtained by Zirin et al. (1990) and that found from RATAN-600 observations are in reasonably good agreement. The differences, at wavelengths longer than 10 cm are probably due to a nonuniform brightness distribution at these wavelengths; the high spatial resolution of the RATAN-600 radiotelescope in this case is important. This explanation is consistent with the fact that the averaged brightness temperatures are in much closer agreement. The centimeter wavelength part of the two spectra agree within a few percent using the calibration data of Pelushenko (1982). We also note (see Fig.1b) that other measurements made at separate frequences with large radio telescopes (such as in Bonn, Crimea, Toyokawa, Fleurs, Ottawa) have shown good agreement. Results compiled by Trottet and Lantos (1978) and Grebinsky (1987) also do not contradict the above results.

The interpretation of the central brightness at microwaves, though essential for model computations, are not very critical to a model. In fact, even a homogeneous model can easily account for the spectrum. An example of such a computation was made by Borovik et al. (1992) and is shown in Fig 1a,b. On the other hand the brightness distribution at the limb is very critical to most of the model parameters and in most cases the results are not compatible with widely assumed models of the solar atmosphere.

In this paper we have presented data on the radio radius of the quiet Sun and the center-to-limb brightness distribution at a large number of wavelengths in the range of 2-32 cm gained with same instrument of high spatial resolution (RATAN-600). We have shown also that these data do not contradict some episodical results obtained by other authors (see Fig 1 and review by Fürst 1979). Optical, UV and X-ray observations of the Sun, together with microwave radio observations carried out with high spatial resolution, all show the inhomogeneous structure of the quiet solar atmosphere: chromospheric network, spicules, macrospicules, small bipolar structures and HeII dark points (X-ray bright points), coronal loops. The averaged radio characteristics of the quiet Sun at centimeter and decimeter wavelengths depend on the physical parameters of these structures and are essential in order to develop realistic inhomogeneous models of the quiet solar atmosphere, consistent with both multi-wave radio observations and UV-data.

Acknowledgements

The author thanks Gelfreich G.B, Livshitz M.A., Korzhavin A.N. and Bogod V.M. for discussions. It is pleasure to thank the Local Organising Committee of the Catania Conference for financial support.

References

Bachurin, A.F. and Eriyshev, N.N. 1976, *Izv. Crimea Obs.* **54**, 241

Bachurin, A.F.: 1982, *Izv. Crimea Obs.* **65**, 71

Borovik, V.N. : 1979, *Astrofiz. issled. (Izv. SAO)* **11**, 107

Borovik, V.N. : 1981, *Astrofiz. issled. (Izv. SAO)* **13**, 17

Borovik, V.N., Kurbanov, M.Sh., and Makarov, V.V.: 1992, *Sov. Astron.* **36**, 656

National Geophys. Data Center: 1992, *Catalogue of CH 1970-1991*, Boulder, USA

Dulk, G.A., Sheridan, K.V., Smerd, S.F., and Withbroe, G.L.: 1977, *Solar Phys.* **52**, 349

Fürst, E.: 1979, in *Radio Physics of the Sun*, M.R. Kundu and T.E. Gergely (eds.), IAU Symp. **86**, p. 25

Gelfreich, G.B., and Peterova, N.G.: 1972, Radiastron. nabl. soln. zatm., Nauka, Moscow 5

Grebinsky, A.S.: 1987, *Sov. Astr. Letters* **13**, 4

Ishiguro, M., Enome, S. and Shibasaki, K.: 1980, *Publ. Astron. Soc. Japan* **32**, 533

Korzhavin, A.N.: 1979, *Thesis*, Pulkovo, Leningrad

Krotikov, V.D., Porfirev, V.A., and Troitzkii, V.S.: 1961, *Radiofizika* **4**, 1004

Pelushenko, S.A.: 1982, *Radiofizika* **25**, 977

Tanaka, H. et al.: 1973, *Solar Phys.* **29**, 243

Trottet, G., and Lantos, P.: 1978, *Astron. Astrophys.* **70**, 245

Zirin, H., Baumert, B.M., and Hurford, G.: 1990, *Ap.J.* **370**, 779

Frequency Spectra of Solar Microwave Bursts Associated with Coronal Mass Ejections

I.M.Chertok, A.A.Gnezdilov

IZMIRAN, Troitsk, Moscow Region, 142092, Russia

Abstract: Data for the period 1979-1982 on Coronal Mass Ejections (CME), observed by the SOLWIND white light coronagraph aboard the P78-1 satellite and on associated microwave bursts, are analyzed. It is shown that in events with coronal mass ejections a soft radio spectrum (maximum frequency $f_m \leq$ 5-7 GHz) is mainly observed in the longest (duration $d \geq 5$ min) microwave bursts of moderate and strong intensities, i.e. for a combination of parameters typical of flares associated with large CMEs. It is supposed that in such events, both long duration and soft microwave bursts are caused by prolonged energy release at high coronal altitudes, when the magnetic field above the active region, strongly disturbed by the CME, relaxes via magnetic reconnection in vertical current sheets and the formation of a post-flare loop system.

1 Introduction

Usually, when speaking about flare energy release, one refers mainly to the initial (primary) or impulsive energy release. This occurs in a region of fast magnetic reconnection in current sheets in the corona where a non-potential magnetic energy component, stored in processes of emergence of new magnetic flux, in the interaction of magnetic loops, and so on, is present (Somov 1987). However, there is another source of much more prolonged energy release, which arises in the late phase of eruptive flares at high altitudes in the corona, after the coronal mass ejection (CME) (Kopp and Pneuman 1976; Martens and Kuin 1989). Here the magnetic field above an active region, strongly disturbed in the CME eruption process, relaxes via magnetic reconnection in vertical current sheets and the formation of a post-flare loop system.

There are many manifestations of this post-eruptive energy release, ranging from high-altitude sources of the so-called delayed or gradual microwave and hard X-ray bursts (Cliver et al. 1986; Kai et al, 1986) up to the high-energy gamma-ray emission lasting several hours after the impulsive phase (Akimov et al. 1993).

In the present work we shall give more evidence of the importance of the post-CME energy release. In a previous paper (Chertok et al. 1992), we considered the distribution of CMEs having different parameters on diagrams of "absolute peak intensity (S) versus the effective duration at the half peak flux level (d)" of microwave bursts. There are definite zones on the S–d-plane where events with different CMEs and events without CMEs are concentrated (Fig. 1). Consideration of near-limb events showed that large, massive, high-speed CMEs of complex form are observed mainly in coincidence with the most intense and long microwave bursts. CMEs having intermediate characteristics and comparatively simple forms are identified with moderate non-impulsive bursts as well as with relatively weak, but long-duration "gradual rise and fall" (GRF) radio bursts. The majority of impulsive bursts are not accompanied by CMEs; only the most intense of them may be associated with small CMEs of simple shape, moderate speed and mass.

Besides the peak intensity and effective duration, one more important parameter characterizing the microwave bursts is the spectral maximum frequency f_m. In the framework of the gyrosynchrotron mechanism of radio emission, this frequency depends on the magnetic field strength of the source and on the energy spectrum of the accelerated electrons (Dulk and Marsh 1982). The totality of the parameters (peak intensity, effective duration, and spectral maximum frequency) gives a OAsufficiently complete description of the microwave burst.

In this paper our previous analysis is supplemented by considering the characteristics of the microwave burst frequency spectrum in events with CMEs.

2 Analysis and results

We analyzed several tens of SOLWIND CMEs which, according to Sheeley et al. (1984,1985), Kahler et al. (1984), Cane et al. (1987), and others, are associated with definite flares on the disk. For comparison, we considered also about twenty events in the near-limb zone (heliolongitudes $\mid l \mid \geq 45°$) which have not been positively accompanied by CMEs. The spectral maximum frequency of the microwave bursts f_m was established from data published in Solar-Geophysical Data (1979-1982). The effective duration d was measured from time profiles obtained at a number of observatories (Chertok et al. 1992). In radio bursts with a complicated profile consisting of several components, d was the sum of the duration of the individual components at the common half peak level for the entire burst. The GRF-events have been omitted because, in the majority of such events, the determination of f_m was not possible.

In Fig. 2 a standard S–d-diagram is shown where events (a) with CMEs and (b) without CMEs are marked by different signs depending on the maximum frequency of microwave burst. In particular, bursts with hard ($f_m \geq 15$ GHz), moderate ($f_m \sim$ 8.4-12 GHz), intermediate ($f_m \sim 7$ GHz), and soft ($f_m \leq 5$ GHz) radio spectra are separated. Boundaries of zones of intense, moderate, impulsive, and GRF bursts are marked by dashed lines. The solid vertical line at $d \sim 5$ min separates the regions of long and comparatively short-term radio bursts.

The main feature of the frequency spectrum in flares with CMEs is the following. The soft and intermediate radio spectra ($f_m \leq 5$-7 GHz) are mainly observed in microwave bursts of large effective duration. About 92 % of events with such a spectrum are concentrated in the region of long-duration bursts with $d \geq 5$ min, and a soft radio spectrum ($f_m \leq 5$ GHz) was observed in about 73 % of events. The percentage of bursts with soft and intermediate spectra in the $d \geq 5$ min region is ~ 70 %, and for moderate radio flux density ($S \sim 100 - 1000$ s.f.u.) it achieves 92 %.

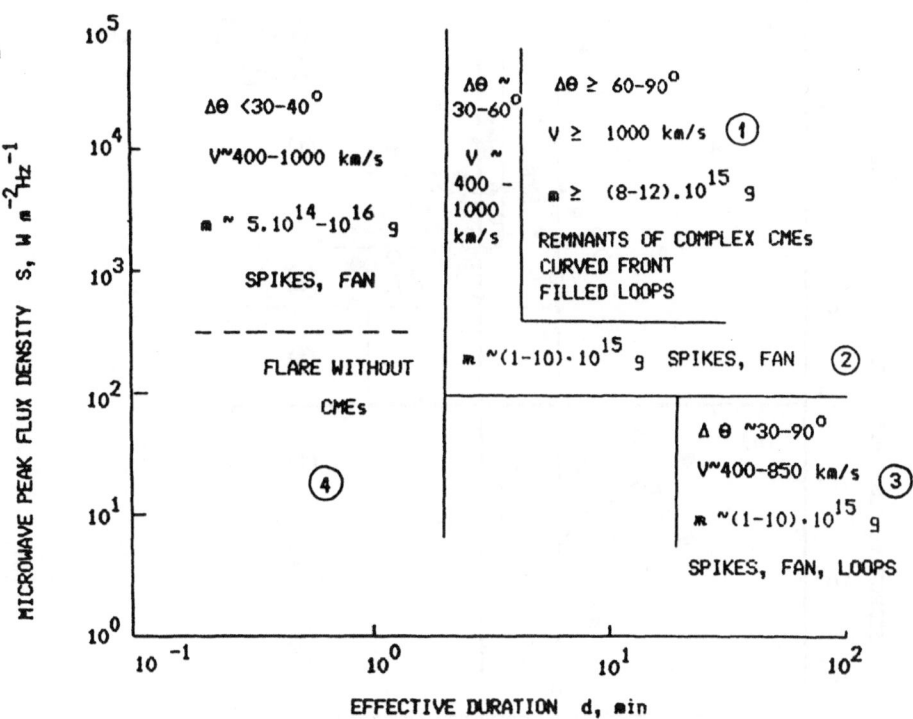

Fig. 1. Summary of the correlation between the microwave bursts and the characteristics of CMEs. The zones of intensive (1), intermediate (2), GRF (3), and impulsive (4) radio bursts are indicated

On the other hand, the majority ($\sim 87\ \%$) of shorter-term ($d < 5$ min) microwave bursts accompanied by CMEs, show moderate or hard radio spectra with $f_m \geq 8.4$-12 GHz. It should be added that about 69 % of events without CMEs, which mainly belong to a category of impulsive bursts, have also moderate or hard spectra, while soft and intermediate spectra are observed only in impulsive bursts of moderate intensity $S \leq 500$ s.f.u.

Taking into account the previous regularities describing the distribution of CMEs with different parameters on the S–d-diagram (Chertok et al, 1992), the data show that for many flares associated with CMEs the following scenario is typical: a long-duration ($d \geq 5$ min) microwave burst with a soft or intermediate radio spectrum ($f_m \leq 5$-7 GHz), and a sufficiently powerful (large, high-speed) CME.

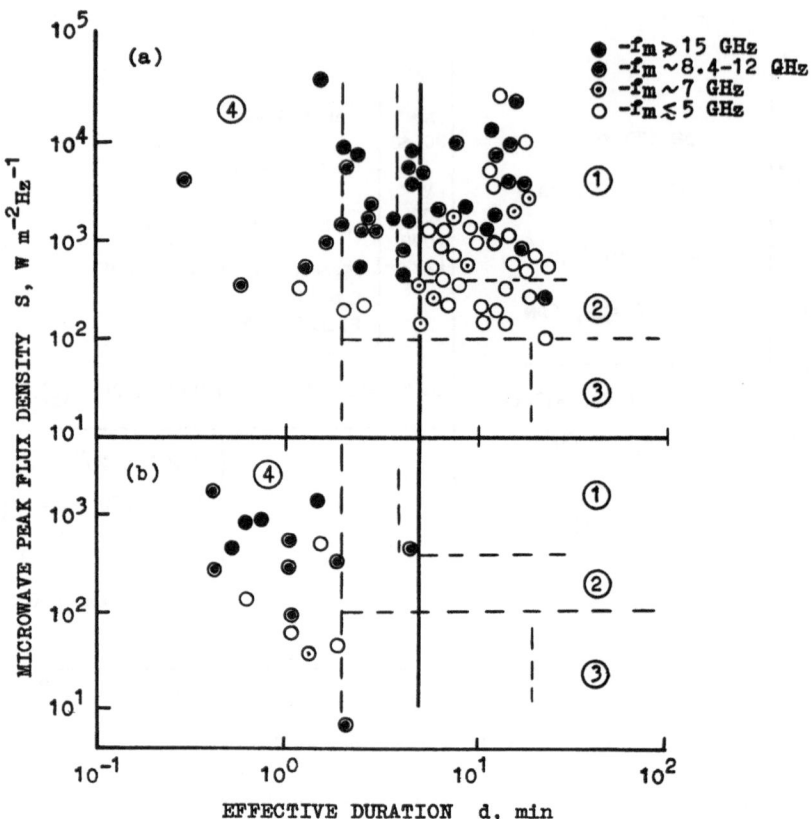

Fig. 2. Distribution on the S–d-diagram of microwave bursts with different frequency spectra for events (a) with CMEs and (b) without CMEs; dashed lines indicate the boundaries of (1) intense, (2) intermediate, (3) GRF and (4) impulsive radio burst zones; the solid vertical line at $d \sim 5$ min separates the regions of long and comparatively short radio bursts

3 Discussion and conclusions

To discuss the results outlined above, let us return to the concept of sources of initial and post-eruptive energy release outlined above. From many observational data it is evident that both sources of energy release are accompanied by particle acceleration, including electrons with an energy of $\sim 10^2$ keV, and manifest themselves in a definite way with electromagnetic emission, in particular with microwave bursts. The majority of the energetic electrons accelerated in the primary energy release turn out to be captured in low, compact arches which have a strong magnetic field. These electrons generate an impulsive component of the microwave burst with a relatively hard radio spectrum. The long-duration (gradual) microwave component, corresponding to prolonged particle acceleration at the post-eruptive stage, appears to be generated by electrons which are trapped in high coronal loops with lower density and magnetic field strength, and are characterized by a duration od tens of minutes - hours, and a softer spectrum.

Generally speaking, both sources of energy release and the corresponding impulsive and gradual microwave components in flare events may occur independently and can be observed in different combinations including the ones with a significant time delay (Cliver et al. 1986; Kai et al. 1986). In relatively short-duration flares without CMEs the impulsive component predominates, and the gradual component is either quite absent or manifests itself poorly. Underneath a large CME, on the contrary, due to the strong disturbance of the magnetic field above an active region, the coronal post-eruptive energy release at the final phase of a flare appears to become more significant and may control both the duration and the spectrum of the microwave bursts. In the context of present analysis, there are arguments to show that the first variety occurs in the majority of flares with $d < 5$ min, and the second one is characteristic of events with $d \geq 5$ min.

By way of illustration, let us estimate some parameters corresponding to the hard and soft radio spectra in the framework of the gyrosynchrotron emission mechanism of microwave bursts. According to Dulk and Marsh (1982), for a number of emitting electrons on the line of sight $NL \sim 10^{16} \text{cm}^{-2}$, an index of the power-law energy spectrum $\delta \sim 3.5$ at the range $E \sim$ 10-1000 keV, and a viewing angle $\sim 45°$, the soft spectrum with $f_m \leq$ 3-5 GHz corresponds to a magnetic field strength $H \leq 150 - 300$ G and the hard one with $f_m \geq$ 9-12 GHz corresponds to $H \geq$ 700-1000 G. These parameters are typical of microwave bursts associated with gradual and impulsive flares respectively.

Of course, this picture of flare energy release and the conclusions outlined above should be considered as an idealized working hypothesis. The spectral maximum frequency depends in a rather complicated way on a number of flare characteristics: localization and structure of the sources of radio emission, the electron energy spectrum, number of accelerated particles and their dynamics in the loop-like magnetic traps, and so on. However, it becomes clear that the eruption of CMEs and the resulting strong disturbance of the coronal magnetic field, plus subsequent relaxation of the magnetic field to its initial state and the associated long-term energy release, all play an important part in the flare and flare-like events. These processes determine to a considerable extent many of the observed characteristics, in particular the frequency spectrum and effective duration of microwave bursts discussed in the present paper.

Acknowledgements

I.M.Ch. is grateful to the Organizing Committee of the Catania Meeting for providing him with financial support. This work has been supported from a Soros Foundation Grant awarded by the American Physical Society and from a grant of the Russian Fundamental Research Foundation.

References

Akimov, V.V., Leikov, N.G., Belov, A.V., Chertok I.M., Kurt, V.G., Magun, A., Melnikov, V.F. (1993): "Some Evidences of Prolonged Particle Acceleration in the High-Energy Gamma-Ray flare of June 15, 1991", in Proc. of the Meeting "High Energy Solar Phenomena: A New Era of Spacecraft Measurements", Waterville Valley, New Hampshire, March 2-5, 1993 *(in press)*

Cane, H.V., Sheeley, N.R., Jr., Howard, R.A. (1987): Energetic interplanetary shocks, radio emission and coronal mass ejections J. Geophys. Res. Vol. 92, No. A9, pp. 9869–9874

Chertok, I.M., Gnezdilov, A.A., Zaborova, E.P. (1992): Microwave and Soft X-Ray Emission from Solar Flare Events Associated with Coronal Transients. Sov. Astron. Vol. 36, No. 3, pp. 301–306

Cliver, E.W., Dennis, B.R., Kiplinger, A.L., Kane, S.R., Neidig, D.F., Sheeley, N.R., Jr, Koomen, M.J. (1986): Solar Gradual Hard X-Ray Bursts and Associated Phenomena. Astrophys. J., Vol. 305, No. 2, pp. 920–935

Dulk, G.A., Marsh, K.A. (1982): Simplified Expressions for the Gyrosynchrotron Radiation from Mildly Relativistic, Non-thermal and Thermal Electrons, Astrophys. J., Vol. 259, No. 1, Part 1, pp. 350–358

Kahler, S.W., Sheeley, N.R., Jr., Howard, R.A., Koomen, M.J., Michels, D.J., McGuire, R.E., von Rosenvinge, T.T., Reames, D.V. (1984): Associations between Coronal Mass Ejections and Solar Energetic Proton Events, Astrophys. J., Vol. 89, No. A11, Part 1, pp. 9683–9693

Kai, K., Nakajima, H., Kosugi, T., Stewart R.T., Nelson, G.J., Kane, S.R. (1986): Radio Evidence for a Delayed Acceleration Process in Solar Flares. Solar Phys., Vol. 105, No. 2, pp. 383–398

Kopp, R.A., Pneuman, G.W. (1976):Magnetic Reconnection in the Corona and the Loop Prominence Phenomenon. Solar Phys., Vol. 50, No. 1, pp. 85–100

Martens, P.C.H., Kuin, N.P.M. (1989):A circuit model for filament eruptions and two-ribbon flares, Solar Phys., Vol.122, No. 2, pp. 263–302

Sheeley N.R., Jr., Stewart, R.T., Robinson, R.D., Howard, R.A., Koomen, M.J., Michels, D.J. (1984): Associations between Coronal Mass Ejections and Metric Type II Bursts, Astrophys. J., Vol. 279, No. 2, Part 1, pp. 839–847

Sheeley N.R., Jr., Howard, R.A., Koomen, M.J., Michels, D.J., Schwenn, R., Müthlhäuser, K.H., Rosenbauer H. (1985): Coronal Mass Ejections and Interplanetary Shocks, J. Geophys. Res., Vol. 90, No. A1, pp. 163–175

Solar-Geophysical Data (1979-1982): NOAA, Boulder

Somov, B.V. (1987): "Solar Flares", in Results of Science and Technics: Astronomy, ed. by I.S.Shcherbina-Samoilova (VINITI, Moscow), Vol. 34, pp. 78–135 *(in Russian)*

Observations of High-Energy (E≥10 MeV) Gamma-Rays with the PHEBUS Instrument

N. Vilmer [1], G. Trottet [1], C. Barat [2], J.P. Dezalay [2], R. Talon [2],

R. Sunyaev [3], O. Terekhov [3], A. Kuznetsov [3]

[1] Observatoire de Paris, Section de Meudon, DASOP, F-92195 Meudon,
France
[2] Centre d'Etude Spatiale des Rayonnements, BP 4346, F-31029 Toulouse,
France
[3] Space Science Institute, Profsoyouznaya 84/32, 117810 Moscow, Russia

Abstract: The PHEBUS instrument aboard GRANAT has observed several high energy
(E ≥ 10 MeV) gamma-ray bursts since 1990. We describe here the characteristics of these
high energy bursts with respect of the longitude of the associated Hα flare. We discuss our
results in the context of the transport of high energy particles in the solar atmosphere.

1 Introduction

Measurements of hard X-ray (HXR) and gamma-ray (GR) emissions represent the
most direct diagnostics of energetic particles accelerated during solar flares. High
energy radiation (≥10 MeV) has been observed in solar flares since Cycle 21 with
the GR spectrometer aboard SMM (Rieger et al. 1983; Vestrand et al. 1987). It
is produced by bremsstrahlung radiation of ultrarelativistic electrons or by decay
radiation of pions produced through the interaction of high energy ions (e.g. ≥200
MeV protons) with the ambient solar atmosphere (Ramaty and Murphy 1987) or
by a combination of both processes. This high energy radiation is thus a good probe
of the most energetic particles accelerated in flares. Given the sensitivity of existing
detectors, high energy events are rather unfrequent since less than 30 events with
≥10 MeV emission have been detected during 10 years of continual observations
with SMM (e.g. Vestrand et al. 1991). The production of pion decay radiation was
reported so far only for 6 events (Mandzhavidze and Ramaty, 1993). In most cases,
high energy radiation is dominated by ultrarelativistic electrons. The high energy
events of Cycle 21 have shown a tendency to be associated with Hα flares at the
limb (Vestrand et al. 1987). Such a "limb brightening" has been interpreted in terms
of the transport of the emitting high energy electrons in converging magnetic fields,
provided that these fields are nearly normal to the photosphere (MacKinnon and
Brown 1989; Kocharov and Kovaltsov 1990; Ramaty et al. 1988, Mc Tiernan and

197

Petrosian 1991). However, during the early phase of Cycle 22, SMM has detected several events associated with flares located well on the disk (Vestrand et al. 1991) raising the question whether Cycle 22 behaves differently from Cycle 21. More recently, the experiments aboard GAMMA1 (Akimov et al. 1991) and COMPTON (Kanbach et al. 1993) have also observed high energy events associated with disk flares. In this paper, we summarize the observations of high energy GR bursts detected from January 1990 to June 1991 by the PHEBUS instrument aboard GRANAT and discuss them in the context of the previous observations and interpretations.

2 Phebus Observations of $\geq$10 MeV Events

The PHEBUS instrument (Barat et al. 1988; Talon et al. 1993) recorded 8 HXR/GR events with significant continuum emission above 10 MeV between January 1990 and June 1991. These solar flares are listed in Table 1 which also indicates for each event the burst trigger time as well as some characteristics of the associated soft X-ray and Hα signatures taken from Solar Geophysical Data. Some of these bursts have also been observed up to 15 MeV by the Anticoincidence shield of the SIGMA telescope aboard GRANAT (Pelaez et al. 1991; Pelaez 1993). The longitudinal distribution of the Hα flares associated with the events of Table 1 do not show any evidence for a clustering towards the limb. However, the limited number of events do not allow to draw a strong conclusion on their center-to-limb distribution. Furthermore, as indicated in the last columns of Table 1, the events present different spectral behaviour at high energies indicating that different components of the high energy radiation (electron bremsstrahlung or pion decay radiation) are predominant. As these components are expected to present different anisotropies, we shall discuss in the following the observations of flares with highly different spectral characteristics that occurred either close to the disk center or close to the limb.

Figure 1 shows the time profile of the 25 January 1991 burst in 5 energy bands. This event is associated with an H α flare close to the limb. The emission from the first hardest peak lasts for about 20 seconds and extends up to about 25 MeV with a significant counting rate. The sharp spectral steepening above 25 MeV suggests that the $\geq$10 MeV flux is predominantly radiated by ultrarelativistic electrons. Furthermore, GR line emission is very weak during this first peak indicating that this event has characteristics similar to "electron dominated events" (Rieger and Marschhaüser 1990). Figure 2 shows the time profile of an event of the same type associated with an Hα flare close to the disk center, e.g. the 30 June 1991 event. Another similar event (11 June 1990) had previously been reported in Talon et al. (1993). Both events exhibit a sharp spectral steepening at energies around 40 MeV and weak, if any, GR line emission during the hardest peak.

On the contrary, the two events of 24 May 1990 and 22 March 1991 present strong emission up to the highest energy bands available for each event (resp 80-110 MeV and 60-80 MeV). Moreover, they are both associated with neutron events detected at ground levels (Pyle et al. 1991; Pyle and Simpson 1991). The 24 May 1990 flare associated with an intense neutron monitor increase due to neutrons (Pyle et al. 1991) has been described in details in Talon et al. (1993). At energies above 10 MeV it comprises essentially two peaks followed by a long-lasting tail. The spectrum of the second peak exhibits some hardening above 60 MeV suggesting that

Table 1. ≥10 MeV events observed by PHEBUS from January 1990 to June 1991

Date	Time UT	GOES	Hα Imp	Hα Pos	comments
11/05/90	20:58:09	X3.6	2B	N29E86	flux up to 85 MeV no spectr. steep.
24/05/90	20:46:23	X9.3	1B	N33W78	spectr. harden. ≥ 80 MeV neutron at earth strong 2.2 MeV line
11/06/90	09:43:01	M4.5	2B	N10W22	spectr. steep. at 40 MeV
25/01/91	06:29:38	X10.0	SF	S16E78	spectr. steep. at 25 MeV
12/03/91	12:43:02	X1.7	2B	S07E59	flux up to 100 MeV no spectr. steep.
22/03/91	22:42:51	X9.4	3B	S26E28	spectr. harden. at high en. neutron at earth strong 2.2 MeV line
31/03/91	19:08:38	X1.0	SF	S22W88	flux up to 20 MeV no spectr. steep.
30/06/91	02:56:07	M5.0	1N	S06W19	spectr. steep. at 40 MeV

the dominant process is emission from neutral pion decay. The long-lasting tail is observed in rough coincidence with the detection of neutrons at ground level and is probably the signature of neutrons reaching the satellite. Figure 3 shows the time profile above 420 keV of the 22 March 1991 event which is associated with a less intense neutron event at ground level (Pyle and Simpson 1991). The photon emission is so intense below 420 keV that it saturates the detector. The event consists in the 420-600 keV band of essentially 3 peaks. Some hardening of the spectrum at high energies is observed after 22:44:10 UT. Furthermore, as for the 24 May 1990 event, a strong 2.22 MeV line is observed in the decay phase after 22:44:20 UT. On the other hand, no significant high energy tail is observed for that flare by PHEBUS. The whole set of GR and ground-level neutron observations for these two events, associated respectively with disk and limb Hα flares, indicate that a large number of ions with energies above a few hundred MeV has been accelerated.

Figure 4 shows the time profile of the 12 March 1991 event. The emission from the hardest peak between 12:43:40 and 12:44 UT shows strong emission up to the highest energy band around 100 MeV. However, no clear hardening of the spectrum is observed at energies above 60 MeV. As no neutron detection has been reported

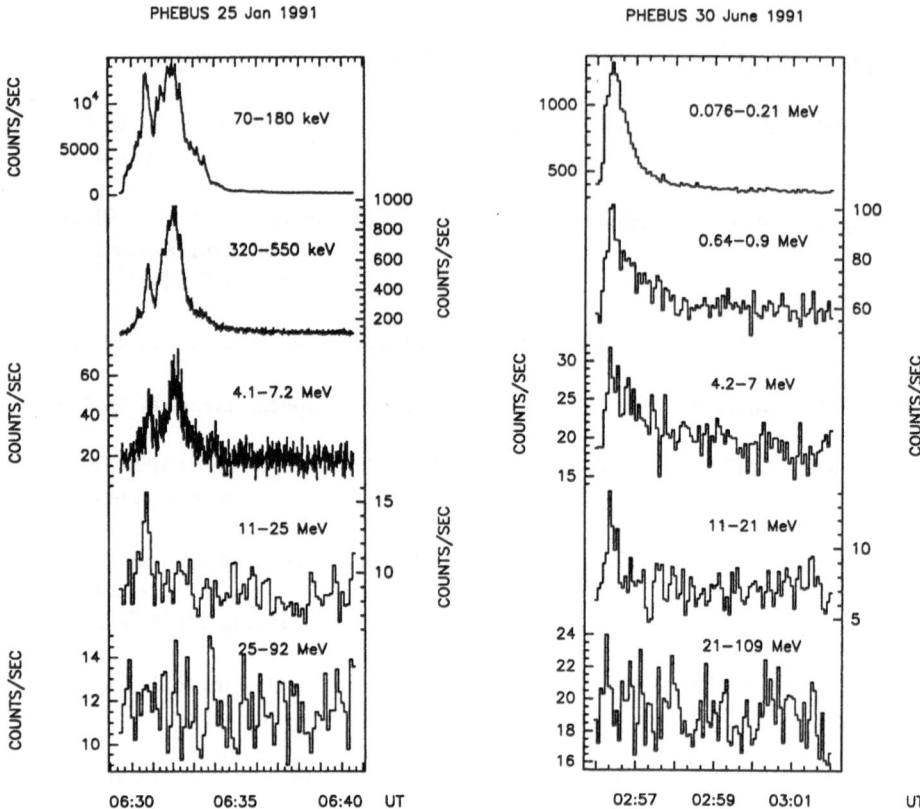

Fig. 1. Time profiles of the 25 January 1991 event observed with PHEBUS with an integration time of 1s below 10 MeV and of 8s above 10 MeV.

Fig. 2. Same as Figure 1 for the 30 June 1991 event with an integration time of 4s in all the energy bands.

for that flare, a complete study is necessary to determine which physical process is predominant at high energies. This is also the case of the 11 May 1990 event discussed in more details in Talon et al (1993).

3 Discussion

The PHEBUS observations presented above tend to show that the spectral properties of GR bursts with $\geq$10 MeV emission are not strongly dependent on the location of the Hα flare. Both events with characteristics similar to "electron dominated events" and events in which a large number of high energy ions are produced (neutron detection at ground level and probable pion radiation) are observed for limb as well as disk flares. The finding based on the analysis of the SIGMA observations of some of these bursts (Pelaez 1993) that the spectral index of the photon

PHEBUS 22 March 1991

PHEBUS 12 March 1991

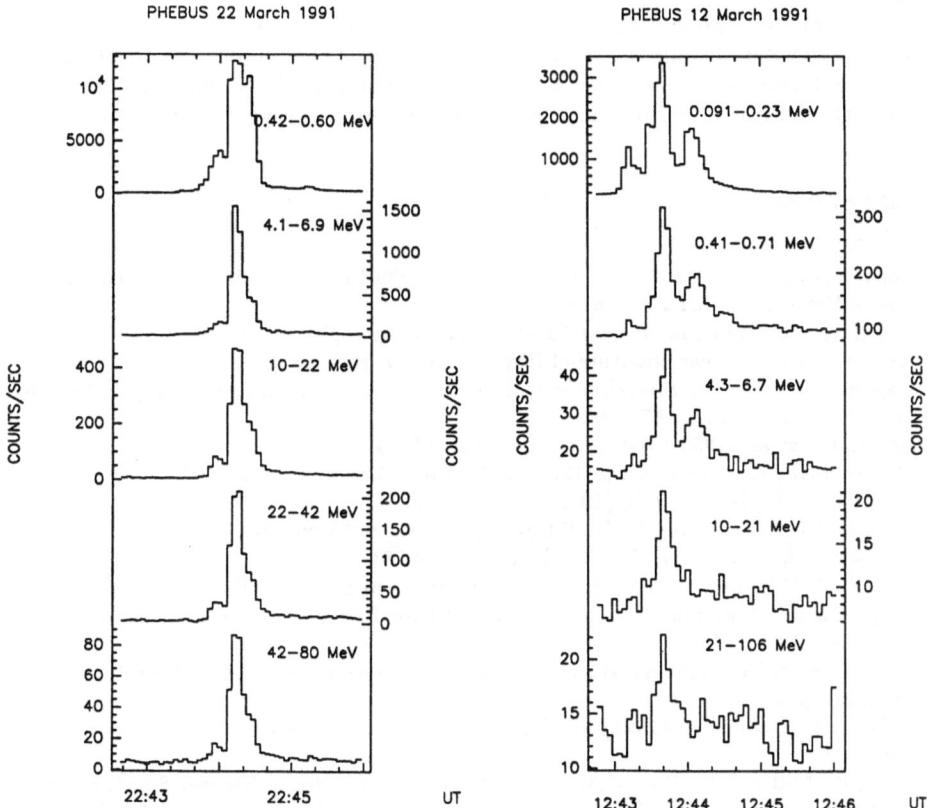

Fig. 3. Same as Figure 2 for the 22 March 1991 event.

Fig. 4. Same as Figure 2 for the 12 March 1991 event.

continuum in the 0.3-10 MeV band has no correlation with the heliographic longitude of the Hα flare is in agreement with the PHEBUS observations at higher energies. The present results are however based on a too small sample of events to conclude that high energy events have an isotropic distribution for the present solar cycle. The intercalibration of the different instruments in operation during this cycle as well as new PHEBUS observations of high energy bursts will allow to achieve a more complete heliocentric angle distribution for this cycle and to compare it with the one obtained in Cycle 21.

The PHEBUS observations of two events located close to the disk center for which the ≥10 MeV emission is dominated by electron bremsstrahlung confirm the SMM observations of similar events in the present cycle (Rieger and Marschhaüser 1990; Chupp et al. 1993). In the frame of the models developed so far to predict the high energy radiation anisotropy, the observations of events close to the disk center indicate that electron bremsstrahlung arises from upwardly moving electrons, probably after their reflection at mirror points below the transition region (Vestrand et al. 1991). Another promising possibility, suggested by several HXR/GR, radio and/or Hα coordinated studies (Wülzer et al. 1990; Trottet et al. 1993; Chupp et al. 1993) is that the high energy emission is produced in complex magnetic structures which may evolve during the flare and with a varying configuration from flare to flare. Apart from studying their heliocentric distribution it is thus also crucial to analyse the magnetic environment in which these high energy events occur.

References

Akimov, V.V. and the GAMMA1 Team (1991): in 22nd International Cosmic Ray Conference Proceedings SH 2.3-6, p73

Barat, C., et al. (1988): in Nuclear Spectroscopy of Astrophysical Sources, ed G.H Share, N. Gehrels (American Institute of Physics, New York), p395

Chupp, E.L., Trottet, G., Marschhäuser, H., Pick, M., Soru- Escaut, I., Rieger, E., Dunphy, P.P. (1993): Astr. Ap. **275** 602

Kanbach, G., et al. (1993): Astr. Ap. Supp. Ser. **97** 349

Kocharov, L.G., Kovaltsov, G.A. (1990): Solar Phys. **125** 67

MacKinnon, A.L., Brown, J.C. (1989): Astr. Ap. **215** 371

Mandzhavidze, N., Ramaty, R. (1993): Nuclear Phys. B **33** Nos 1-2 141

McTiernan, J.M., Petrosian, V. (1991): Ap. J. **379** 381

Pelaez, F. (1993):Thèse de doctorat de l'université Paul Sabatier, Toulouse, France

Pelaez, F., et al. (1991): in 22nd International Cosmic Ray Conference Proceedings SH2.3-11, p89

Pyle, K.R., Shea, M.A., Smart, D.F. (1991): in 22nd International Cosmic Ray Proceedings SH2.2-7, p57

Pyle, K.R., Simpson, J.A. (1991): in 22nd International Cosmic Ray Proceedings SH2.2-6, 53

Ramaty, R., Miller, J.A., Hua, X.M., Lingenfelter, R.E. (1988): in Nuclear Spectroscopy of Astrophysical Sources, ed by G.H. Share, N. Gehrels (American Institute of Physics, New York), p217

Ramaty, R., Murphy, R. (1987): Space Science Reviews 45 213

Rieger, E., Marschhäuser, H. (1990): in Proceedings of the 3rd MAX'91/SMM Workshop on Solar Flares: Observations and Theory, ed by R.M. Winglee, A.L. Kiplinger, p68

Rieger, E., Reppin, C., Kanbach, G., Forrest, D.J., Chupp, E.L., Share, G.H. (1983): in 18th International Cosmic Ray Conference Proceedings **10** 338

Talon, R., Trottet, G., Vilmer, N., Barat, C., Dezalay, J.P., Sunyaev, R., Terekhov, O., Kuznetsov, A. (1993): Solar Phys. **147** 137

Trottet G., Vilmer, N., Barat, C., Dezalay, J.P., Talon, R., Sunyaev, R., Terekhov, O., Kuznetsov, A. (1993): Adv. Space Res. (COSPAR), Vol. 13 No 9, 285

Vestrand, W.T.,Forrest, D.J., Chupp, E.L., Rieger, E., Share, G.H. (1987): Ap. J. **322** 1010

Vestrand, W.T., Forrest, D.J., Rieger, E. (1991): in 22nd International Cosmic Ray Conference Proceedings SH2.3- 5, p69

Wülzer, J.P., Canfield, R.C., Rieger, E. (1990): in Proc. of the 3rd MAX'91/SMM Workshop on Solar Flares: Observations and Theory, ed by R.M. Winglee, A.L. Kiplinger (Univ. Colorado, Boulder), p149

V

Small-Scale Dynamics of the Corona

Observational Characteristics
of Explosive Events

J.-C. Hénoux [1], K.P. Dere [2]

[1]Observatoire de Paris, DASOP, UA326, 92195 Meudon Principal Cedex,
FRANCE
[2]E.O. Hulburt Center for Space Research, Naval Research Laboratory,
Washington, D.C., USA

Abstract: The characteristics of dynamic phenomena observed in the EUV with the HRTS instruments, chromospheric jets and explosive events, are reviewed here. Most of the review is devoted to explosive events that are short duration explosions involving a plasma at 10^5 K. These events are characterised by strong Doppler shifts rather than by brightness enhancements, contrary to EUV impulsive brightenings observed by SMM. They are presumably due to magnetic reconnection involving unresolved fine structures of the magnetic field near the network. The rôle of the density of concentrated magnetic fluxtubes in a reconnection scenario has still to be understood.

1 Introduction

Explosive events can be defined as events in which energy is released impulsively and where some parameters vary discontinuously in space and time. Solar flares are explosive events; however this paper is not devoted to solar flares but to explosions observed in transition zone lines. Despite the fact that these events involve a limited amount of energy, they may be produced by a physical process similar to the one or one of the ones that take place in solar flares. Their study may help to understand the solar flare phenomena. Explosive events were discovered with the NRL High Resolution Telescope and Spectrograph (HRTS) during its first rocket flight in 1975. All HRTS observations clearly showed that these high velocity events are quite common.

2 Instrument and Observing Techniques

The NRL High Resolution Telescope and Spectrograph (HRTS) experiment consists of a Cassegrain telescope, which focusses an image of the Sun onto the slit-jaw of a tandem-Wadsworth spectrograph (Brueckner and Bartoe 1983). The spectrograph slit is a solar radius long and has an equivalent width of 0.5-1 arc sec, producing a spectral resolution of 0.05-0.1 Å. The wavelength interval in successive flights was within the 1175 Å to 1710 Å domain and the stigmatic spectrum was recorded on photographic film. In addition to the spectrograph, the HRTS mission carried a broadband UV spectroheliograph and an Hα slit-jaw imaging system. From 1975 to 1992 there have been 8 rocket flights of the HRTS and a space shuttle Spacelab 2 mission. A series of papers describes the data reduction techniques (Dere *et al.* 1984, 1986a,b). An Atlas of the Sun between 1190 and 1730 Å using data obtained during a rocket flight in 1978 and covering a variety of different solar regions and an explosive event has been published by Brekke (1993).

3 Classification of High Resolution Spectra

Brueckner and Bartoe in their first report (Brueckner and Bartoe 1983) gave a classification of the phenomena observed in high resolution spectra. They distinguished
    - flows and oscillations with associated regular line profiles of full width at half maximum between 10 and 40 km s^{-1},
    - jets, where the short wavelength wing is enhanced, observed either in the chromospheric C I λ 1560 line or in the C IV λ 1548 transition zone line,
    - turbulent events, later called explosive events, that show strong enhancement in both long and short-wavelength wings in the C IV lines.
    We shall briefly recall here some of the results derived from the study of jets and then devote most of the paper to a review of the characteristics of explosive events.

4 Jets

The jets described below have been observed in EUV lines. In Hα, jet-like phenomena have also been observed (Koutchmy and Loucif 1990) in a polar coronal hole with speeds higher than 50 km s^{-1} . These jets may be macro-spicules.

4.1 Coronal Jets

Coronal jets were observed in C IV lines. The line profiles showed a rapid velocity increase up to 400 km s^{-1}. From the analysis of data accumulated from 1975 to now, it appears (Dere 1992, 1993) that coronal jets are relatively rare events. Apparently their observation in the first two rocket flights was simply fortuitous, and it seems unlikely that they can play a major rôle in the solar corona mass and energy balance.

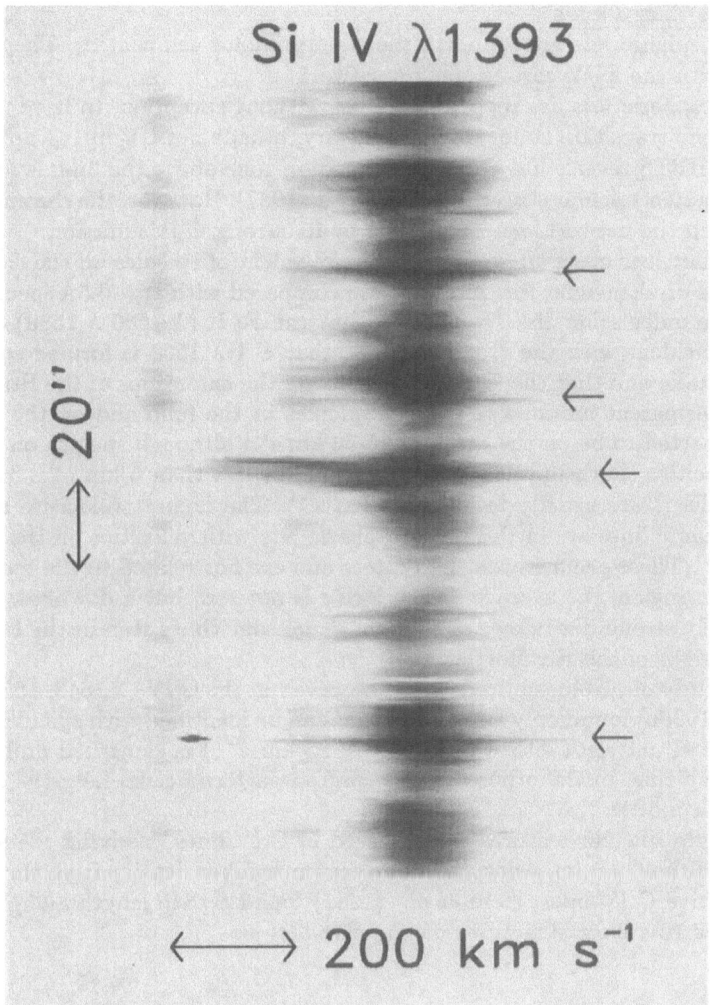

Fig. 1. Si IV spectra of explosive events

4.2 Chromospheric Jets and the UV Counterpart of Spicules

Chromospheric jets observed in the chromospheric C I λ 1560 line were reported by Dere *et al.* (1983, 1984). They are observed as up and down velocity structures of amplitude up to 20 km s^{-1} in supergranulation cell centers. The observed properties of chromospheric jets resemble to the properties of visible light spicules. The only difference is in the lifetime which is much smaller than for spicules.

However, a one-to-one correspondence between chromospheric jets and spicules seems to be ruled out. Since Hα dark mottles are presumed to be the disk manifestation of spicules, a correlation between these mottles and chromospheric jets was studied with a negative result:

- The dark Hα mottles are correlated with nearly symmetric C I profiles with velocities less than 3 km s^{-1}.

- Chromospheric jets are found in the supergranular cell centers. They are not associated with the EUV chromospheric network.

- Chromospheric jets are reported, almost without exception, to have no manifestation in the transition zone. On the contrary, broadband UV filtergraph images obtained by HRTS reveal that the C IV transition zone above the limb is formed in discrete elongated spicular structures (Dere *et al.* 1987). However, the chromospheric jet emission in the network can be masked by its strong C IV emission.

There is no clear observation of the UV equivalent of spicules on the solar disk. When images of ultraviolet line intensity are compared with Hα -0.5Å spectroheliograms on the quiet solar disk, it is found that the Fe II (λ 1560-λ 1580) emission is nearly coincident with the Hα structures, that C I λ 1560 is formed somewhat lower in altitude, and that the C IV structures are the extensions of the Hα mottles or spicules. Apparent motions of the Hα spicules at the limb and on the disk are generally reported to be on the order of 25-50 km s^{-1} although spectra on the disk indicate velocities (both upward and downward) of less than 6 km s^{-1}. Velocities observed in Fe II are usually less than 3 km s^{-1}. The highest velocities observed in chromospneric lines are in the chromospheric jets with velocities on the order of 10-20 km s^{-1}. These occur in the cell centers and are not related to the mottles. In the transition region, the average flow velocity is not zero but a downflow of 5 km s^{-1}. Relatively strong downflows of 10 km s^{-1} are sometimes seen in the transition region extensions of the Hα mottles.

From a study of 11 blue-shifted events observed in the C IV (λ 1548, 1550) lines, Dere *et al.* (1986b) found only two events that can be identified with spicules. In one the blue-shifted pattern (with a blue-shift $\approx$ 25 km s^{-1}) is elongated and parallel with an Hα spicule. In the other the blue-shifted pattern occurs near the end of a large dark Hα mottle.

On the limb spicular features are observed in C IV lines. Deriving the emission measure from the C IV intensity and the electron number density from the ratio of density sensitive O IV lines, Dere *et al.* (1987) found a path length along the line of sight below 10 km for structures on the disk.

5 Explosive Events

5.1 Definition of Explosive Events

As seen in Fig. 1, explosive events are characterized by strong intensity enhancements in both the long and short-wavelength wings of lines. Explosive events are typically observed in lines formed at a temperature of about 1×10^5 K - transition region lines (typically C IV and O IV lines). They are rarely seen in chromospheric lines such as C I λ1560 line. Phenomena very similar to the transition region explosive events have also been observed in coronal and upper transition zone lines with the NRL slitless spectroheliograph on Skylab (Brueckner *et al.* 1976, Cheng and Kjeldseth-Moe 1991). These events are more energetic than the events observed in transition region lines. However, they are rare and occur mostly in active regions. Therefore their characteristics are not so well known as the ones of the transition region explosive events, and they will not be reviewed here.

The Doppler shifts in transition region explosive events are large (100 km s^{-1}) compared with typical Doppler shifts in the quiet and active sun, which are less than 20 km s^{-1}. The velocities in explosive events are clearly supersonic but the velocities are lower than the Alfvén speed for magnetic fields higher than $\approx$ 20 to 50 Gauss. The presence of wide wings often spatially separated makes the term Rexplosive" to be more appropriate than the term Rturbulent" to describe the observed profiles. A detailed analysis of six events selected among 82 observed during the third HRTS rocket flight was published by these authors. A solar area of 10 arc sec × 800 arc sec was covered by 10 complete and one partial rasters, with 6 positions of the spectrograph slit separated by 2 arc sec in each raster. The exposure time and the total observing time were respectively 3 s and 200s. The conclusions of this study (see also Dere 1992, 1993 and Cook and Brueckner 1991) are reviewed here.

5.2 Properties of Explosive Events

We review below in detail the main characteristics of explosive events wich are recapitulated at the end of this subsection in Table I.

5.2.1 C IV lines profiles and Doppler shifts

In most explosive events the profile is dominated by the intense, slowly evolving component at low velocity. The example of profiles given in Fig. 2 and Fig.3 show the presence of both blueshifted and redshifted emission at the same location. The profiles are labelled according to the raster sequence from wich the data were taken. Since there is a 20 s interval between two sequential exposures at the same spatial location, the figures indicate a rise time of both the red and blue wings in about 100 s and a decay time of 50 s. However, the red wing is often different from the blue wing and in many examples one of the components is quite weak. This is illustrated on Fig.4 that shows the line profile of an intense but short lived burst where the maximum velocity increased from 80 km s^{-1} to 160 km s^{-1} during 20 s.

On average the line of sight velocity appears to be independent of the viewing angle. In a statistical sense the horizontal velocities observed at the limb are roughly equal to the vertical velocities observed at disk center. Histograms of the maximum blue-shift and red-shift velocity in explosive events give an average maximum blue shift of 108 km s^{-1} and an average maximum red shift of 114 km s^{-1}. In general the profiles are not symmetric. There is a large number of events with only one measurable wing. The events with dominant blue wings are found to outnumber the events with dominant red wings by about 3/2.

Over time periods of 200 s, no apparent motion of C IV emitting matter along the slit is observed. Either the plasma velocity perpendicular to the line of sight is always lower than 5 km s^{-1} or, if the plasma ejection perpendicular to the line of sight takes place with a velocity of 100 km s^{-1}, this implies a significant change in the temperature of the plasma in less than 5s. Since the absence of apparent motions is independent of the positions of the explosive events on the solar disk the second hypothesis is presumably the best.

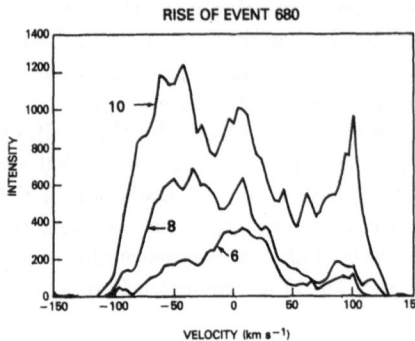

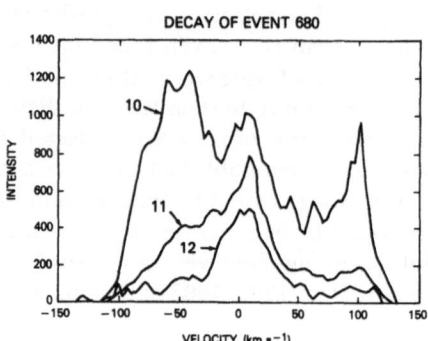

Fig. 2. C IV line profiles during the rise phase of an explosive event (Dere *et al.* 1989, Fig.2)

Fig. 3. C IV line profiles during the decay phase of the same explosive event as in Fig. 2 (Dere *et al.* 1989, Fig.3)

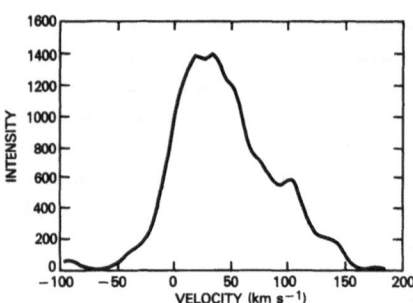

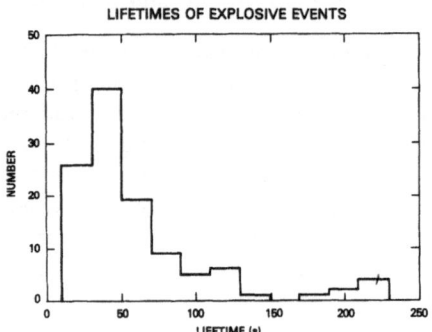

Fig. 4. C IV line profile showing a strong blueshifted wing at maximum (Dere *et al.* 1989, Fig.11)

Fig. 5. Histogram of the lifetime of explosive events taken from Dere *et al.* (1989), Fig. 22

5.2.2 Size of explosive events

The positions of the blue and red wing emission along the slit are often displaced. The displacement is very close to the size of the events measured along the slit, which is about 1500 km. Structures of smaller sizes are evident on the images.

5.2.3 Time-scales

Explosive events evolve on time-scales between 20 and 200 s. A histogram of the lifetimes taken from Dere *et al.* (1989) is presented in Fig. 5. The average life-time is 60 s. No differences in life-time are observed between the red and the blue wings.

Dere reports evidence for repeatability of explosive events at the same location with some residual high nonthermal velocity between events.

5.2.4 Birthrate

Defining the birthrate as the ratio between the observed number of explosive events and the product of the field of view by the observing time Dere *et al.* (1989) found a birthrate of 10^{-20} cm^{-2} s^{-1} on the quiet sun from a sequence of 11 rasters. The birthrates in a coronal hole on the disk and in a polar coronal hole were found respectively equal to or lower than the birthrate on the quiet sun.

5.2.5 Density in explosive events

The derivation of the density from the determination of the emission measure ($\int n^2 dV$) fails since the filling factor of the emitting elements in the observed structure is not known. For an assumed filling factor of unity, the average electron density is 1×10^9 cm^{-3}. A more accurate measurement uses the intensity ratio of lines that are differently sensitive to the electron number density, such as the O IV lines near 1400 Å (Dere *et al.* 1982). One event was found bright enough in the HRTS Spacelab 2 data to derive the electron number density from the O IV line ratio giving a value of 7×10^{10} cm^{-3}. The ratio of the square of these two density values indicates a volumetric filling factor f of 2×10^{-4}. dM/dt (the mass loss rate) and $d\mathcal{E}/dt$ (the energy flux) in explosive events can be estimated as the following:

$$\frac{dM}{dt} = M \times \text{Birthrate} \times \text{Area on the solar surface,}$$

where M the mass ejected per event of observed volume V is

$$M = m_p\, n_e\, f\, V$$

A numerical estimate of dM/dt gives 6 10^9 g s^{-1} which is small compared to a mass loss rate in the solar wind of $\approx$ 3 10^{11} g s^{-1}.

$$\frac{d\mathcal{E}}{dt} = \frac{1}{2}\, M\, v^2 \times \text{Birthrate} \approx 10 \,\text{ergs cm}^{-2}\, \text{s}^{-1}$$

As a consequence of the low value of the filling factor, the explosive events can not play a significant rôle in the mass and energy balance of the corona.

5.3 Sites of the Explosive Events

Images of the intensity in C IV lines have been reconstructed from spectra taken over a region of mostly quiet features, including the limb, during the fifth HRTS rocket flight (HRTS5). The comparison of these images with photospheric magnetograms (Dere 1992, 1993) indicates that a number of explosive events occur at a height less than 2000 km just above the photosphere and below the limb brightening in the C IV transition region lines. Following Dere, the space distribution of explosive events appears to be nearly random with regard to the X-ray bright points. A number

Table 1. Properties of Explosive Events as recapitulated in Dere (1993)

Properties of Explosive Events	
Velocities	100 km s^{-1} Asymmetric profiles Apparent motions < 5 km s^{-1} No center-limb variation: statistically isotropic
Spatial scales	1500 km Evidence for smaller scales
Time scales	Average lifetime = 60 s Evidence for repeatability Residual high nonthermal velocities
Height	1000-2000 km above photosphere Below quiet transition zone
Density	7 x 10^{10} cm^{-3} (one event) Low fill factor
Temperature	Maximum brightness at 10^5 K Rarely seen in chromospheric lines Occasionally seen in coronal lines
Birthrate	10^{-20} cm^{-2} s^{-1} (quiet sun and coronal holes)

of explosive events occur inside He dark points but there is no general correlation between the two phenomena.

The fact that the motions observed in explosive events are directed suggests that they are driven by electromagnetic forces. The association between the location of explosive events and the magnetic field characteristics has been studied by Porter and Dere (1991) and Dere *et al.* (1991) and reviewed in Dere (1992, 1993).

5.3.1 Explosive events and emerging magnetic flux

5.3.1.1 Active Regions. HRTS spectra obtained during the Spacelab 2 mission provided the first evidence of explosive events as being the result of emerging magnetic flux (Brueckner *et al.* 1988). Dere *et al.* (1991) pointed out that the C IV profiles of this explosive event are of the most spectacular even seen. Systematic and random motions at velocities up to 300 km s^{-1} were observed on a fraction of a sunspot perimeter where an Hα filament indicated a newly emerging magnetic flux. The origin of the strong line broadening observed is unclear and this event is not a typical explosive event. Other examples were reported by Dere *et al.* (1990, 1991) of explosive events associated with emerging loops in an active region or with the formation of an active region.

5.3.1.2 Quiet Sun. A concentration of explosive events in a quiet sun area where numerous C IV loops were seen is also reported by Dere *et al.* (1991). The UV continuum was showing several bright points, and the HeI λ 10830 spectroheliogram area was dark suggesting enhanced coronal EUV emission. These observational features led these authors to suggest a relation with the emergence of a bipolar magnetic region . However, no bipolar magnetic region was observed in the two magnetograms of that day and the following day. Only a weak bipolar structure was reported two hours before the observations. Therefore the association between explosive events and emerging flux is clearly demonstrated only for active regions.

5.3.2 Explosive events in the quiet sun

Dere (1992, 1993) stress the point that at first glance the explosive events appear to be randomly distributed over the quiet sun, and that most events are scattered over the solar surface in places where there appear to be no observable magnetic field. The location of the explosive events in the quiet sun was studied by Porter and Dere (1991) using Spacelab 2 observations, simultaneous magnetograms and near-simultaneous HeI λ 10830 spectroheliograms. Generally λ 10830 spectroheliograms are dark in places of bright coronal emission. Consequently in λ 10830 spectroheliograms, the magnetic network appears as dark-grey lanes, and the interiors of supergranulation cells are the brightest regions. Porter and Dere found that the explosive events are not correlated with either the bright inner cell area or the darkest part of the network lanes. Rather the events occur in regions of intermediate darkness along the lanes of the network.

Comparison of Spacelab 2 observations with Kitt Peak magnetograms confirmed the results of the comparison with HeI λ 10830 spectroheliograms. The explosive events are not found in fairly strong bipolar regions. Instead, they appear in low field areas on the borders of high field regions. Since the apparent strength or weakness of the magnetic field is presumably the observed signature of a high or low concentration of structures with high field strength, the explosive events appear to occur in regions were the number of magnetic field concentrated flux tubes per unit area is low.

5.4 A Qualitative Magnetic Reconnection Scenario for Explosive Events

The confinement of most of the transition region explosive events to the network lanes led Porter and Dere (1991) and Dere *et al.* (1991) to suggest a magnetic interaction - magnetic reconnection - between intra-network magnetic loops and funnel-like open field stuctures of the network as illustrated in Fig. 6.

Dere (1992, 1993) compared the apparent size scales (1500 km) and time-scales (60 s) of explosive events to the size and time-scale of the photospheric magnetic flux cancellations, that are respectively 6000 km and several hours. He concluded that the reconnection process occurs in short bursts at discrete locations along the interaction boundary.

5.4.1 Checking the magnetic reconnection hypothesis

In the Petschek model, τ the reconnection time is given by $\tau = L/(\alpha V_A)$, where L is a characteristic length, V_A is the Alfvén velocity and α is between 0.01 and 0.1. Taking the observed maximum velocity of 100 km s^{-1} as the Alfvén velocity and 1000 km for L, Dere *et al.* (1991) derived time-scales between 100 and 1000s. (If the observed convection velocity of flux elements through the supergranular cell interior of 0.3 km s^{-1} is used instead of αV_A, then the estimated reconnection time-scale becomes 50 hours).

The time-scales expected from a reconnection process are slightly higher than the observed characteristic time-scale. However, any attempt to estimate the value of the magnetic reconnection hypothesis must be taken with care since explosive events may consist of spatially unresolved elements. From electron density measurements

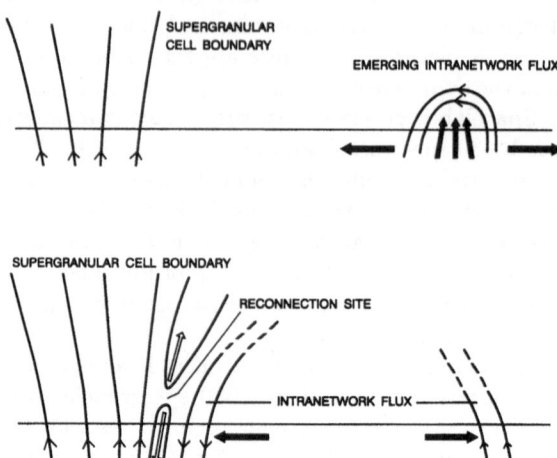

Fig. 6. A schematic representation of intranetwork motion and reconnection, taken from Dere *et al.* (1991), Fig. 2

(see subsection 5.2.5), a volumetric filling factor of 2×10^{-4} was derived which could be interpreted as resulting from the presence of fine structures of transverse size about 20 km. The corresponding expected time-scale is close to 4 s, and therefore below the time resolution of HRTS observations which is of 20 s. Observations with high time resolution together with electron number density measurements would help to check both the existence of unresolved high density fine structures and the occurrence of magnetic reconnection.

5.4.2 Additional requirements

Since the lifetimes, spatial scales, frequency of occurrence and spectral characteristics of explosive events and EUV impulsive brightenings (microflares) observed by SMM (Solar Maximum Mission) on neutral lines of small magnetic elements in the network are similar, Porter *et al.* (1987) suggested that they were members of the same class of events. As a matter of fact the transition region line emission in explosive events was reported to be dimmer than the enhanced emission coming from the stronger field concentration of the network. Moreover, the UVSP (Ultra Violet Spectro Polarimeter) elements were selected for brightness, not for high velocity.

Therefore, as suggested by Porter and Dere (1991), explosive events could just be one end of a range of phenomena generated by magnetic reconnection. The other end could be enhancements in brightness like the EUV impulsive brightenings. Consequently a convincing magnetic field reconnection model must include scenarios leading either to plasma ejection or to local heating. Since this model must reproduce the observed characteristics of explosive events, it must be able to generate not only upward or downward motions but also horizontal motions.

6 Conclusion

Explosive events are presumably due to magnetic reconnection. However, a quantitative model of generation of these events requires additional observations both of transition region line profiles and of the magnetic field:

- Higher time resolution in observations of explosive events must be achieved. If explosive events are the signature of flux cancellation lasting several hours, then extended time series of observations with high time resolution are also needed.

- Due to their occurrence in regions with a low density of magnetic flux tubes, the measurements of the local magnetic field requires special techniques that are not currently used in routine magnetic field observations.

Since explosive events occur in regions of low concentration of magnetic flux-tubes, in contrast to impulsive EUV brightenings, the rôle of the density of these fluxtubes, or of the magnetic field strength, on the magnetic reconnection scenario has yet to be understood.

References

Brekke, P. (1993): ApJS, 87, 443

Brueckner, G.E., Patterson, N.P., Scherrer, V.E. (1976): Solar Phys., 47, 127

Brueckner, G.E., and Bartoe, J.-D. (1983): ApJ, 272, 329

Brueckner, G.E., Bartoe, J.-D., Cook, J.W, Dere, K.P., Socker, H., Kurokawa, H., and McCabe, M. (1988): ApJ, 335, 986

Cheng, C.C., and Kjeldseth-Moe, O. (1991): Dynamics of solar flares, eds. B. Schmieder and E. Priest, p.101

Cook, J.W., and Brueckner, G.E. (1991): in Solar Interior and Atmosphere, ed. by A.N. Cox, W.C. Livingston and M.S. Matthews (The University of Arizona Press, Tucson), pp. 996-1028

Dere, K.P. (1992): in Electromagnetical coupling of the solar atmosphere, ed. by D.S. Spicer and P. MacNeice (AIP, New York), pp. 63-70

Dere, K.P. (1993): Adv. Space Res., in press

Dere, K.P., Bartoe, J.-D., and Brueckner, G.E., (1982): ApJ, 259, 366

Dere, K.P., Bartoe, J.-D., and Brueckner, G.E. (1983): ApJL, 267, L65

Dere, K.P., Bartoe, J.-D., and Brueckner, G.E. (1984): ApJ, 281, 870

Dere, K.P., Bartoe, J.-D., and Brueckner, G.E. (1986a): ApJ, 305, 947

Dere, K.P., Bartoe, J.-D., and Brueckner, G.E. (1986b): ApJ, 310, 456

Dere, K.P., Bartoe, J.-D., Brueckner, G.E., Cook, J.W., and Socker, D.G. (1987): Solar Phys., 114, 223

Dere, K.P., Bartoe, J.-D., and Brueckner, G.E. (1989): Solar Phys., 123, 41

Dere, K.P., Schmieder, B., Alissandrakis, C.E. (1990): A&A, 233, 207

Dere, K.P., Bartoe, J.-D., Brueckner, G.E., Ewing, J., Lund, P. (1991): JGR, 96, 9399

Koutchmy, S. and Loucif, M.L. (1990): in Mechanisms of Chromospheric and Coronal Heating, ed. by P. Ulmschneider, E. Priest and R. Rosner (Springer), pp. 152-158

Porter, J.G., and Dere, K.P. (1991): ApJ, 370, 775

Porter, J.G., Moore, R.L., Reichmann, E.J., Engvold, O. and Harvey, K.L. (1987): ApJ, 323, 380

Plasma Physics of Explosive Phenomena

V.V. Zaitsev

Applied Physics Institute, Russian Academy of Sciences,
603600 Nizhnii Novgorod, Russia

Abstract: Some physical aspects of explosive type phenomena in the solar atmosphere are reviewed. Since the time of the Skylab experiments, we know that at least a considerable fraction of solar flares take the form of heated magnetic loops. In order to explain these observations two types of models with current-carrying magnetic loops are discussed. In the first type of model there are a limited number of loops with regular magnetic field structure and a neutral current sheet as a key element (e.g. isolated flaring loop model, loop coalescence model, emerging flux model, the central helmet streamer configuration, et al.). The main problem with these models is the high magnetic Reynolds number in the solar corona plasma and the resulting long time for neutral sheet evolution. In the second type of model there is an ensemble of the twisted flux tubes, isolated from each other, which are driven by turbulent photospheric motion. In this case the "flare" is a statistical flare and it is an ensemble of microflares that are triggered almost simultaneously either by emerging flux or large scale flows. The possibility of "statistical flare" is supported by the observations of decimetric radio spikes which indicate that there are multiple particle accelerations during the flare. Mm-wave and hard X-ray emission of solar flares sometimes show that the energy content in the fast electrons accelerated in the coronal part of flaring loop is insufficient for chromospheric plasma evaporation. It cannot be excluded that in this case the main energy release process occurs in the chromosphere. In this connection we discuss the "circuit model" of solar flares which can provide the main energy release at the footpoints of the flaring loop.

1 Introduction

In this short review we shall emphasise some kinetic aspects of explosive phenomena in the solar atmosphere which are important for plasma heating and particle acceleration during a solar flare. Of course, all the main kinetic processes in magnetized plasma are well known. However, their relative role in active regions strongly depends on the general aspects of the energy release during the solar flare, which are not yet understood. For example, we don't know where the main energy release of the flare takes place: in the corona or in the chromosphere? We also don't know twhat form the main energy release takes. Is it transmitted by Joule heating or by fast particles? The answers to these questions depend on the flare model. So,

when we speak about the plasma physics of explosive phenomena in the solar atmosphere we must keep in mind some definite models of flares and their observational
manifestations.

2 Flare Models

Let us now discuss the best-known flare models which play an important role in
the comprehension of the main physical processes ooccurring in solar active regions.
Since the time of the Skylab mission it is common knowledge that at a considerable
fraction of flares take the form of heated magnetic loops. In order to explain these
observations, two type of current-carrying magnetic loops have been discussed. The
first type are loops where the structure of magnetic field inside them is regular due
to the smooth distribution of the electric current across the loop. The second type
constitute loops, or their groups, with a chaotic distribution of the magnetic field,
which results from turbulent photospheric motion. Magnetic loops of the first type
were used for the development of the dynamical models of flares and loops of the
second type were used in the models of the "statistical flare".

Let us consider first the most important dynamical flare models.

2.1 The Isolated Flaring Loop Models

Spicer (1977) has assumed that a flux tube emerges from the photosphere and
becomes twisted due to photospheric motion. It is well known that such a structure
is unstable with respect to the tearing - mode instability (Furth et al., 1963). As a
result some part of the magnetic field energy is converted into energy of fast particles
and into heated plasma. The site of the energy release is in the coronal part of the
loop. Some fraction of the energetic ions and electrons are trapped in the flaring loop,
whereas the remainder precipitate into the chromosphere and generate the γ-ray, X-
ray, optical and microwave emission. Just a very small number of energetic particles
will escape into the upper atmosphere. This well-developed model is however, not
without problems. In particular, the tearing-mode instability cannot yet explain the
observed rise-time of solar flares. That is because the shortest time scale of the
resistive tearing-mode instability (Furth et al., 1963),

$$\tau = \tau_A R_M^{1/2}, \quad \tau_A = \frac{R}{V_A} \approx 1 - 10 \ s, \tag{1}$$

is proportional to the Alfvénic time τ_A and the square root of the magnetic Reynolds
number, which is rather large in the corona ($R_M = 10^{10} - 10^{12}$) if we use a classical
Spitzer conductivity. Moreover, the thickness of the current sheet required for the
explanation of the observed rate of the energy release must be very small ($10^3 -
10^4 \ cm$). It does not agree with the usual estimates of the large volume of the
energy release region $\sim 10^{27} \ cm^3$. R is the characteristic dimension of the region of
instability.

2.2 The Loop Coalescence Model

Sweet (1958) was the first to consider the coalescence of magnetic loops as a possible flare model. Later on Tajima et al. (1982, 1987) have presented a more refined loop coalescence model based on the conception of explosive reconnection which arises from nonlinear simulation of the coalescence instability.

According to Tajima et al. (1982) explosive reconnection takes place if the maximum current in the magnetic islands (or magnetic flux tubes) exceeds some threshold value. In this case the reconnection of the two magnetic islands occurs during a finite time which is not dependent on the magnetic Reynolds number. The driving force in their explosive reconnection model is the attractive force between two islands or current filaments. The characteristic time of the energy release for this model is the Alfvénic time $\tau_A \approx 1 - 10$ s. The energy release takes place in the corona and the energetic particles are transported into both loops.

Two main problems connected with this flare model can be pointed out. First, what is the exact condition leading to the independence of the reconnection process on the magnetic Reynolds number? Second, is the explosive reconnection a universal phenomenon? There are other important questions. For example, we don't know the behaviour of the explosive reconnection in three-dimensions or in the case of coalescence of asymmetric loops.

2.3 The Emerging Flux Model

Another popular model of the flare is the emerging flux model (Heyvaerts, Priest and Rust, 1977). In this model the flare is produced by magnetic reconnection of the emerging loop with the overlying pre-existing coronal magnetic field. The main force driving the emerging flux tube is magnetic buoyancy. The emerging magnetic flux creates the neutral current sheet in the region of interaction of "old" and "new" magnetic fields. The main problems in this model are the same as those of the basic reconnection theory.

2.4 The Central Helmet Streamer Model

Sturrock (1968) considered the central helmet streamer configuration as a possible source of the solar flare. In this case a large neutral current sheet is formed above the coronal loop forming a source of hot plasma and fast particles.

So, all these flare models have presented a method for the formation of one neutral current sheet. All existing theories on the stability and evolution of the current sheet suggest that when its thickness is sufficiently small its evolution is fast enough to explain the rise time of the impulsive phase of flares. However, it is not clear yet whether a single current sheet can provide the acceleration characteristics compatible with observational data. Another question is how does the current sheet fill the entire volume during the short time required by the observations? In this case we need on efficient cross-field transport theory. Recently, some new models were proposed which are free from the difficulties connected with unit current sheet formation. We shall now consider two of these flare models.

2.5 The Statistical Flare Model

It is supposed in the statistical model of a flare (Parker, 1988; Vlahos, 1989) that subphotosheric and photospheric turbulence continuously splits the large scale magnetic field in millions of twisted flux tubes isolated one from another. The continuous interaction of the large number of these flux tubes or fibrils by means of coalescence, or by continuous twisting with tearing or kink instabilities, will produce sporadic activity in the form of very weak nano-flares. This spontaneous process can contribute to coronal heating. What is needed in a statistical flare is some form of coherence in the spontaneous reconnection process. The coherence could result from mutual interaction of current sheets and current filaments, or flux emergence or large scale organised motion. A flare produced by the coherent interraction of nano-flares is usually called a statistical flare.

Formation of the thousands of coherently interacting fibrils solves the majority of the problems noted above. Very small current sheets will evolve rather quickly. The current sheets are distributed over large areas and they will fill effectively the flare loop with hot plasma and energetic particles in a very short time. The electric field and MHD-waves occurring during the interaction process favour particle acceleration. Finally, almost all the flare models considered above are incorporated in the statistical model as a separated parts of general scenario. The statistical model of the flare is not complete yet. We have discussed briefly the general scenario but without a quantitive analysis. The main problem here is the description of the process of collective interaction (self-organisation) of a large number of spontaneously occurring current sheets.

2.6 The Circuit Model of a Flare

Just as with a statistical flare, the circuit model of a flare is not connected with the existence of a unit current sheet in the corona. This model was was proposed by Alfvén and and Carlqvist (1967) and since then has been developed and modifyed by Zaitsev and Stepanov (1992).

The general scenario of the flare in a circuit model is as follows (Zaitsev and Stepanov, 1992). Photospheric convection generates an electric current $I = 10^{11} - 10^{12}$ A in a magnetic loop. The characteristic rise time of the current is of the same order as the inductive rise time $\tau_L \approx L(dL/dt)^{-1}$ of the equivalent electric circuit along the magnetic loop and photosphere, i.e. of the order of the time scale of the magnetic flux emergence. This time scale usually ranges from several hours to several days. A flare can be triggered by a prominence or fibril which lies above the magnetic loop or loop system. If the thickness d, of the prominence becomes sufficiently large,

$$d > \frac{B^2}{10\pi\rho g},$$
(2)

(where B is magnetic inductance, ρ is the prominence density and g is gravity) the fluid instability of the ballooning mode becomes important (Pustilnik, 1973). As a result, the cold, partly ionized prominence plasma penetrates the current channel of the magnetic loop with a characteristic time

$$\tau_{fl} \approx \frac{r}{V_{Ti}} = 1 - 10 \ s,$$
(3)

where r is the radius of the loop and V_{Ti} is the thermal velocity of the ions. The penetration of the cold plasma into the current channel switches on an effective dissipation mechanism which is connected with ion-atom collisions in a non-stationary plasma (Cowling 1957). As a consequence, the nonlinear resistance of the channel

$$R(I) = \frac{2\pi F^2 I^2 d}{c^4 n m_i \nu_{ia} S} \tag{4}$$

grows to a magnitude $R(I) \approx 3 \times 10^{-2} \Omega$ corresponding to a heating rate of $W = RI^2 \approx 10^{27} - 10^{28} \; erg/s$. In formula (4) $F = n_a/(n_a + n)$ is the relative number density of the neutral component of the plasma, n_a and n are the number densities of atoms and ions respectively, ν_{ia} is ion - atom collision frequency, S is the cross-section of the current channel and m_i is the ion mass. This nonlinear resistance is 8 to 10 orders of magnitude greater that the classical Spitzer resistance.

Since electric field components parallel to the magnetic field are present in the magnetic loop, some fraction of the electron population is accelerated by the runaway effect. Thus, in the framework of this model the energy of the accelerated electrons is only a part of the total flare energy.

The duration of the flare process is determined by the characteristic time scale of the electric current dissipation in the equivalent circuit after the current disruption,

$$\tau_0 \approx \frac{L}{R(I)} = 500 - 5000 \; s, \tag{5}$$

where we have used the induction $L \approx 10 \; H$ calculated by Alfvén and Carlquist (1967) for a slender semi-circular flux tube with a length $l \sim 10^9 \; cm$ and thickness $\sim 2 \times 10^8 \; cm$. This time is of the same order as the duration of the flaring process taking place in a magnetic loop or loop system. However, the energy release can be fragmented in the form of "subflares", if the prominence experiences quasiperiodic oscillations. The oscillation period, and consequently the repetition time of "subflares" can vary from 10^2 to $10^3 \; s$ depending on the parameters of the loop and the prominence.

In addition to a long range of time scales from 10 s to 1 hour in this model, a hierarchy of short time scales can be noted which may be related to the fine-structure of the flare process. Alfvénic and fast magneto-acoustic oscillations of the magnetic flux tube excited during the impulsive flare phase can modulate the flow of energetic electrons in the flux tube due to the modulation of the ratio B_{max}/B_{min} and it can also modulate the electron streams penetrating the chromosphere (Roberts et al 1984; Zaitsev and Stepanov, 1989). The characteristic time scales of such modulations are $\tau_A \approx l/V_A \approx 10 \; s$ and $\tau_{MS} \approx r/V_A \approx 1 \; s$, respectively. Oscillating plasma instabilities can also lead to modulation both of the electron stream and of the energy density of the plasma wave with a time scale (Zaitsev 1971) $\tau \approx \Lambda/\nu_{ei} - \Lambda/\gamma \approx 10^{-1} - 10^{-4} \; s$, where ν_{ei} is the collision frequency between ions and electrons and γ is the increment of plasma instability for Langmuir waves; $\Lambda \simeq 10$ is the logarithmic factor. These short time-scales can result in the fine time structure of the flare emission in different frequency bands. Thus, the observed fragmentation of energy release during the flare can appear not only as a result of the multiple energy release process but also due to nonlinear dynamics of all the flaring loops.

The circuit model of a flare opens up the possibility of a more powerful energy release in the chromosphere, compared with the coronal part of the loop, since the current channel cross-section S decreases in the chromosphere (see formula (4)).

3 Comparison with Observations

It is widely accepted that the electrons accelerated or heated in a flare are most directly detected through their emission in different frequency bands from radio waves up to the γ-ray continuum. These emissions supply important information on the underlying processes of energization of the radiating particles, in particular on the different time scales of the flare process. By using flare microwave burst data, Sturrock et al. (1984) described four characteristic time scales of sub-seconds, a few seconds, a few minutes, and tens of minutes which were found to be present in temporal structures. In addition some solar flares are accompanied by thousands of decimetric spikes with typical durations of 100 ms (Benz 1985, 1986), which appear clustered over a broad frequency range.

Sakai and Ohsawa (1987) and Tajima et al. (1982, 1987) have applied the coalescence model of solar flares to specific observations to try to understand the time-scales and the dynamical processes which were occurring. They succeeded in explaining the fast rise-time ($\sim$ 1-3 s) of the impulsive phase of flare, and the rapid amplitude oscillations found in hard X-ray and γ-ray emissions during some solar flares. However, the coalescence model has difficulties with explaining longer time scales (with durations $\geq$ 10 s) observed in solar flares. In this case "coalescence events" can be included in the statistical model of flares as a part of general scenario (Vlahos, 1989).

The statistical model of flare is supported by observations of decimetric millisecond spikes which may accompany the flare and indicate the multiple processes of particle acceleration. It must be mentioned, however, that recent studies of the evolution of decimetric spikes and hard X-ray emission during solar flares (Aschwanden and Güdel, 1992) shows that the bulk of the radio spike emission is always delayed with respect to the correlated hard X-ray emission, typically by about 2-5 s. This observational fact rules out a simple model where the radio spikes are considered as direct tracers of a fragmented accelerator. However, at least some common origin seems to be required to explain the striking correlations of both emissions (Aschwanden and Güdel, 1992). It cannot be excluded that the radio spikes reflect only the runaway process for some fraction of the fast electrons which are accelerated in the region of the primary energy release.

In connection with the observed time delay between the hard X-ray emission and radio spikes the next essential question arises in our discussion of different flare models. Where is the main energy release situated? Is it in the coronal part of flaring loop where the radio spikes are generated or in the chromosphere where the thick-target hard X-ray emission is produced? Recently Bachareva et al. (1993) used mm-wave and X-ray diagnostics to determine the total energy of fast electrons injected into the chromosphere during the flare and the energy of evaporated chromospheric plasma arising as a result of chromospheric heating. For five flares investigated by these authors the energy of the evaporated chromospheric plasma was large compared with the total energy of the fast electron beams injected into the chromosphere. They concluded that additional energy release in the chromosphere is needed to account for the evaporation. Such a possibility can be achieved in the circuit model for the solar flare.

It is difficult to explain the short rise times of the impulsive phase of flares in the framwork of the emerging flux model, the isolated flaring loop model, and the central helmet streamer configuration. However, these models can play the important role in pre-flare heating of plasma in active regions (Kundu and Vlahos, 1982), in the origin of extended flares with durations greater than about 10 min (Crannell et al., 1988) and in the processes of post-flare long-lived acceleration of particles (Akimov et al., 1993). For example, Akimov et al. (1993) analyzed data from the solar flare on June 15, 1991 for which the γ-ray emission in the continuum and nuclear lines was observed for at least two hours after the impulsive phase of the flare. these authors suggested that the prolonged energy release in this case was connected with a vertical current sheet formed above the active region due to the coronal mass ejection. Relaxation of the disrupted magnetic field above the active region to its initial configuration through reconnection in a vertical current sheet can be accompanied by effective particle acceleration for a long time after the main flare has died away.

4 Conclusions

1. The solar flare models considered above which are connected with the formation of one single neutral current sheet have certain difficulties related to the long time-scale of the energy release in the current sheet. So these models cannot give a satisfactory explanation for the impulsive phase of a flare. However, these models can play an important role in understanding pre-flare heating of active regions and may also account for extended flares and post-flare long lived particle acceleration following the impulsive phase.
2. Recently, the idea of a statistical flare which overcomes a number of the difficulties of previous models was suggested as an explanation of the impulsive phase of flares. This idea requires a detailed theoretical investigation. One critical element is the explanation of a flare trigger which can stimulate the switching-on of thousands of spontaneously-reconnecting regions.
3. Mm-wave and X-ray diagnostics of the flares show that at least in some cases the main energy release takes place in the chromospheric part of flaring magnetic loop. This fact shows that the circuit model of flares may be a possible explanation of certain events.

Acknowledgements

The author acknowledges fruitful discussions with L. Vlahos. The present work was supported by a grant from the Russian Foundation for Fundamental Research.

References

Akimov, V.V., Belov, A.V., Chertok, I.M., Kurt, V.G., Leikov, N.G., Magun, A., Melnikov, V.F. (1993): *Proc. of the 23th ICRC*, Calgary, Canada
Alfvén, H. and Carlquist, P. (1967): Solar Phys. 1, 120
Aschwanden, M.J. and Güdel, M. (1992): Ap.J., 401, 736
Bakhareva, N.M., Zaitsev, V.V., Dennis, B.R., Stepanov, A.V. and Urpo, S. (1993): Solar Phys., (submitted)
Benz, A.O. (1985): Solar Phys., 96, 357
Benz, A.O. (1986): Solar Phys., 104, 99
Cowling, T.G. (1957): *Magnetohydrodynamics* (New York: Interscience), p. 107
Crannell, C.J., Dulk, G.A., Kosugi, T. and Magun, A. (1988): Solar Phys., 118, 155
Furth, H., Rosenbluth, M and Kileen, Y. (1963): Phys. Fluids, 6, 459
Heyvaerts, J., Priest, E.R. and Rust, D.M. (1977): Ap.J., 216, 123
Kundu, M.R. and Vlahos, L. (1982): Space Sci. Rev., 32, 405
Parker, E. (1988): Ap.J., 330, 474
Pustilnik, L.A. (1973): A. Zh., 50, 1211
Roberts, B., Edwin, P.M. and Benz, A.O. (1984): Ap.J., 279, 857
Sakai, J. and Ohsawa, Y. (1987): Space Sci. Rev., 46, 113
Spicer, D.F. (1977): Solar Phys., 53, 305
Sturrock, P.A. (1968): Astron. J., 73, S. 79
Sturrock, P.A., Kaufmann, P., Moore, R.L. and Smith, D.F. (1984): Solar Phys., 94, 341
Sweet, P.A. (1958): in: *Electromagnetic phenomena in cosmical physics*: Cambridge Univ. Press, p. 123
Tajima, T., Brunel, F. and Sakai, J. (1982): Ap.J, 258, L45
Tajima, T., Sakaj, J., Nakajima, T., Kosugi, T., Brunel, F. and Kundu, M.R. (1987): Ap.J., 321, 1031
Vlahos, L. (1989): Solar Phys., 121, 431
Zaitsev, V.V. (1971): Solar Phys., 20, 95
Zaitsev, V.V., and Stepanov, A.V. (1989): Soviet Astron. Letters, 15, 154
Zaitsev, V.V., and Stepanov, A.V. (1992): Solar Phys., 139, 343

A Search for Small Solar Flares with BATSE

D.A. Biesecker [1], J.M. Ryan [1], G.J. Fishman [2]

[1] University of New Hampshire, Space Science Center, Durham, NH 03824
[2] NASA / Marshall Space Flight Center, ES-62, Huntsville, AL 35812

Abstract: The Burst and Transient Source Experiment on board the Compton Gamma Ray Observatory is being used to measure solar flares. The Large Area Detectors are well suited for studies of small flares. We are searching the BATSE LAD data for solar flares with an automated algorithm. In this search we have detected solar flares an order of magnitude smaller than those detected by a visual search of the data. The results are consistent with the differential distribution of peak flare rates measured by other researchers. These results also show that the frequency of the smallest flares observed by BATSE can be predicted from the frequency of larger flares. We also present results on the factors limiting the sensitivity of the search.

1 Introduction

The study of solar flares is important in understanding energy release and the acceleration and transport of particles on the sun. One of the difficulties in studying solar flares is their complexity. By studying small flares the interpretation of the data is much simpler. A *microflare* is an ideal event for investigating the basic features of the flare process. Microflares have been proposed as the basic component of solar flares (van Beek *et al.* 1974; de Jager and de Jonge 1978). They have also been suggested as a source for heating the corona (Parker 1981).

The importance of the flare size distribution has been reviewed by Hudson (1991). Earlier researchers (Datlowe *et al.* 1974; Dennis 1985; Crosby *et al.* 1993) have found the differential distribution of peak flare rates to be well fit by a power-law with a slope of ~ -1.8. If microflares follow the same distribution and the flare energy available to heat the corona scales linearly with peak X-ray flux, then microflares cannot heat the corona. It is not known, however, if the distribution of microflares below the threshold of earlier instruments follows the same power-law as larger flares. If microflares below the threshold of earlier instruments are frequent enough, they could provide enough energy to heat the corona.

The Burst and Transient Source Experiment (BATSE) on-board the Compton Gamma Ray Observatory (CGRO) is being used to measure solar flares. The BATSE instrument consists, in part, of eight uncollimated Large Area Detectors (LADs), placed on the corners of the spacecraft. The LADs are arranged so that the detector faces are parallel to the faces of a regular octahedron. Every direction in space is viewed by four detectors. The relative count rates, a function of the geometry of the detectors, is used to locate the Sun.

The LAD data have four energy channels: 25–50, 50–100, 100–300, and > 300 keV, with a time resolution of 1.024 seconds. Only the first energy channel is used in this study in order to maximize the signal-to-noise ratio. The LADs provide for the most sensitive long-term study of microflares ever conducted.

We present preliminary results of a search for small solar flares. In order to provide insight in the role of flares in coronal heating, it is necessary to detect the smallest flares possible. Searching for small flares with BATSE requires that instrumental effects must be understood. We report the detection of flares down to the instrumental threshold.

2 Solar Flare Search

Events are identified by a computer algorithm that searches the data for count-rate increases. The algorithm then uses the geometry of the instrument to independently identify events of solar origin. The result of the search is a large database of events for studying the characteristics of microflares.

The algorithm calculates the second difference in time of the count-rate data, thereby locating peaks or bursts (Mariscotti 1967). The four solar-facing detectors are searched simultaneously. The ratio of the second difference at a local maximum to the statistical error in the second difference yields the significance of the peak in units of standard deviations. A significance of at least 3σ in one BATSE detector or of at least 2.5σ in two detectors identifies peaks for further analysis. There is no characteristic time scale or structure that identifies all flares, so the data are smoothed over neighboring data points on four different time scales. This ensures that the algorithm detects the weak, rapid events as well as the long, smooth events.

Once a significant count-rate increase has been detected, it is necessary to determine if the Sun is the source of the increase. Using the total counts above background and knowing the Sun's position, the count rate in each detector is corrected for angular response. A solar flare should have equal corrected counts among the solar-facing detectors. The solar identification process is tested using flares measured by the GOES X-ray monitors. A correction for the non-ideal detector response is calculated using these known flares. This correction is applied to unidentified events to determine if the events are of solar origin.

All significant count-rate increases are stored in a database. The information contained in the database includes the time, peak rate, total counts in the search window, background rate, and the second difference of each peak. Data from all four solar-facing detectors are included.

The database includes many events that have been counted more than once. This can happen for two reasons. Since a flare may have many local count-rate maxima, the algorithm may detect a single flare several times. In addition, each day is searched with four different smoothing averages, resulting in many events being detected by the algorithm at several smoothing sizes. Data from all four searches are combined and all of the events that fall within 2 minutes of the previous event are considered to be part of the same flare, minimizing double counting of events. The largest events are checked individually.

3 Results

The differential distribution of peak flare rates is plotted in Fig. 1 for the period 11 May to 26 Dec 1992. The peak rates are taken from the most solar facing detector and only corrected for angular response.

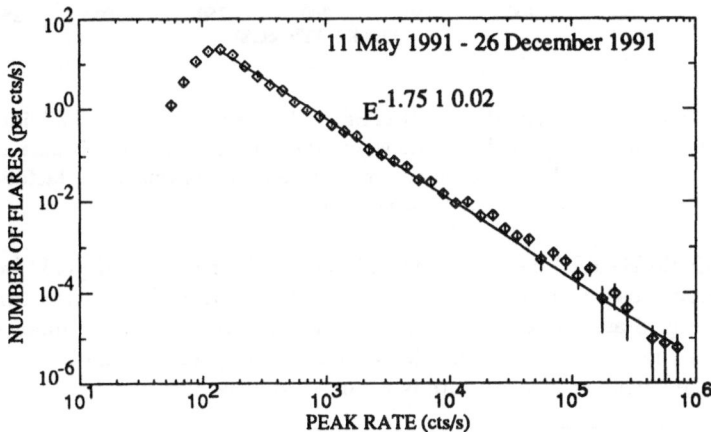

Fig. 1. The differential frequency distribution of BATSE peak flare rates is plotted.

The solid line is a power-law fit with spectral index of -1.75 ± 0.02. This result is consistent with earlier researchers' findings (Schwartz *et al.* 1991; Crosby *et al.* 1993). However, the search extends the distribution down to peak flare rates a factor of three smaller than earlier analyses. The instrumental turnover occurs at about $150\,s^{-1}$. There are several factors which affect the ability of BATSE to detect small flares. By identifying the factors that limit the sensitivity of BATSE, it is possible to extend the distribution even further. The two most important factors are the background rate and $\cos\theta$, i.e., the angle between the sun and the normal to the face of the most solar facing detector.

From 11 May 1991 to 26 Dec 1991 the Compton spacecraft underwent 22 different orientations. This resulted in four different detectors being the most solar facing. In an earlier work a difference was reported in the location of the peak rate distribution turnover for each detector (Biesecker *et al.* 1993). Figure 2 shows that this difference between detectors is due mostly to differences in $\cos\theta$ of the detectors. Figure 2 includes the same data as Fig. 1, except that they are now plotted as a function of BATSE detector module. When the data for each detector are restricted to a small range of $\cos\theta$, the difference in the distribution turnover vanishes.

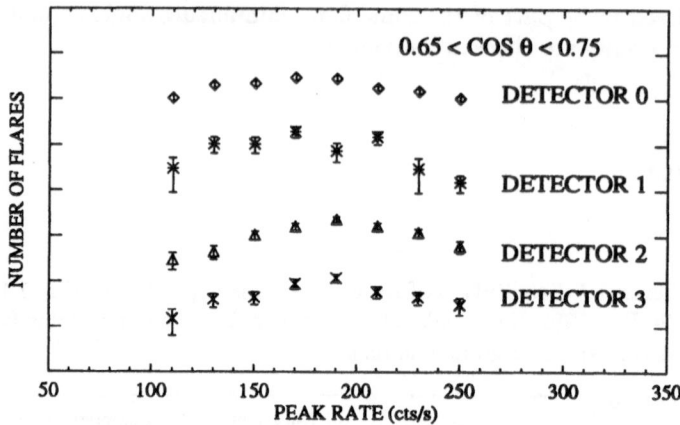

Fig. 2. The differential distribution of BATSE peak flares rates for the individual detector modules. The data are restricted to cases where the sun-detector angle is between 0.65 and 0.75. The y-axis tick labels have been suppressed because multiplicative factors have been applied to the data to make the plot easier to read.

For a single BATSE detector, the differential distribution of peak flare rates is plotted versus $\cos\theta$ values. The peak rate at which the distribution turns over is plotted in Fig. 3 as a function of $\cos(\theta)$. The solid line is a least squares fit of the data. When the most solar-facing detector is directly facing the Sun ($\theta = 0$), it is seen that the turnover in peak rate is reduced to only $85\,\mathrm{s}^{-1}$.

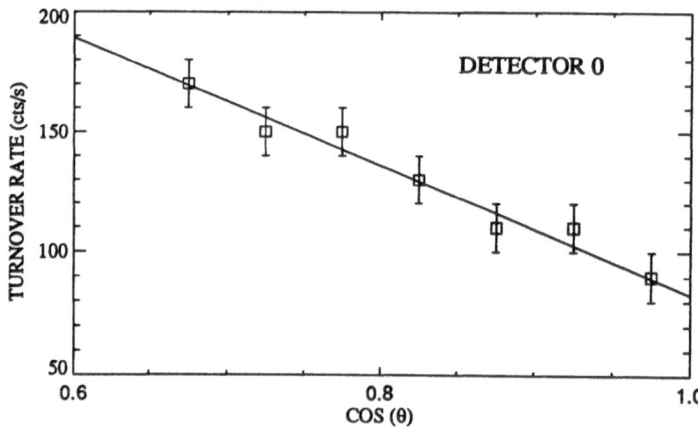

Fig. 3. The peak rate distribution turnover is plotted for different $\cos\theta$.

The sensitivity of BATSE is also limited by the background rate, which varies from 800 to $2200\,\mathrm{s}^{-1}$. With a $3\,\sigma$ detection threshold, only flares with peak rates $> 140\,\mathrm{s}^{-1}$ are detected for all levels of background. The minimum background rate corresponds to a $3\,\sigma$ detection of about $85\,\mathrm{s}^{-1}$.

If $\cos\theta$ and the background rate are restricted to optimal levels the differential distribution can be extended down to $85\,\mathrm{s}^{-1}$. The amount of time that the spacecraft is oriented in this configuration is small. By partially relaxing these conditions, the turnover can still be at a low level, while retaining enough flares for reasonable statistics. In Fig. 4 the data are restricted to background rates $< 1700\,\mathrm{s}^{-1}$ and $\cos\theta > 0.85$. The data are still fit with a power-law, but the distribution now extends down to $\sim 100\,\mathrm{s}^{-1}$, rather than $\sim 150\,\mathrm{s}^{-1}$ as seen in Fig. 1. The duty cycle for solar viewing is reduced by $\sim 65\%$ with these restrictions. The reduction in duty cycle, as well as the varying nature of solar activity results in only 20% of the data from Fig. 1 being included in Fig. 2.

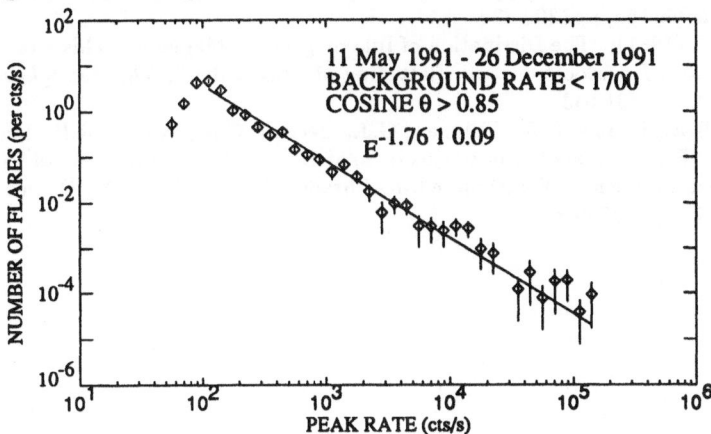

Fig. 4. The differential frequency distribution of BATSE peak flare rates is plotted for events with low background rates and favorable $\cos\theta$.

4 Conclusions

The BATSE instrument is a valuable tool in the search for small solar flares. We have shown that the differential distribution of peak flare rates is well fit by a power-law extending to the limiting sensitivity of the instrument. No departure from a power-law needed for significant coronal heating is observed. With long term observations, BATSE will prove useful in trying to determine if there are differences in the characteristics of the small flares as compared to large flares.

References

van Beek, H.F., de Feiter, L.D., de Jager, C. (1974): "Hard X-Ray Observations of Elementary Flare Bursts, and Their Interpretation", *Space Research* XIV, pp. 447-452
Biesecker, D.A., Ryan, J.M., Fishman, G.J. (1993): "Observations of Small Solar Flares with BATSE", to be published in AIP Conference Proceedings "High Energy Solar Phenomena"

Crosby, N.B., Aschwanden, M.J., Dennis, B.R. (1993): "Frequency Distributions and Correlations of Solar X-Ray Flare Parameters", *Solar Physics* Vol. 143, pp. 275-299

Datlowe, D.W., Elcan, M.J., Hudson, H.S. (1974): "OSO-7 Observations of Solar X-Rays in the Energy Range 10-100 keV", *Solar Physics* Vol. 39, pp. 155-174

Dennis, B.R. (1985): "Solar Hard X-Ray Bursts", *Solar Physics* Vol. 100, pp. 465-490

Hudson, H.S. (1991): "Solar Flares, Microflares, Nanoflares, and Coronal Heating", *Solar Physics* Vol. 133, pp. 357-369

de Jager, C., de Jonge, G. (1978): "Properties of Elementary Flare Bursts", *Solar Physics* Vol. 58, pp. 127-137

Mariscotti, M.A. (1967): "A Method for Automatic Identification of Peaks In the Presence of Background and its Application to Spectrum Analysis", *Nuclear Instruments and Methods* Vol. 50, pp. 309-320

Parker, E.N. (1981): "The Dissipation of Inhomogeneous Magnetic Fields and the Problem of Coronae. I. Dislocation and Flattening of Flux Tubes", *The Astrophysical Journal* Vol. 244, pp. 631-643

Schwartz, R.A., Dennis, B.R., Fishman, G.J., Meegan, C.A., Wilson, R.B., Paciesas, W.S. (1992): "BATSE Flare Observations in Solar Cycle 22", in Proceedings of The Compton Observatory Science Workshop, ed. by Shrader, C.R., Gehrels, N., Dennis, B. (NASA CP-3137) pp. 457-468

Solar Flares and Laboratory Experiments

Sergey Yu. Bogdanov, Anna G. Frank

and Natalya P. Kyrie

General Physics Institute Russian Academy of Sciencies,
117942 Vavilov st. 38, Moscow, Russia.

Abstract: We report on a laboratory experiment where we have simulated in three dimensions the magnetic configuration which might be a scaled version of that existing in some solar flares. The results of current sheet (CS) evolution and plasma dynamics in nonuniform 2-D and 3-D magnetic configurations which have both lines and points of zero field are presented. The CS formation seems to be a general property of 3-D magnetic fields with neutral points. The evolution of the CS is in qualitative agreement with the principal features of solar flares.

1 Introduction

We have set up special laboratory experiments which allow us to investigate flare-type phenomena in magnetized plasmas which can be useful to solar flare physics. For this to be realistic, theoretical models (Syrovatskii, 1981) have shown that the following criteria must be satisfied:
 (i) a nonuniform magnetic field with a neutral line;
 (ii) a highly-conductive plasma; (i) and (ii) form the initial state;
(iii) a perturbation of the initial state which excites both plasma motion and the electric current.
We first produce the initial state, introduce the perturbations, and then let the system evolve naturally.

At present we have both theoretical (Syrovatskii, 1981) and experimental (Frank, 1989) evidence that production of an electric current along the neutral line of a two-dimensional (2–D) magnetic field (which is the simplest singular line), results in the formation of a plane pinch current sheet (CS) which separates opposing magnetic fields and stores the magnetic energy. Three typical stages of CS evolution are usually identified: the formation, the metastable stage, and the impulsive phase. In comparison with solar flares, it is natural to consider the CS metastable stage as the pre-flare situation, while the impulsive phase is associated with the flare itself (Syrovatskii, 1979). The spatial scales of laboratory installations differ from those in the Sun by 9-10 orders of magnitude. Nevertheless, if the other parameters of the laboratory experiment are correctly chosen, it is possible to reproduce processes which are similar to those in flares.

Compared with 2–D configurations with zero lines, the 3–D configurations with singular lines and zero points are both more general and much more probable in the natural state. We have shown for the first time that the excitation of electric current results in CS formation in various 3–D configurations with zero points. The CS development seems to be a general property of nonuniform magnetic fields.

2 Current Sheet Evolution in 2–D Magnetic Fields with Zero Lines

The main results of our studies have been published (Frank, 1989; Bogdanov et al., 1990, 1992; Kyrie et al., 1988, 1992; Frank et al., 1992). The experimental arrangement is shown in Fig. 1.

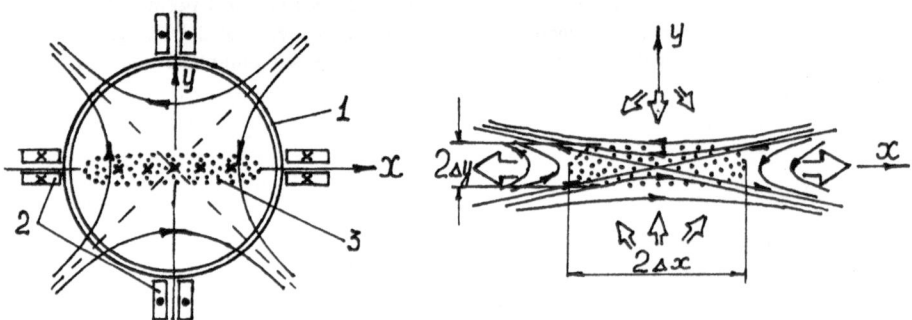

Fig. 1. (*left*) The cross-section of the device CS-3. CS-3 produces the impulsive discharge (I_z current up to 70 kA at $t = 2.3$ ms) in a 2–D magnetic field with gradient $h \leq 3$ kG cm^{-1} and zero line Oz at the axis of the vacuum chamber (1), 10 cm in diameter. The initial magnetic configuration (solid lines) is formed by the electric currents of external conductors (2). The initial plasma parameters are: $N_{e0} \simeq 10^{15}$ cm^{-3}, $T_{e0} \simeq 1 \div 3$ eV. (3) is the position of the plane current sheet. (*right*) Magnetic configuration (solid lines) after the CS formation. The CS cross-section dimensions are: $2\,\Delta x \simeq 6 \div 9$ cm; $2\,\Delta y \simeq 0.6 \div 1.5$ cm; $B_x \simeq 4 \div 6$ kG; $j_z = 5 \div 10$ kA/cm^2. The short arrows show the directions of plasma flows.

The initial state perturbation generates a magnetoacoustic wave converging to the zero line and gives rise to plasma flows and electric currents. The process results in the CS formation in $0.3 \div 0.8$ ms. Plasma is compressed into a plane sheet, the final electron density exceeds the initial density and the surrounding density more than 10 times: $N_e^{sh} = (1 \div 2) \cdot 10^{16}$ cm^{-3}. The CS electron and ion temperatures are: $T_e \simeq 10$ eV, $T_i \simeq (20 \div 50)$ eV. The high-temperature dense CS plasma is balanced by the magnetic pressure outside, so $\beta = 8\pi N_e (T_e + T_i/Z_i)/B_x^2 \simeq 1$.

The CS proved to be rather stable relative to the tearing-mode instability, although the elongation of the CS cross-section exceeded the instability threshold value: $\Delta x/\Delta y = 10 \div 15 > 2\pi$. The time interval of the CS metastable existence, $1 \div 3$ ms, was more than 10 times longer than the specific tearing-mode rise-time.

We varied the initial conditions of CS formation by changing the magnetic field gradient h, the initial perturbation value, etc.

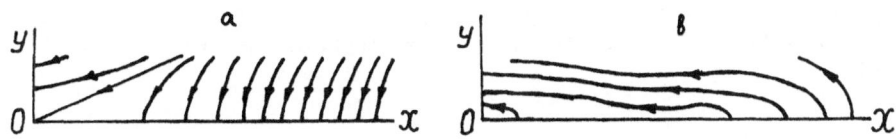

Fig. 2. The two types of CS magnetic field structure: open (a) and closed (b). Only one quadrant of a CS cross–section is shown; the pictures are symmetrical relatively to the planes $(x=0)$ and $(y=0)$.

The internal magnetic structure of the CS at the metastable stage depends significantly on the initial conditions, so we can produce an open magnetic configuration with a single zero-line of the X-type, (Fig. 2a), or a closed configuration with zero-lines of both X- and O-types, (Fig. 2b); the CS can be almost neutral as well. At the open configurations the CS plasma is accelerated by electrodynamical forces $F_x = j_z \cdot B_y/c$ along the CS surface, from the middle to the edges. Plasma flow velocities increase with time and are also nonuniform along the CS surface: the most energetic superthermal flows have been observed at the CS edges. The increase in the thermal plasma energy dominates in closed magnetic fields.

The long-lived metastable stage can be destroyed by the impulsive phase of magnetic reconnection. The process causes the change in the magnetic field topology, the electric current density redistribution, and the excitation of the nonlinear wave, which propagates along the CS surface with a super-Alfvén velocity, $v_x \simeq 10^7$ cm s^{-1}, while $v_a \simeq 1.5 \cdot 10^6$ cm s^{-1}.

The inductive electric fields result in particle acceleration $(\mathcal{E}_e \geq 10 \, \text{keV})$; the process has a burst-type time dependence with the appearance of the most energetic particles at the beginning of the burst. Fast electrons leave the small nonadiabatic region near the zero line, propagating over large distances across strong magnetic fields; they can go away from the CS if the magnetic configuration is open (Fig. 2a).

We have observed that a magnetic island formation inside the CS and the superfast plasma heating are the most essential processes for explosive CS transverse equilibrium disruption and the start of the impulsive phase of magnetic reconnection, which are followed by the total CS destruction. Thus, the evolution of the laboratory-produced CS displays qualitative agreement with the principal features of solar flares. The comparison of physical conditions typical for solar corona and laboratory experiments with CS has been made on the basis of dimensionless parameters. We conclude that the physical phenomena in both cases have to be similar.

3 Current Sheet Formation in 3–D Magnetic Configurations with Zero Points

Up to now flare-type phenomena have been studied experimentally in 2–D magnetic configurations. The magnetic field of translational symmetry with the zero line (Fig. 3a), is an example: $B_q \equiv \{hx; -hy; 0\}$, where h is the magnetic field gradient; the zero line, Oz, is the intersection of two separatrix planes (SP): $(x=0)$ and $(y=0)$.

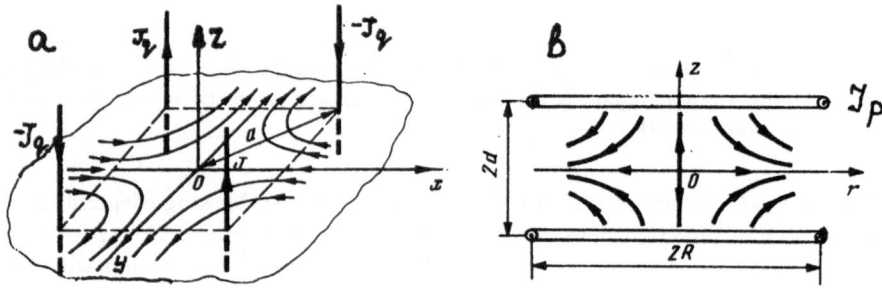

Fig. 3. Two magnetic configurations: (a) the quadrupole 2–D field with zero-line (Oz-axis), produced by 4 straight conductors with electric currents I_q of alternated directions; (b) the axial symmetrical "picket-fence" field with zero-point O, produced by 2 coaxial coils with electric currents I_p of alternated directions.

The variety of 3–D magnetic configurations with singular lines, passing through the zero points, can be produced by the superposition of two fields (Bulanov and Frank, 1992): the first is the field with the zero line, (Fig. 3a), the second is the field with the zero point O and SP $(z=0)$, (Fig. 3b): $B_p \equiv \{h_r x; h_r y; -2h_r z\}$, where h_r is the radial field gradient. The sum $B_{\text{total}} \equiv B_q + B_p = \{(h+h_r)x; -(h-h_r)y; -2h_r z\}$ results in the new 3–D configuration, which can be changed by varying the h_r–h correlation (see the table). The theoretical analysis of the plasma behaviour in the vicinity of the magnetic zero points was performed on the basis of self-similar solutions of the MHD equations (Bulanov and Ol'shanetskij, 1984).

The experimental arrangement CS–3D includes two separate systems to produce the initial 3–D magnetic configuration: (i) 2–D magnetic field with the zero line along the Oz-axis and $h \leq 1.5$ kG cm^{-1}; (ii) axial (Oz) symmetrical magnetic field, produced by 4 coils, with $h_r \simeq 1.0$ kG cm^{-1}. Thus, the configurations with 1, 2 or 3 zero points at the Oz-axis are possible. The total number of configurations is more than 20. Among them, there is a straight analog of Sweet's configuration with four sunspots, as well as a magnetic field with uniform B_z -component up to 24 kG. The value of plasma electric current, directed along Oz, is $I_z \leq 80$ kA.

The first experimental investigation of electric current and plasma configurations has been carried out in a 3–D magnetic field with a single zero point. The most impressive result is that the CS formation occurs in 3–D configurations as well;

Table. Magnetic configurations produced by the superposition of two magnetic fields, $B_{\text{total}} = B_q + B_p$, $h > 0$; $h_r > 0$.

h_r/h	B_x/x	B_y/y	B_z/z	SP	Comments
0	1.0 h	−1.0 h	0	(x=0); (y=0)	2-D quadrupole field; $B_z=0$; Oz-zero line; $h_x=-h_y$
0.1	1.1 h	−0.9 h	−0.2 h	(x=0)	B_z appearance; destruction of the second SP; $|h_y|>|h_z|$
0.33	1.33 h	−0.67 h	−0.67 h	(x=0)	"picket-fence" with Oz-axis; $h_x = 2|h_y| = 2|h_z|$
0.67	1.67 h	−0.33 h	−1.34 h	(x=0)	$|h_y| < |h_z|$
0.9	1.9 h	−0.1 h	−1.8 h	(x=0)	$|h_y| \ll |h_z|$
1.0	2.0 h	0	−2.0 h	(x=0); (z=0)	2-D quadrupole field; $B_y=0$; Oy-zero line; $h_x = -h_z$
1.1	2.1 h	0.1 h	−2.2 h	(z=0)	B_y sign change; SP position change; $|h_x| \gg |h_y|$
∞ (h=0)	1.0 h_r	1.0 h_r	−2.0 h_r	(z=0)	"picket-fence" with Oz-axis; $|h_z| = 2h_x = 2h_y$

contrary to the 2–D configurations, the CS is no more plane, but twisted and, in the vicinity of a zero point, the CS position angle depends on the h–h_r correlation (Fig. 4).

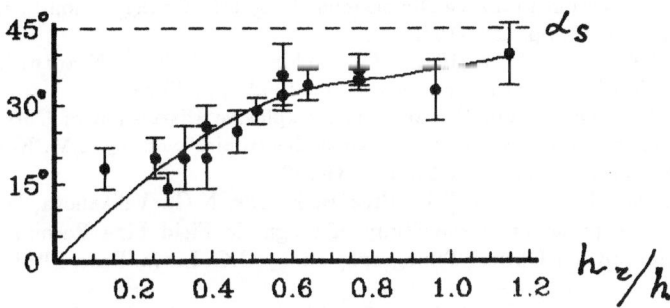

Fig. 4. The CS position angle α at the $z=0$ plane vs. h_r/h ratio: $\alpha=0$ if $h_r=0$; α increases with the h_r/h ; $\alpha \to \alpha_s = 45°$ asymptotically.

It should be emphasized that in, the whole range of the 3–D configurations with zero points we have investigated experimentally, we have always observed the formation of CS. On the basis of this result we may assume that CS formation is a general property of 3–D fields with zero points.

4 Conclusions

1. The CS evolution in a 2–D magnetic field with zero line can ensure the occurrence of flare-type phenomena showing the main features of solar flares. The variety of flare-type phenomena can be realized by the CS formation under different initial conditions: magnetic field gradient, plasma electric current, etc.
2. The plasma ejection from CS and the impulsive phase of magnetic reconnection (the flare) are two different manifestations of the same fundamental process, that is the CS formation and evolution.
3. The superfast plasma heating inside a magnetic island can result in the start of the impulsive stage of magnetic reconnection and the total CS destruction.
4. The CS formation occurs in 3–D magnetic configurations with zero points as well as in 2–D ones. The CS position depends on the specific type of the 3–D configuration. Thus, the CS formation seems to be a general property of nonuniform magnetic fields with zero points and singular lines.

References

Bogdanov, S.Yu., Dreiden, G.V., Kirii, N.P., Komissarova, I.I., Markov, V.S., Ostrovskaya, G.V., Ostrovskii, Yu.I., Phylippov, V.N., Frank, A.G., Khodzhaev, A.Z., Shedova, E.N. (1992): "Plasma Dynamics in Current Sheets", in Russ. J. Plasma Phys. Vol. 18, pp. 1269–1295

Bogdanov, S.Yu., Frank, A.G., Markov, V.S. (1990): "Observation of Accelerated Particle Fluxes Emerging from a Current Sheet across a Strong Magnetic Field", in JETP Lett. Vol. 51, pp. 638-64

Bulanov, S.V., Frank, A.G. (1992): "About one Approach to the Experimental Study of Magnetic Reconnection in Three-Dimensional Magnetic Configurations", in Russ. J. Plasma Phys. Vol. 18, pp. 1535-1544

Bulanov, S.V., Ol'shanetskij, M.A. (1984): "On the Magnetic Collapse Near the Zero Points of the Magnetic Field", in Physics Letters Vol. 100A, pp. 35-38

Frank, A.G. (1989): "Formation, Evolution and Explosive Disruption of Current Sheets in Plasma", in: "Plasma Physics and Plasma Electronics", ed. by L.M. Kovrizhnykh. Nova Science Publ. Commack, N-Y., pp. 131-169

Frank, A.G., Kiselev, D.T., Kyrie, N.P., Preobrazhensky, N.G., Velikanova, L.G. (1991): "Tomographic Approaches in the Study of Magnetic Field Line Reconnection", in: Analytical Methods for Optical Tomography ed. by G.G. Levin. Proc. SPIE. Vol. 1843, pp. 2-30

Kyrie, N.P., Markov, V.S., Frank, A.G. (1988): "Flash of Emission from Multiply Charged Ions in a Current Sheet", in JETP Lett. Vol. 48, pp. 459-463

Kyrie, N.P., Markov, V.S., Frank, A.G. (1992): "Local Pulsed Plasma Heating and Destruction of a Current Sheet", in JETP Lett. Vol. 56, pp. 82-87

Syrovatskii, S.I. (1979): "Key Problems of Flare Theory", in Izv. Akad. Nauk SSSR. Vol. 43, pp. 695-707 (in Russian)

Syrovatskii, S.I. (1981): "Pinch Sheets and Reconnection in Astrophysics", in Ann. Rev. Astron. Astrophys. Vol. 19, pp. 163-229

Particle Acceleration and Radiation Generation by Nonlinear Mode-Mode Coupling Processes in the Solar Corona

A.C.-L. Chian[1,2], J.R. Abalde[2], M.V. Alves[2] and S.R. Lopes[1,2]

[1]Department of Applied Mathematics and Theoretical Physics, University of Cambridge, Silver Street, Cambridge CB3 9EW, UK
[2]National Institute for Space Research - INPE, P.O. Box 515, 12227-010 São José dos Campos-SP, Brazil

Abstract: We discuss various nonlinear mode-mode coupling processes, involving the interaction of high-frequency Langmuir waves with low-frequency Alfvén waves, which may play a role in particle acceleration and radiation generation in the solar corona. Intense Langmuir waves can generate solar radio waves such as whistler waves or electromagnetic electron cyclotron waves, near the electron plasma frequency, via ponderomotive coupling with Alfvén waves. These induced high-frequency cy clotron waves and low-frequency Alfvén waves may contribute toward the energization of coronal particles through wave-particle interaction processes.

Langmuir turbulence can be an effective mechanism for accelerating particles in the solar corona (Benz 1977; Hoyng et al. 1980; Heyvaerts 1981; Karlicky and Jungwirth 1989). Previous theoretical works have considered Langmuir turbulence as the result of nonlinear interaction of Langmuir waves with low-frequency plasma density fluctuations such as ion-acoustic waves (Goldman 1983; Chian and Alves 1988; Chian 1991; Melrose 1991; Rizzato and Chian 1992). These works have shown that electromagnetic waves near the fundamental and higher harmonics of the electron plasma frequency are emitted in the presence of Langmuir turbulence.

In this paper, we show that Langmuir turbulence can also evolve as the result of nonlinear interaction of Langmuir waves (L) with low-frequency magnetic field fluctuations, namely, Alfvén waves (A). Moreover, we show that such nonlinear interaction can produce electromagnetic electron cyclotron waves such as whistler waves (W) near the fundamental electron plasma frequency. The electromagnetic electron cyclotron waves and Alfvén waves induced by Langmuir turbulence can contribute toward particle acceleration in the solar corona.

We treat the case wherein all the interacting waves are aligned with the ambient magnetic field. Consider a large-amplitude Langmuir wave travelling along the ambient magnetic field $\mathbf{B}_0 = B_0\hat{z}$, with wave electric field given by

$$\mathbf{E}_L(\mathbf{x}, t) = \frac{1}{2}\mathcal{E}_0 \exp i(k_0 z - \omega_0 t) + c.c. \tag{1}$$

237

where the pump wave frequency ω_0 and wave number k_0 obey the dispersion relation

$$\omega_0^2 = \omega_{pe}^2 + 3v_{th}^2 k_0^2 \tag{2}$$

where $\omega_{pe} = (n_o e^2/m_e \epsilon_0)^{1/2}$ and $v_{th} = (KT_e/m_e)^{1/2}$ is the electron thermal velocity. The total electric field is given by $\mathbf{E} = \mathbf{E}_L + \mathbf{E}_W + \mathbf{E}_A$, where the fields of induced whistler wave $\mathbf{E}_W$ and Alfvén wave $\mathbf{E}_A$ are represented by

$$\mathbf{E}_W(\mathbf{x},t) = \frac{1}{2}\mathcal{E}_\mp \exp i(k_\mp z - \omega_\mp t) + c.c. \tag{3}$$

$$\mathbf{E}_A(\mathbf{x},t) = \frac{1}{2}\mathcal{E}_A \exp i(kz - \omega t) + c.c. \tag{4}$$

The parametric interaction is triggered when the Langmuir pump amplitude exceeds a certain threshold and the interacting waves satisfy the following matching conditions for wave frequencies and wave vectors

$$\omega_\mp = \omega_0 \mp \omega , \quad \mathbf{k}_\mp = \mathbf{k}_0 \mp \mathbf{k} . \tag{5}$$

Two types of three-wave parametric processes are possible: $L \to W + A$ (decay) and $L + A \to W$ (fusion). In the decay process the Stokes whistler wave $\mathbf{E}_-(\omega_-, \mathbf{k}_-)$ is excited, whereas in the fusion process the anti-Stokes whistler wave $\mathbf{E}_+(\omega_+, \mathbf{k}_+)$ is excited. In addition to the matching conditions (5), each wave-triplet must also satisfy the conservation of wave helicity (Tajima 1977). Since the whistler wave has right-hand circular polarization, the Alfvén wave must be left-hand circularly polarized (i.e. shear Alfvén mode) in the decay process and right-hand circularly polarized (i.e., fast magnetosonic mode) in the fusion process.

We divide the interacting wave fields into two time scales: high-frequency fields, $\mathbf{E}_L$ and $\mathbf{E}_W$, oscillating near the electron plasma frequency; and low-frequency field, $\mathbf{E}_A$, oscillating below the ion cyclotron frequency $\omega_{ci} = eB_0/m_i$. It is evident from (5) that the whistler waves generated by $L \to W \pm A$, acquire frequencies close to ω_{pe} since $\omega < \omega_{ci} \ll \omega_{pe}$. By introducing the above time scales into the momentum and continuity equations for electrons and ions, and the Maxwell's equations, we obtain the following set of Fourier-transformed coupled wave equations for the parametric decay process $L \to W + A$:

$$D_- E_- = \alpha E_0 E_A^* \tag{6}$$

$$D_A E_A = \beta < E_0 E_-^* > \tag{7}$$

with the dispersion functions for the Stokes whistler waves and Alfvén waves given, respectively, by

$$D_- = c^2 k_-^2 - \omega_-^2 + \frac{\omega_{pe}^2 \omega_-}{\omega_- - \omega_{ce}} \tag{8}$$

$$D_A = c_A^2 k^2 - \omega^2 \tag{9}$$

and the coupling coefficients

$$\alpha = -\frac{ie\omega_{pe}^2 \omega_-}{2m_e \omega_0(\omega + \omega_{ce})}\left[\frac{k_0}{\omega_0} + \frac{k\omega_{ce}}{\omega(\omega_- - \omega_{ce})}\right] \tag{10}$$

$$\beta = -\frac{ie\omega_{pe}^2 \omega}{2m_e \omega_0(\omega_- - \omega_{ce})}\left[\frac{k_0}{\omega_0} - \frac{k_-\omega_{ce}}{\omega_-(\omega + \omega_{ce})}\right] \tag{11}$$

where $\omega_{ce} = eB_0/m_e$ is the electron cyclotron frequency, $c_A = B_0/(n_0 m_i \mu_0)^{1/2}$ is the Alfvén velocity, the angular bracket denotes averaging over the fast time scale, and the asterisk denotes the complex conjugate. The right-hand side of (6) is related to the nonlinear current density arising from the beating of Langmuir and Alfvén waves. The right-hand side of (7) is related to the low-frequency ponderomotive force resulting from the beating of Langmuir and whistler waves. Thus, the parametric interaction is produced by the combined action of ponderomotive force and nonlinear current density. Likewise, a set of coupled wave equations similar to (6) and (7) can be derived for the fusion process $L + A \rightarrow W$.

From (6) and (7), we obtain the nonlinear dispersion relation

$$D_A D_-^* = \alpha^* \beta |E_0|^2 .$$ (12)

For the resonant decay instability, we let $\omega = c_A k + i\Gamma$. It then follows from (12) that the growth rate is

$$\Gamma = \frac{|E_0|}{2} \left(\frac{\alpha^* \beta}{c_A k \omega_-} \right)^{\frac{1}{2}} \left[1 + \frac{\omega_{ce} \omega_{pe}^2}{2\omega_- (\omega_- - \omega_{ce})^2} \right]^{-\frac{1}{2}} .$$ (13)

The threshold condition can be determined by the inclusion of wave damping in (12), yielding

$$|E_0|^2 = \frac{4\Gamma_W \Gamma_A c_A k \omega_-}{\alpha^* \beta} \left[1 + \frac{\omega_{ce} \omega_{pe}^2}{2\omega_- (\omega_- - \omega_{ce})^2} \right]$$ (14)

where Γ_W and Γ_A are the damping frequencies of whistler and Alfvén waves, respectively. Hence, we have shown that for Langmuir pump intensity exceeding (14), both whistler and Alfvén waves are parametrically excited with the growth rate given by (13).

For oblique propagation the whistler mode is limited to the frequency range below the minimum of f_{pe} or f_{ce}, whichever is lower. However, for parallel propagation considered here the condition for whistler waves becomes $f < f_{ce}$. Since the Langmuir wave frequency is close to f_{pe}, $L \rightarrow W \pm A$ can operate only inside the density depletion region of the solar corona where $f_{pe} < f_{ce}$.

It is important to point out that the electromagnetic dispersion relation for whistler waves given by (8) is applicable only if the wave vectors are strictly parallel to the background magnetic field. For wave propagation at oblique angles to the background magnetic field the whistler waves become quasi-electrostatic, hence modifications to the dispersion relation of whistler waves would be necessary. In this case, direct mode conversion of Langmuir waves into whistler waves becomes possible provided $\omega_{pe} < \omega_{ce}$ (Budden 1985).

Note that in addition to the processes $L \rightarrow W \pm A$ discussed above the whistler-mode radiation, with $f \leq f_{pe}$, can also be driven by Langmuir waves along the ambient magnetic field lines through other parametric processes such as $L \rightarrow W + \ell$ (Hasegawa 1974), where ℓ is the left-hand circularly polarized electromagnetic electron cyclotron wave. An interesting feature of the process $L \rightarrow W + \ell$ is the possibility of its occurrence either inside or outside of the density cavity (i.e., either $f_{pe} < f_{ce}$ or $f_{pe} > f_{ce}$), provided the frequency of the ℓ-mode exceeds its cut-off frequency ω_ℓ which is a function of ω_{pe} and ω_{ce}. The induced whistler emission covers a wide range of frequencies, from $f \leq (f_{ce}, f_{pe})$ down to $f \ll (f_{ce}, f_{pe})$. However, this parametric instability does not involve the excitation of Alfvén waves. It is worth mentioning that a Langmuir wave can also parametrically decay into a left-hand circularly polarized electromagnetic electron cyclotron wave near ω_{pe} and a right-hand circularly polarized Alfvén wave (Shukla and Sharma 1982); alternatively, a Langmuir wave can parametrically combine with a left-hand circularly polarized Alfvén wave to generate a left-hand electromagnetic electron cyclotron wave near ω_{pe}. These processes can occur either inside or outside of the plasma density cavity..

In summary, we have discussed various nonlinear mode-mode coupling processes, involving the interaction of Langmuir waves with Alfvén waves, which may play an important role in the energization of coronal particles. Traditionally, it is believed that the nonlinear coupling of Langmuir waves with low-frequency density fluctuations (e.g., ion-acoustic waves) is the dominant mechanism for Langmuir turbulence to dissipate energy in a plasma. We have shown, in this paper, that the nonlinear coupling of Langmuir waves with Alfvén waves may also be an efficient dissipation mechanism of Langmuir turbulence. The high-frequency whistler and electromagnetic electron cyclotron waves emitted by Langmuir–Alfvén turbulence may account for part of the spectrum, near the fundamental plasma frequency, of solar radio emissions. Hence, they provide the electromagnetic signatures of the energization processes taking place in the solar corona. Furthermore, these high-frequency waves and the low-frequency Alfvén waves induced by Langmuir turbulence may accelerate coronal particles through various wave-particle interaction processes.

Acknowledgements

The authors are grateful to CEC, CNPq and FAPESP for financial support.

References

Benz, A.O.: 1977, *Astrophys. J.* **211**, 270
Budden, K.G.: 1985, *The Propagation of Radio Waves*, Cambridge Univ. Press, p. 542
Chian, A.C.-L.: 1991, *Planet. Space Sci.* **39**, 1217
Chian, A.C.-L., and Alves, M.V.: 1988, *Astrophys. J. Lett.* **330**, L77
Goldman, M.V.: 1983, *Solar Phys.* **89**, 403
Gurnett, D.A., and Frank, L.A.: 1975, *Solar Phys.* **45**, 477
Hasegawa, A.: 1974, *Phys. Rev. Letters* **32**, 817
Heyvaerts, J.: 1981, in *Solar Flare Magnetohydrodynamics*, ed. by E.R. Priest, Gordon and Breach, New York, p.429
Hoyng, P., Duijveman, A., van Grunsven, T.F.J., and Nicholson, D.R.: 1980, *Astron. Astrophys.* **91**, 17
Karlicky, M., and Jungwirth, K.: 1989, *Solar Phys.* **124**, 319
Lin, R.P., Levedahl, W.K., Lotko, W., Gurnett, D.A., and Scarf, F.L.: 1986, *Astrophys. J.* **308**, 954
Melrose, D.B.: 1991, *Ann. Rev. Astron. Astrophys.* **29**, 31
Rizzato, F.B., and Chian, A.C.-L.: 1992, *J. Plasma Phys.* **48**, 71
Shukla, P.K. and Sharma, R.P.: 1982, *Phys. Rev.* **A25**, 2816
Tajima, T.: 1977, *Phys. Fluids* **20**, 61

A Fast Mechanism for the Acceleration of Solar Energetic Particles in Solar Flares

G. Fiorentini [1], S.S. Gershtein [2]

[1]Dipartimento di Fisica and INFN, I-44100 Ferrara, Italy
[2]Institute for High Energy Physics, Protvino, Russia

Abstract: It is pointed out that the mechanism of collective ion acceleration by means of electron beams may occur in solar flares. This mechanism remarkably provides the following features in agreement with the observational data: i) proton acceleration up to GeV energies within a $10^{-1} - 10^{-2}$ s time interval, ii) proton and electron spectrum similarity and a scaling law that relates the number of the accelerated protons and electrons, iii) acceleration of nuclei, whose spectra are similar to that of electrons and protons.

1 The Acceleration Mechanism

Experiments on satellites – Solar Maximum Mission and Hinotori – have shown that acceleration of electrons up to energies of hundreds of keV and of ions up to hundreds of MeV takes place almost simultaneously in times shorter than $1 - 2$ s (Forrest and Chupp, 1983; Nakajima, 1983; Yoshimori et al., 1983; Chupp, 1984; Yoshimori et al., 1984; Kane et al., 1986; Chupp et al., 1982). This is in drastic contrast to widely accepted theories, where acceleration of protons up to GeV energies is due to the stochastic Fermi mechanism, with typical time scales of the order of minutes. In this paper we show that the mechanism of collective ion acceleration, presented by one of the authors (Gershtein, 1981), can naturally explain fast acceleration of protons and heavier nuclei up to GeV energies in times shorter than 10^{-2} or 10^{-1} s. This mechanism also provides a common origin for the Solar Energetic Particles (SEP), that produce various phenomena in the solar atmosphere.

The mechanism we consider is similar to the "smokotron" acceleration mechanism of nuclei in laboratory accelerators, proposed by V.I. Veksler and V.P. Sarantsev in Dubna and by A.M. Sessler in Berkeley, (Sarantsev and Perelshtein, 1979; Olson and Schumacher, 1979) by means of which ions are accelerated with the help of electron bunches. At the first stage of this process, rings of relativistic electrons are produced in a longitudinal magnetic field between two magnetic mirrors. The gas entering the ring gets ionized and a small concentration of positive ions is trapped by the electron bunch. When one of the mirrors is switched off, the electron rings are pushed towards weaker magnetic field regions and the positive ions are carried

241

together with the electron bunch. In this case, due to the conservation of the adiabatic invariant p_t^2/H (where H is the strength of the magnetic field and p_t is the electron momentum transverse to the field direction) and of the electron energy, the momentum is transferred from the transverse to the longitudinal direction. For ions co-moving with the electron ring with velocity v_l along the magnetic field, the total energy becomes:

$$E = \frac{M\varepsilon}{m_t} \qquad (1)$$

where M is the ion mass, ε is the total electron energy and

$$m_t = \sqrt{m^2 + p_t^2} \quad ; \qquad p_t^2 = p_{t0}^2 \frac{H}{H_0} \qquad (2)$$

where p_{t0} and H_0 are the initial values of the transverse momentum and the magnetic field respectively (hereinafter the speed of light is set equal to 1 and m is the electron mass). From the above equations it follows that, if the electron bunch can transport the ions to the region where $H \ll H_0$, then the ion energy will be larger than that of the electrons by a factor of M/m. In the subrelativistic limit the kinetic energy of ions, $T = E - M$, and that of electrons, $T_e = \varepsilon - m$, are related as follows:

$$T = \left(\frac{M}{m}\right) T_e \left(1 - \frac{H}{H_0}\right) \qquad (3)$$

Thus, in order to accelerate ions up to hundreds of MeV, it is sufficient to accelerate electrons up to energies of the order of hundreds of keV only.

Of course, Eqs. (1) and (3) are valid only when the number N of ions in the bunch is much smaller than that of electrons N_e, i.e. for ion concentration:

$$C = \frac{N}{N_e} \leq \frac{m}{M} \qquad (4)$$

In the general case $T = M(\varepsilon - m_t)/(m_t + CM)$ and, for subrelativistic energies, $T \simeq T_e(1 - H/H_0)M/(m + CM)$.

The complex magnetic field configuration during solar flares favours this mechanism of acceleration. The intense flares develop in active solar regions characterized by the presence of a group of sunspots with different polarities. The sunspot magnetic field, which can reach a strength of about 10^3 gauss, provides both open structures and "arc-like" configurations and can act as a magnetic mirror. If electrons are accelerated near a sunspot in the direction orthogonal to the magnetic field by any impulsive mechanism, e.g. breaking of current layers, then, due to the magnetic field gradient, they will reach the velocity v_l along the field direction. The existence of such electron bunches moving along open field lines is well-known. They are the source of microwave bursts of type III and are detected in satellite experiments. In the same way, bunches of electrons moving along magnetic arcs can also be generated. The existence of these bunches, which is inferred from the detection of U- and V-microwave radiation, is also well-known. Such bunches can trap nuclei and keep them during the acceleration along the magnetic field (Gershtein, 1979).

Small values of ion concentration (see Eq. (4)) appear to be the natural condition for a fast electron bunch. Once the electron bunch is produced in a neutral plasma, it must have a large velocity in order to escape from positive ions. On the other hand, this large velocity can be attained only when the concentration of the trapped ions is small. In other words, Eq. (4) can also be viewed as a condition for the electron bunch to form and propagate.

For the small value of the concentration given by Eq. (4), we get from the conservation of energy and the adiabatic invariant the following equation:

$$\frac{dp_l}{dt} = -\frac{p_{t0}^2}{2H_0\varepsilon}\frac{dH}{dz} \tag{5}$$

This equation implies an upper limit to the time of nuclei acceleration up to the maximum energy, such as:

$$\tau = \frac{2}{v_{t0}}\frac{H_0}{|dH/dz|} \tag{6}$$

where v_{t0} is the initial electron velocity. The typical values of the magnetic field strength and its gradient during solar flares are $H_0 \simeq 10^2 - 10^3$ G and $dH/dz \simeq 0.1$ G/km. Protons are accelerated up to energies of hundreds of MeV by 100 keV electrons in less than $10^{-2} - 10^{-1}$ s, which is consistent with the satellite data already quoted. Note that this time is much shorter than the characteristic time of energy damping by synchrotron radiation, $\tau_{syn} \simeq 2.6 \cdot 10^8 (1 \text{ gauss}/H)^2$ s.

For $C \simeq m/M$ we expect the following scaling law which involves both number and energy of accelerated electrons and protons:

$$N_p(> T) \simeq 10^{-3} N_e(> 10^{-3}T) \tag{7}$$

where $N_i(> T)$ is the number of particles with kinetic energy larger than T. This relationship is well confirmed by observations. It was shown by Lin (1974) that the flux ratio is $I_p(T > 10 \text{ MeV})/I_e(T > 45 \text{ keV}) \simeq 3 \cdot 10^{-4}$ and more recent data for impulsive gamma flares indicate that this ratio is smaller than 10^{-3} (Kurt et al., 1981). The differential spectra should be approximately identical and satisfy the relationship:

$$\frac{dN}{dT}(T) \simeq 10^{-6}\frac{dN_e}{dT_e}(T_e \simeq 10^{-3}T) \tag{8}$$

This relationship is in surprisingly good agreement (at least for energies $T \leq 200$ MeV, $T_e \leq 200$ keV) with the spectra obtained by Ramaty et al. (1975) from the experimental data on the yields of bremsstrahlung and nuclear γ lines observed in solar flares. For power extrapolation of spectra, $dN/dT \sim T^{-s}$, the spectral index s varies weakly in different flares and, for protons with few hundreds of MeV, $s = 3 \pm 0.5$ (Kocharov, 1988). The spectral index for the electron spectra is within the same limits. It is interesting to note that the kink in the electron spectrum at $T_e \simeq 200$ keV, corresponding to the transition to a higher spectral index, $s = 3.5$ (Ramaty et al., 1975), is reproduced in the spectrum of protons at energies of some hundreds of MeV. The similarity between the accelerated electron and proton spectra together with the approximate relationship between their numbers can be considered as a strong argument in favour of this mechanism.

In Gershtein (1981) the collective mechanism was considered only for the solar cosmic rays, i.e. particles accelerated in open configurations of magnetic fields. Obviously, the same process can work in closed arc-like structures, because in both cases, energetic electron bunches are generated. Thus, the same mechanism is responsible for solar cosmic rays and energetic particles. Protons accelerated along open configurations escape from the Sun, whereas those accelerated in arc structures remain trapped. In the latter case, the density and path length can be sufficient for yielding nuclear collisions. These produce excited nuclear levels with subsequent emission of nuclear γ-rays, high-energy neutrons ($E_n > 50$ MeV), which can reach the Earth, and also neutral and charged pions, which by their decay generate energetic γ-rays, ultra-relativistic electrons and positrons.

In arc structures nuclear reactions can occur when protons are moving almost parallel to the Sun's surface. In this case, neutrons are also produced primarily parallel to the Sun's surface and thereby have better chances to reach the Earth if the flare develops at the solar limb. The observed enhancement of energetic neutrons in flares at the solar limb (Breneman and Stone, 1985; Van Hollebeke et al., 1985) favours the idea that these particles are produced in arc structures.

One of the principal features of this acceleration mechanism is that not only protons, but also nuclei of heavier elements can be accelerated in a natural way. In our case there is no need of a pre-acceleration mechanism: the well-known difficulty of the stochastic acceleration mechanism for the case of heavy nuclei is avoided. This is an important feature, since it appears that accelerated heavy nuclei are also present among SEP (Breneman and Stone, 1985; Van Hollebeke et al., 1985; Murphy et al., 1985; Ramaty et al., 1977; Ibragimov and Kocharov, 1977; Yoshimori et al., 1985). According to the mechanism proposed, all the nuclei spectra should be identical.

To explain the intensity of the nuclear γ-lines and the flux of neutrons observed during solar flares, one has to assume that the interaction of the accelerated ions with nuclei occurs in a region where the number density is $n \simeq 10^{11} \div 10^{12}$ cm^{-3}. In this regard we would like to note that the acceleration of nuclei by electron bunches in arc structures can principally yield fluxes of nuclei directed towards the chromosphere and photosphere. This can happen if the electron bunch is generated near a big sunspot and then, following the magnetic field lines, travels towards smaller sunspots, where the magnetic mirror is located closer to the Sun's surface. In this way the accelerated ions can penetrate the deeper layers of the solar atmosphere. An additional possibility for accelerated ions to reach larger densities is to get through prominences, that usually exist in active solar regions, placed on the top of magnetic arcs.

2 Discussion

In this paper we do not discuss the mechanism of initial acceleration of electrons up to MeV energies and their spectrum formation. Clearly, this difficult problem is beyond the aim of our paper and should be solved on the basis of the solar plasma physics. However, although the generation of energetic electron beams is a well-known experimental fact, the related mechanism of collective ion acceleration, that we suggested above, seems fairly plausible. It provides a quite natural and simple explanation of a wide class of phenomena in solar flares: the fast acceleration of protons with a spectrum similar to that of electrons and the acceleration of nuclei. It is worth noting that the connection between the acceleration of electrons and nuclei is also confirmed by the approximately constant (within a factor of two) ratio of the bremsstrahlung radiation flux produced by electrons to the nuclear γ-lines flux produced in nuclear collisions, even in flares having quite different intensities and time profiles (Forrest, 1983).

Notice that in our mechanism one should assume that electrons must be initially accelerated in a very short time. Data on very fast variations of the microwave (Slottie, 1978) and hard X-rays emission (Kiplinger et al., 1983; Brown et al., 1985) seem to support this hypothesis. Presumably, the impulsive process yielding the initial acceleration of electrons in a solar flare resembles the repeated electric discharges occurring during a thunderstorm in the Earth's atmosphere.

It is important to note that our mechanism also explains the observed delay of microwaves compared to hard X-rays (Cornell et al., 1984; Kaufman et al., 1983; Takakura et al., 1983). According to observations the hard X-ray sources are located at the footpoints of magnetic loops, whereas the microwaves are essentially emitted from the top of these loops (Marsh and Hurford, 1980; Hoyng et al., 1983). This is in full agreement with the commonly accepted picture in which the hard X-rays are produced by accelerated electron bremsstrahlung in dense chromospheric layers, while the origin of microwaves is due to synchrotron or plasma radiation by electrons in the solar corona. In the conventional scenario, the electron acceleration occurs at the top of magnetic loops and the microwave emission begins immediately, while the hard X-ray emission starts after the electrons have propagated down to the chromosphere. However, the conventional scenario contrasts with the evidence that microwaves are observed to follow, not precede, the X-rays, with a mean delay of about 0.2 s (Cornell et al., 1984). On the contrary, in the mechanism suggested in the present paper the "wrong" sign of the time delay is accounted for. The electron acceleration starts at the footpoints of loops, where the magnetic field is large, and is accompanied by X-ray emission. Then, due to the field gradient, the accelerated electrons are pushed up to the solar corona and provide the microwave radiation. Moreover, for reasonable values of the magnetic field and its gradient, Eq. (6) gives a typical time scale $\tau \sim 0.1$ s, in agreement with the observed time delay.

It is clear from Eqs. (1) (2) that for electron energies $\varepsilon \gg m$ the relationship $E \simeq (M/m)\varepsilon$ is valid only if the electron bunch is able to transport trapped protons to regions with substantially low magnetic fields, $H/H_0 \ll m^2/\varepsilon^2$. In the opposite case, the proton energy has the upper limit $E \simeq M\sqrt{H_0/H}$. Therefore, the suggested mechanism seems to be effective up to energies of some GeV per nucleon.

Our mechanism, intended for the impulsive phase of solar flares, can also help to provide the proper initial conditions (the pre-acceleration of ions) for the subsequent inclusion of other acceleration mechanisms (namely, for the stochastic one). These mechanisms could provide further acceleration of ions and also explain the emission of energetic γ-rays in the late phase of solar flares (few hours after the impulsive phase) detected by EGRET on board the Compton Gamma-Ray Observatory (Kanbach et al., 1992). However, as was shown by Mandzhavidze and Ramaty (1992), the γ-ray production by the particles accelerated during the impulsive phase and subsequently trapped in magnetic loops, can be sufficient by itself, while the possibility of a continuous (e.g. stochastic) acceleration is disfavoured by the observational data (Kanbach et al., 1992).

Acknowledgements

The authors are grateful to Z.G. Berezhiani, A.E. Chudakov and M.A. Livshits for useful discussions.

References

Breneman, H., Stone, E.C.: 1985, *Astrophys J. Lett.* **299**, L5
Brown J. et al.: 1985, *Astron. Astrophys.* **147**, L10
Chupp E.L.: 1984, *Ann. Rev. Astr. Ap.* **22**, 359
Chupp E.L. et al.: 1982, *Astrophys. J. Lett.* **235**, L95
Cornell M.E. et al.: 1984, *Astrophys. J.* **279**, 875
Forrest D.J.: 1983, *Positron-Electron Pairs in Astrophysics*, M.L. Burns et al. (eds), New York, AIP, 3
Forrest, D.J., Chupp, E.L.: 1993, *Nature* **305**, 291
Gershtein S.S.: 1979, *Geomagnetism and Aeronomy* **19**, 202
Gershtein S.S.: 1981, Proc. XVII Int. *Cosmic Ray Conf.* **3**, 419
Hoyng P. et al.: 1983, *Astrophys. J.* **268**, 865
Ibragimov, I.A., Kocharov G.E.: 1977, *Pis'ma Astr. J.* **3**, 412
Kanbach G. et al.: 1992, *Astron. Astropys.*, in press
Kane S.R. et al.: 1986, *Astrophys. J. Lett.* **300**, L95
Kaufman P. et al.: 1983, *Solar Phys.* **84**, 311
Kiplinger A.L. et al.: 1983, *Astrophys J. Lett.* **265**, L99
Kocharov G.E.: 1988, *Sov. Sci. Rev., sect. E; Astrophys. and Space Phys. Rev.*, **6**
Kurt V.G. et al.: 1981, *Proc. XVII Int. Cosmic ray conf.* **3**, 69
Lin R.P.: 1974, *Space Sci. Rev.* **16**, 189
Mandzhavidze, N., Ramaty, R.: 1992, *Astrophys. J. Lett.* **396**, L111
Marsh, K.A., Hurford G.J.: 1980, *Astrophys. J. Lett.* **240**, L111
Murphy R.L. et al.: 1985, *Proc XIX Int. Cosmic Ray Conf.* **4**, 249
Nakajima H.: 1983, *Nature* **305**, 292
Olson, C.L., Schumacher U.: 1979, *Collective Ion Acceleration*, Springer Tracts in Modern Physics, **84**
Ramaty, R., Kozlovsky, B., Lingehfelter, R.E.: 1975, *Space Sci. Rev.* **18**, 341
Ramaty R. et al.: 1977, *Astrophys. J.* **214**, 617
Sarantsev, V.P., Perelshtein E.A.: 1979, *The collective acceleration of ions by electron rings*, Moscow, Atomizdat
Slottie C.: 1978, *Nature* **275**, 520
Takakura T. et al.: 1983, *Nature* **302**, 317
Van Hollebeke M.A.I. et al.: 1985, *Proc XIX Int. Cosmic Ray Conf.* **4**, 209
Yoshimori M. et al.: 1983, *Solar Phys.* **86**, 375
Yoshimori M. et al.: 1984, *Proceedings of Phys. Soc. Japan* **53**, 4499
Yoshimori M. et al.: 1985, *J. Phys. Soc. Japan* **54**, 4462

VI

Propagation of Energetic Particles in the Corona

VI

Acceleration and Storage of Energetic Particles in the Solar Corona

K.-P. Wenzel

Space Science Department of ESA,
ESTEC, 2200 AG Noordwijk, The Netherlands

Abstract: Observations of solar energetic particles (SEP) in interplanetary space have considerably advanced our understanding of particle acceleration near the Sun in recent years. Based on particle abundances, ionisation states, time evolution and longitude distributions, two distinct classes of SEP events have been identified. They are usually described as "impulsive" and "gradual", with reference to the time scale of the accompanying soft x-ray flare. It is now widely accepted that these classes carry the signature of different acceleration processes: impulsive heating in the flare process and shock acceleration, respectively. It is argued that coronal particle transport in large events (gradual events) can be understood in terms of shock propagation and acceleration eliminating the ad hoc concept of coronal diffusion. Finally, new results of solar flare gamma-ray emission at high energies are discussed which shed new light on particle storage and acceleration in the solar corona.

Keywords: Acceleration, solar energetic particles.

1 Introduction

Solar energetic particles (SEP) are manifestations of solar activity associated with the evolution of the solar magnetic field. SEP (ions > 1 MeV/n and electrons > 100 keV) are an important element in many solar flares producing a complex variety of electromagnetic emissions and carrying information on plasma processes at work in flaring regions. SEPs are also the result of energisation processes related to the transient ejection of material from the Sun into interplanetary space. To try to understand SEP acceleration, one can observe the escaping particles directly or infer particle properties from the electromagnetic radiation that the interacting particles emit. In this paper the subject is approached via the first method: SEP observations in interplanetary space. Electromagnetic signatures of particle acceleration are discussed in a companion paper by Klein (1993).

The exploitation of solar particle observations at 1 AU for the study of acceleration has been hampered by the difficulty in disentangling effects of various processes in different regions that the particles have to traverse on their way from the source to the observer. In a very simplistic view one may divide these processes into five stages: acceleration, storage and transport in the solar corona, injection into the interplanetary medium, propagation along interplanetary magnetic field (IMF) lines and interplanetary acceleration.

In this paper, aspects of the first two stages are discussed: acceleration and transport near the Sun. In the next section examples of observations are presented that have led to the classification of SEP into impulsive and gradual events. In section 3 recent models for particle acceleration are outlined. Section 4 deals with coronal transport. In section 5 the significance of recent high-energy gamma-ray observations in solar flares is discussed.

2 Observational Properties and Classification of Solar Particle Events

Major advances in our understanding of particle acceleration on the Sun have been made in the last decade through improved instrumentation and extensive analysis, combining SEP observations with those of neutral radiation (x-rays, gamma-rays, neutrons) from the same flares. Theorists have responded with interpretations taking into account not only the wealth of new data, but also systematic relationships among the data which in many cases provide severe constraints on relevant models.

Observing with instruments of limited sensitivity, the early measurements of SEP events at 1 AU were obviously confined to large events that we now regard as gradual SEP events or major proton events (MPEs). These events evolve gradually on a time scale of days. Attempts to associate them with flares having durations of only hours led to models with extremely inefficient particle transport. The ad hoc process of "coronal propagation" (see section 4) was proposed to allow particles to diffuse uniformly around the corona from a point source of acceleration, followed by slow leakage into interplanetary space onto field lines connected to the observer.

Only with improvements in instruments, for example, those launched on ISEE-3 in 1978, was the significance of a new type of particle event, the ^{3}He-rich event, recognised. These events were distinguished by values of ^{3}He/^{4}He in the range of 0.01 to 10 (corresponding values in the solar atmosphere or solar wind are about 5×10^{-4}). Enhancements in heavy nuclei, such as Fe by factors of 10, were also observed. The absence of the isotopes ^{2}H and ^{3}H in these events eventually excluded the possibility that the ^{3}He might result from nuclear fragmentation (Mewaldt and Stone 1983).

One of the first great successes in the studies of SEP was the establishment of significant correlations between the escaping energetic particles produced by flares and other properties of these flares. The most dramatic ordering of certain SEP properties, e.g. the slope of the energy spectrum of electrons (Moses et al. 1989), was obtained by considering the duration of soft (<10 keV) x-ray emission from the flare. In 1977, Pallavicini, Serio and Vaiana had observed that the duration of soft x-ray emission was an indicator of the physical parameters at the main site of energy release in solar flares. They found that impulsive soft x-ray events came from compact flares, low ($\approx 10^4$ km) in the corona, while gradual soft x-ray events came from extended sources that occurred high (5×10^4 km) in the corona.

Over the last decade there has been growing and now widely accepted evidence that there are two SEP populations derived from at least two acceleration processes on or near the Sun. These two populations can be divided into two event classes on the basis of the time scale of the soft x-ray emission of their associated flare (if there is one) into impulsive SEP events related to short duration (<1 hr) emission, and gradual SEP events related to long (>1 h) duration x-ray emission. Some SEP events appear to be composites of these basically distinct types (e.g. Mason et al. 1984).

One of the first attempts to distinguish the two acceleration processes was made when Cane et al. (1986) studied the soft x-ray durations of flares producing relativistic electron events. They found that the impulsive events have higher e/p ratios than the gradual SEP events. A recent statistical study of Helios events by Kallenrode et al. (1992) including He expanded those results.

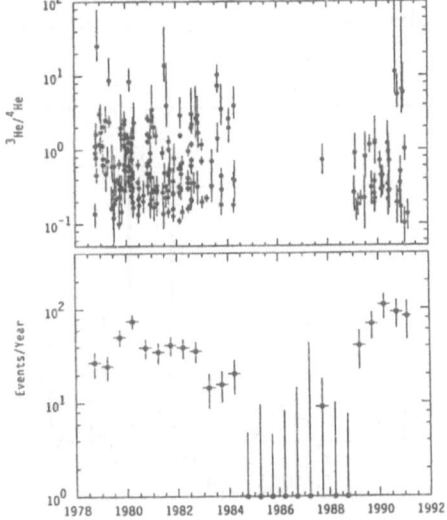
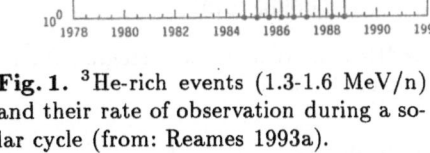
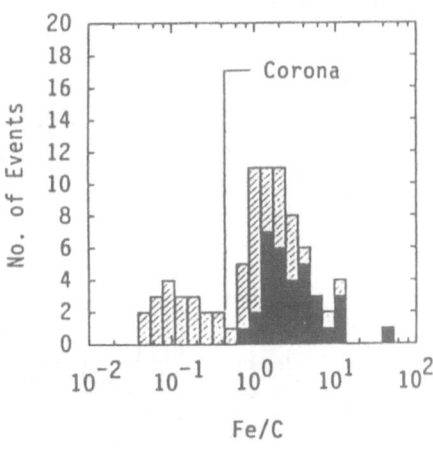

Fig. 1. ^{3}He-rich events (1.3-1.6 MeV/n) and their rate of observation during a solar cycle (from: Reames 1993a).

Fig. 2. Histogram of the Fe/C ratio. ^{3}He-rich events are blackened. The coronal abundance derived by Meyer (1985) is indicated (Reames et al. 1990).

Until a few years ago, it was generally believed that ^{3}He-rich events implied some unusual conditions in the flare plasma where the ^{3}He enhancements originated. It was furthermore assumed that unusual conditions in the interplanetary medium allowed the ions to propagate with minimal scattering, since ^{3}He-rich events were found to be correlated with the numerous non-relativistic electron events (Reames et al. 1985). As more events were observed, however, these beliefs become less attractive. Figure 1 shows statistics of ^{3}He- rich events. Correcting the observed frequency of ^{3}He rich events for their relatively small "opening angle" ($\approx 20^0$), Reames (1993a) *estimates an effective rate of these events* to be about 1000 ^{3}He rich events/year on the visible disk of the Sun at solar maximum.

Energetic particle abundances are probably the best tool to distinguish impulsive events with their enhanced ^{3}He, electron and heavy element abundances from most gradual events that have heavy element abundances near or below those in the corona or solar wind. Figure 2 displays, as an example, the distribution of Fe/C for a sample of 90 impulsive and gradual events (Reames et al. 1990). ^{3}He-rich events, blackened in the figure, clearly show an enhanced Fe/C ratio compared to the other events (shaded) that have ^{3}He/^{4}He < 0.1.

A key to the origin of the two particle populations was the measurement of the charge state of Fe. Luhn et al. (1987) found that Fe ions in ^{3}He-rich events had a mean charge of 20.5±1.2, indicating a source temperature of about 2 × 10^7 K. In contrast, an Fe charge state of 14.1±0.2 was found by averaging over gradual events, indicating a temperature of about 2 × 10^6 K as might be found in the ambient corona or solar wind (Fig. 3). These observations led to the idea that the ^{3}He-rich, Fe-rich material has been heated at the flare site, while the material observed in gradual events originates in the corona and solar wind, and has been accelerated in the corona and the interplanetary medium by a shock wave, driven by a coronal mass ejection (CME).

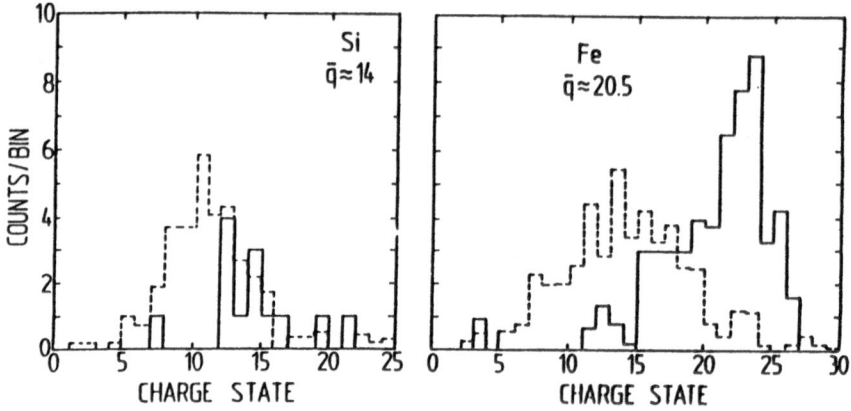

Fig. 3. Measured ionisation states of Si and Fe for ^{3}He-rich events (solid histograms) and large proton events (dashed histograms) (after Luhn et al. 1987).

Understanding gradual events in terms of shock acceleration also provided an explanation for two other distinguishing features, namely, the long time duration of these events and their wide extent in solar longitude (e.g. Cane et al. 1988, Kallenrode et al. 1992). Gradual events can have risetimes as long as a day and typically persist at high intensity levels for several days or more. Impulsive events typically reach maximum quickly and decay over a period of several hours. Whilst impulsive events appear to be restricted a narrow cone of emission clustered around the observer's field line, i.e. connected to ≈ W 50 for near-Earth observations, the gradual events are distributed more uniformly across the Sun. Energetic particles in gradual events are observed on or close to IMF lines that connect to a shock running in front of a fast CME and extending over a longitude range as wide as 180^0. The largest number of accelerated particles are probably produced near the Sun where the CME-driven shock is strongest and the ambient density is highest.

Throughout the outward journey of the disturbance accelerated particles continually leak away from the acceleration region near the shock along the IMF. The detailed SEP intensity time profiles that are observed depend sensitively on the longitude where the CMEs originate relative to the observer.

Event Classes (see Table 1). Impulsive events are frequent, small, electron-, ^{3}He- and Fe-rich and they are, relative to gradual (or MPE) events, proton-poor. Yet none of these properties is correlated on a flare to flare basis. They have a small cone of emission. The high ionisation charge state suggests heating to about 2×10^7 K.

Gradual (or MPE) events, in contrast, are infrequent, large, and proton-rich. They exhibit coronal abundances in heavy ions. They originate over a large longitudinal (latitudinal?) extent of the corona. Their properties can be understood in terms of a CME-driven coronal/interplanetary shock (which may or may not be associated with a flare) that provides an acceleration source that evolves more gradually in space and time. The shock accelerates particles with abundances and ionisation states similar to the ambient coronal or solar wind plasma. This scenario is consistent with the lower charge state of the heavy ions.

Table 1. Characteristics of SEP events (adapted from Reames 1992)

	Impulsive	Gradual
Particles	electron-rich	proton-rich
	^{3}He/^{4}He ~ 1	~ 0.0005
	Fe/O ~ 1.0	~ 0.1
	H/He ~ 10	~ 100
Duration state	Q_{Fe} ~ +20	~ +13
Duration partic.event	hours	days
Longitude cone	<30°	~ 180°
Radio type	III, V (II)	II, IV
Duration soft x-rays	<1h	>1h
Coronagraph	-	CME (96%)
Solar Wind	-	IP shock
Flares/year	~ 1000	~ 10

It is important to note that virtually all of these characteristics of the SEP events and their associated electromagnetic emission form a continuum, with no clean division into a bimodal distribution. The one exception appears to be heavy ion abundances for which a bimodal distribution has been found (Reames et al. 1990). This lack of bimodality may be due to 'hybrid' events that do not fall clearly into either class of events, but exhibit both impulsive and gradual acceleration signatures. Examples are magnetically well-connected events associated with large long-decay soft x-ray events in which an Fe-rich period from the flare particles is followed by an Fe-poor period from shock-accelerated coronal material.

While most impulsive events are small, there are also some large impulsive events (Van Hollebeke et al. 1990). Examples are the unusual 3 June 1982 and 21 June 1980 SEP events that have impulsive (e.g. ^{3}He rich) ion abundances at E>50 MeV/n throughout the duration of the events and that are associated with shocks driven by narrow- angle CMEs, comparable to the SEP emission cone from impulsive flares. Kahler (1993) therefore concluded that these shocks may have extended over only a narrow coronal region through which impulsive flare SEPs were propagating and preferentially accelerated flare particles rather than ambient coronal material. There are also large impulsive events that do not necessarily show an enrichment in ^{3}He. Their smaller ^{3}He/^{4}He ratios may be a simple consequence of the depletion of the available ^{3}He in the flare region (Reames 1992).

About 1/3 of the large particle events are impulsive (Cane et al. 1986, Kallenrode et al. 1992) and are not accompanied by an interplanetary shock. Their energy can extend to rather high values: 7 out of 22 ground level events (GLEs) observed between 1976 and 1990 originated in impulsive flares (Kahler et al. 1991), and strong gamma-emission can result from impulsive flares (Bai 1986). Kallenrode et al. (1992) therefore suggested a more continuous classification between the small ^{3}He events ("pure impulsive") to the large "pure" gradual events. Although acknowledging the progress made and the importance of shocks for SEP acceleration, these authors raised a number of questions regarding the role of shocks, especially their role in the azimuthal transport of particles in large impulsive events (Kallenrode 1993a, see also section 4).

A terminology that was widely accepted for a long time (Wild, Smerd and Weiss 1963) was the existence of two "phases" of the acceleration, based on radio, x-ray and SEP observations. In the first (impulsive or flash) phase, stored magnetic energy would be converted into energetic electrons up to 100 keV. The second phase of acceleration, believed to occur only occasionally during large solar flares, would produce both ions and relativistic electrons, and be delayed by the order of 10 min with regard to first phase. The terminology "phases" of particle acceleration in solar flares was introduced because it was believed that the second phase could not occur without the first. SMM observations subsequently showed that the separation between low-energy (50 keV) electrons and energetic (>10 MeV) ions could be as short as 1 second (Forrest and Chupp 1983) and that gamma-ray lines (indicative of interacting protons) could be generated simultaneously with impulsive hard x-ray bursts (signature of electrons). These observations prompted various modifications of the classic two-phase paradigm. Cane et al. (1986) retained the basic two-phase picture but require that in certain intense events the traditional impulsive phase process now be capable of accelerating electrons to relativistic energies and protons to moderate (<40 MeV) energies. Other approaches invoked a "second step" (Bai 1986) or "intermediate" process (de Jager 1987) to bridge the gap between the traditional (<100 keV) impulsive phase and the shock-associated second phase. Both Bai's "second step" and de Jager's "intermediate" process occur within the flare impulsive phase. Today it may therefore be more appropriate to speak of two acceleration processes rather than phases: impulsive heating in the flare process and shock acceleration.

Whilst the origin of SEP events with moderate energies (i.e. ions up to several 100 MeV/n) appears to be reasonably understood, the question of the SEP acceleration to relativistic (GeV/n) ion energies, as manifested in GLEs is still debated. GLEs are generally large gradual events, but shock acceleration for the very energetic particles is not yet generally accepted. It is interesting to note that the 21 August 1979 GLE lacked a prominent impulsive flare phase, but was associated with a high-speed CME and a shock (Cliver et al. 1983).

3 Acceleration Models

Definitive theories for the origin of the energetic ions and electrons do not exist, primarily because of the complexity and variety of individual solar events and the absence of in situ observations which could discriminate among competing theories. Nevertheless, over the last few years a qualitively consistent picture has begun to emerge for the energetic particles in both impulsive and gradual events. A recent review of mechanisms for the acceleration of energetic particles in the solar corona was given by Melrose (1992).

Impulsive Events. Early models (e.g. Fisk 1978) for the origin of ^{3}He-rich events explained the enhancement as a 2-step process in terms of selected heating by resonant absorption of waves prior to acceleration. Theories of ion enhancement that require rare or unusual conditions of the pre-flare plasma would seem to be ruled out by the high frequency of these events (section 2).

Since particles from impulsive flares originate in the region of primary energy release, abundance variations provide a valuable diagnostic of the acceleration mechanism. The selectivity of this mechanism, especially in regard to He isotopes, points to a resonant process, such as gyroresonance with waves close to the cyclotron frequency of the enhanced ion. Although less spectacular than the ^{3}He enrichment, the Fe, Ne, Mg and Si enhancements must be explained as well.

A promising new model for the production of ^{3}He and other heavy ion enrichments in impulsive flares is the one proposed by Temerin and Roth (1992) since it relates, in a single step process, the ion enhancements to waves produced as a natural consequence of the streaming electrons that produce type III bursts. H$^+$ electromagnetic ion cyclotron waves excited by electron beams in a H- He plasma accelerate ^{3}He and heavy ions to MeV energies. The acceleration would occur in relatively low-density, high-magnetic field regions where the Alfven velocity (V_a) is large (few times 10^4km/s). At the sites of impulsive energy release, where the magnetic field is high, Va is expected to exceed the velocity of mass motions. In such an environment shocks will probably not develop. The authors argue that the presence of accelerated electrons in impulsive solar flares makes the physics of such flares analogous to that of the Earth's aurora.

This model has recently been expanded by Miller and Viñas (1993). They show that not only H$^+$ EM cyclotron waves, but also shear Alfven waves are driven unstable and that Fe ions (Fe^{20+}) will be selectively accelerated to MeV/n energies. The observed enhancements of Ne, Mg, and Si relative to CNO in the energetic particles from impulsive solar flares also appears to be accounted for by this model (Miller et al. 1993). The abundance ratio in impulsive and gradual events, i.e. the abundance enhancement in impulsive events over coronal abundances, is shown as a function of nuclear charge Z in Figure 4. The enhancement seems to be organised into three element groups, ^{4}He-O, Ne-S and Fe, with factor of 3 enhancements between consecutive groups. This grouping of the abundances is interpreted by Reames (1993b) in terms of the temperature of the preflare gas of about 3.5 MK where the elements

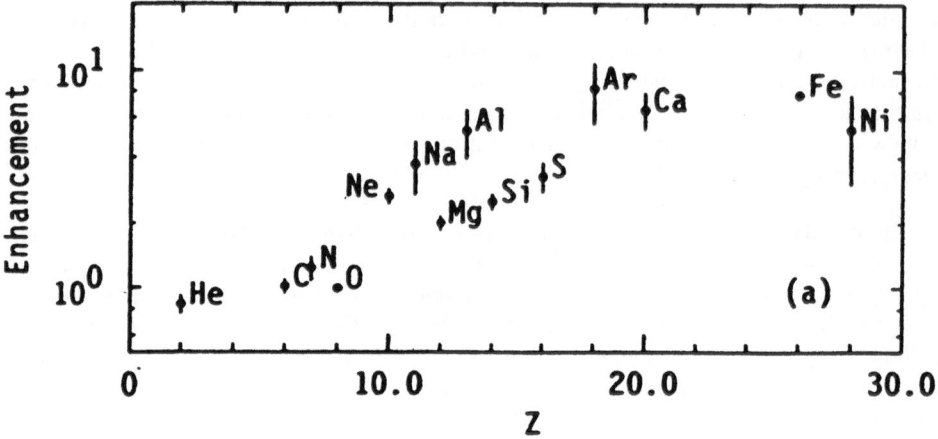

Fig. 4. Element enhancement in impulsive over gradual SEP events (Reames 1993b).

8   K.-P. Wenzel

up to O are essentially fully ionised with Q/A= 0.5, those in the Ne - S group are He-like (2 electrons left) with Q/A about 0.42 and Fe has a charge of +16 so that Q/A about 0.26. Ions with the same Q/A, thus the same gyrofrequency, absorb waves of that frequency and are similarly accelerated and enhanced.

Gradual Events. A consistent picture appears to emerge also for the energetic particles which eventually escape into interplanetary space after having been accelerated at a coronal shock. A fast CME, which may or may not be associated with a flare, drives a shock wave which spreads out through large portions of the corona at speeds of (1-2) × 10^3 km/s. The shock can accelerate ambient coronal plasma or, if there is a flare, energetic particles produced in the impulsive flare phase. While the shock is sufficiently strong it can continually "extract" electrons and ions from the shock-processed plasma along its evolving front and accelerate the electrons to about 1 MeV and the ions to about 1 GeV/n. Eventually the shock encounters open field lines, and a substantial number of accelerated particles escape the confining turbulence into interplanetary space. Coronal shock acceleration of particles resolves the mystery of an ad-hoc mechanisms invented 20 years ago to explain delay times of large events (see section 4).

Lee and Ryan (1986) modelled the acceleration and propagation of SEPs at an evolving coronal shock analytically. Event-averaged spectra and abundance ratios for a variety of ions in large SEP events have been successfully fitted by Mazur et al. (1992) to a stochastic acceleration model which assumes rigidity-dependent diffusion.

4 Coronal Transport

The subject of coronal propagation, or coronal diffusion (e.g. Perez-Pereza 1986, Wibberenz et al. 1989), was a result of the pioneering statistical studies of Reinhard and Wibberenz (1974) and Van Hollebeke et al. (1975) that examined the heliolongitude of optical flares that were associated with large particle events observed at 1 AU. At that time, not knowing about CMEs and not aware of the role of shock acceleration, it was shown that most of these events were associated with western hemisphere flares, and that the event rise times were fastest in a well connected region of about 40^0 width (fast propagation region, FPR) centred at the point (about W50), where the IMF lines connecting the Earth to the Sun intercept the high corona. Theoretical arguments and observations showed that particle populations were not generally able to diffuse across IMF lines and therefore made it unlikely that energetic particles from eastern hemisphere flares could reach the Earth by crossing IMF lines. Therefore the ad hoc concept of transport close to the sun, called coronal propagation, was introduced.

There is little direct evidence for this concept. A number of authors (e.g. Mewaldt and Stone 1983, Mason et al. 1984) have examined composition and spectral data in order to find evidence of propagation of the energetic particles through solar atmospheric material, all with uniformly negative results. The small emission cone of impulsive events imposes constraints on longitudinal diffusion with severe implications for large SEP events. Also, multi-spacecraft studies of the e/p ratio in SEP events show little longitudinal dependence of this ratio for the two species having rigidities differing by a factor of 10^2 - 10^3 (Wibberenz and Cane 1993).

Kallenrode (1993b) recently used the term 'azimuthal transport' independently of whether it is the result of extended (shock) acceleration or a real transport mechanism to refer to the fact that particle related to a flare or other processes (e.g. a shock wave) can be observed at large azimuthal distances. Her analysis supports the association of a FPR with the large-scale polarity cells of the coronal magnetic field, but leaves open the interpretation of "transport" outside the FPR either in terms of acceleration at a shock or some propagation mechanism, possibly involving a coronal shock.

Taking all these arguments, many researchers believe that "there is no process of coronal diffusion" (Reames 1992) and that the subject of coronal propagation has ceased to exist. Instead, the observations of large SEP events over a wide range of heliolongitude are thought to carry information about solar particle acceleration and shock propagation, and not about particle transport (Mason 1990).

5 High-energy Gamma-rays: Evidence for Storage of Energetic Particles in Coronal Loops

Recent spacecraft observations of high-energy neutral emissions from solar flares have renewed the interest in the relative importance of the roles played by SEP storage/trapping versus prolonged acceleration in gamma-ray associated flares. Gamma-rays and neutrons are produced by the interaction of flare accelerated charged particles with deeper layers of the solar atmosphere. Electron interactions produce hard x-ray and gamma-ray continuum which extends from 20 keV to over 100 MeV. Ion interactions produce nuclear gamma-ray lines, neutrons and pions.

The instruments on the Compton Gamma Ray Observatory (GRO) with their high sensitivity, together with the GAMMA-1 telescope, have observed solar flares over a wide range of gamma-ray energies in June 1991. For the 11 June flare event both the COMPTEL (1-30 MeV) and EGRET (50 MeV-2 GeV) instruments observed prolonged gamma-ray emission at energies extending into the GeV range. The emission observed by EGRET for an unusually long period of time (Kanbach et al. 1993), reveals the existence of π^0-decay radiation as late as 8 h after the impulsive flare phase.

Two possibilities are being discussed to explain these long duration gamma-ray emissions and the injection of high energy protons into interplanetary space. The first is continuous acceleration of protons for the duration of the event. The second is acceleration during the impulsive phase of the flare followed by trapping of energetic protons in coronal magnetic loops, for which two models have been suggested. Ryan and Lee (1991) propose proton trapping in a turbulent magnetic loop with second order stochastic acceleration. Mandzhavidze and Ramaty (1992) suggest that the protons responsible for the extended gamma-ray emission arise in the impulsive phase with trapping and precipitation in a relatively quiet coronal loop (see also Lau et al. 1993). In either case it is the diffusion of protons out of the loop which precipitate and produce the gamma-ray emission.

Figure 5 shows the decay time of the gamma-ray emission measured by EGRET and fits using the Mandzhavidze and Ramaty (1992) model with a combination of pion decay radiation and e-bremsstrahlung. The decay is strongly dominated by the pion decay. The fact that the bremsstrahlung decayed faster than the pion radiation is taken by the authors as a strong argument against continuous acceleration, although the latter cannot be excluded.

Extended (>2h) high-energy gamma emission was also detected from the 15 June 1991 flare. The energy spectrum of the extended emission, resulting from pion decay, allowed Mandzhavidze et al. (1993) to derive for the first time the spectrum of the accelerated particles at the Sun for energies up to a few GeV. Whilst these authors explain the prolonged gamma-ray emission by particle trapping, Kocharov et al. (1993) and Akimov et al. (1993) argue for continuous acceleration of protons and electrons in a wide range of energies, supported by time-coincident radio and gamma-ray emission profiles. More analysis of the complete and unique data set for this (and possibly other) event(s) is required to draw firm conclusions about the relative roles of SEP storage and acceleration in the solar corona.

6 Summary/Conclusions

In recent years we have seen major advances in our understanding of the relationship between solar flares and energetic particles and in distinguishing the sources and acceleration mechanisms for SEP associated with solar events of various kinds. This association has led to a significant revision of our understanding of both the origin of the particles and their transport out into the heliosphere.

Our current understanding may be summarised as follows:

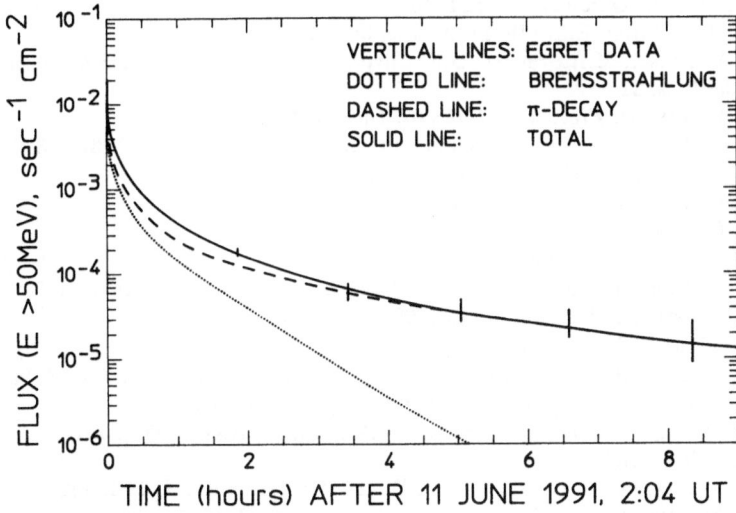

Fig. 5. Comparison of the gamma-ray flux profiles at energies greater than 50 MeV observed by EGRET with pion decay and e-bremsstrahlung calculations (after Mandzhavidze and Ramaty 1992).

1. Energetic ion populations from impulsive solar flares are characterised by ^{3}He-rich and Fe-rich abundances produced by resonant wave-particle interactions in the flare plasma.
2. Particle populations in most gradual events (or major proton events), are accelerated at CME-driven shocks, not in the flares. These populations show coronal abundances.
3. There is no process of "coronal propagation" of particles. Coronal shocks cross field lines to accelerate particles near the shock front.
4. Storage of high-energy ions in coronal magnetic loop structures may occur. Further analysis of combined neutral and charged particle observations is required to understand the relative roles of acceleration and storage of particles.

Acknowledgements. I am grateful to G. Simnett for the challenge of presenting this review at the European Meeting on Advances in Solar Physics. I thank him, as well as my colleagues R.G. Marsden, R. Reinhard, and T.R. Sanderson for comments on the draft manuscript.

References

Akimov, V.V. et al. (1993): *Proc. 23rd Internat. Cosmic Ray Conf.*, **3**, 111
Bai, T. (1986): *Astrophys. J.* **308**, 912
Cane, H.V., McGuire, R.E., von Rosenvinge, T.T. (1986): *Astrophys. J.* **301**, 448
Cane, H.V., Reames, D.V., von Rosenvinge, T.T. (1988): *J. Geophys. Res.* **93**, 9555
Cliver, E. et al. (1983): *Solar Phys.* **89**, 181

de Jager, C. (1987): *Proc. 20th Int. Cosmic Ray Conf.* **7**, 6

Fisk, L.A. (1978): *Astrophys. J.* **224**, 1048

Forrest, D.J. , Chupp, E.L. (1983): *Nature*, **305**, 291

Kahler, S.W., Shea, M.A., Smart, D.F. Cliver, E.W. (1991): *Proc. 22nd Internat. Cosmic Ray Conf.*, **3**, 21

Kahler, S.W. (1993): *Proc. 23rd Internat. Cosmic Ray Conf.* **3**, 1

Kallenrode, M-B., Cliver, E.W., Wibberenz, G. (1992): *Astrophys. J.* **391**, 370

Kallenrode, M-B. (1993a): *Adv. Space Res.* **13**, (9) 341

Kallenrode, M-B. (1993b): *J. Geophys. Res.* **98**, 5573

Kanbach, G. et al. (1993): *Astron. Astrophys. Suppl.* **97**, 349

Klein, L. (1993): this volume

Kocharov, G.E. et al. (1993): *Proc. 23rd Internat. Cosmic Ray Conf.*, **3**, 107

Lau, Y.-T, Northrop, T., Finn, J.M., (1993): *Astrophys. J.* **414**, 908

Lee, M.A., Ryan, J.M. (1986): *Astrophys. J.* **303**, 829

Luhn, A., Klecker, B., Hovestadt, D., Möbius, E. (1987): *Astrophys. J.* **317**, 951

Mandzhavidze, N., Ramaty, R. (1992): *Astrophys. J.* **396**, L111

Mandzhavidze, N., Ramaty, R., Akimov, V.V., Leikov, N.G. (1993): *Proc. 23rd Internat. Cosmic Ray Conf.*, **3**, 119

Mason, G.M., Gloeckler, G., Hovestadt, D. (1984): *Astrophys. J.* **280**, 902

Mason, G.M. (1990): *Proc. 21st Internat. Cosmic Ray Conf.*, **11**, 167

Mazur, J.E., Mason, G.M., Klecker, B., McGuire, R.E. (1992): *Astrophys. J.* **401**, 398

Mewaldt, R.A., Stone, E.C. (1983): *Proc. 8th Internat. Cosmic Ray Conf.*, **4**, 52

Melrose, D.B. (1992): in: *Particle Acceleration in Cosmic Plasma,* AIP Conf. Proc. **264**, p.3

Meyer, J.P. (1985): *Astrophys. J. Suppl.* **57**, 151

Miller, J.A., Viñas A.F. (1993): *Astrophys. J.* **412**, 386

Miller, J.A., Viñas, A.F., Reames, D. V. (1993): *Proc. 23rd Internat. Cosmic Ray Conf.* **3**, 17

Moses, D., Dröge, W., Meyer, P., Evenson, P. (1989): *Astrophys. J.* **346**, 523

Pallavacini, R., Serio, S., Vaiana, G.S. (1977): *Astrophys. J.* **216**, 108

Perez-Pereza, J. (1986): *Space Sci. Rev.* **44**, 91

Reames, D.V., von Rosenvinge, T.T., Lin, R.P. (1985): *Astrophys. J.* **292**, 716

Reames, D.V., Cane, H.V., von Rosenvinge, T.T. (1990): *Astrophys. J.* **357**, 259

Reames, D.V. (1992): in: *Particle Acceleration in Cosmic Plasmas,* AIP Conf. Proc. **264**, p. 213

Reames, D.V. (1993a): *Adv. Space Res.* **13**, (9) 331

Reames, D.V. (1993b): *Proc. 23rd Internat. Cosmic Ray Conf.* **3**, 388

Reinhard, R., Wibberenz, G. (1974): *Solar Phys.* **36**, 473

Ryan, J.M, Lee, M.A. (1991): *Astrophys. J.* **368**, 316

Temerin, M., Roth, I. (1992): *Astrophys. J.* **391**, L 105

Van Hollebeke, M.A.I., Mc Donald, F.B., Meyer, J.P. (1975): *Solar Phys.* **41**, 189

Van Hollebeke, M.A.I., Ma Sung, L.S., McDonald, F.B. (1990): *Astrophys. J. Suppl.* **73**, 285

Wibberenz, G. et al. (1989): *Solar Phys.* **124**, 353

Wibberenz, G., Cane, H. (1993): *Proc. 23rd Internat. Cosmic Ray Conf.* **3**, 274

Wild, J.P., Smerd, S.F., Weiss, A.A. (1963): *Ann. Rev. Astron. Astrophys.* **1**, 291

Electromagnetic Signatures of Particle Acceleration and Propagation

K.-L. Klein

Observatoire de Paris, Section de Meudon, DASOP, F-92195 Meudon,
France

Abstract: The solar atmosphere is a source of suprathermal particles during flares, but also over hours and even days outside flares. Diagnostics based on multi-wavelength and imaging observations (gamma-ray, hard X-ray, optical, radio wavelengths) are reviewed. During impulsive flares, where the most detailed information is available, acceleration occurs on sub-second time scales in extended regions of the corona. They comprise small-scale (some ") structures which are activated simultaneously or successively, produce different types of particle signature and also determine how particles get access to neighbouring or remote regions in the corona and eventually into interplanetary space. Particles accelerated outside the impulsive phase have qualitatively distinct signatures which are attributed to the environment where the acceleration or emissions occur, rather than to basically different acceleration mechanisms.

1 Introduction

The rapid energy release during flares produces particle populations up to relativistic energies. Signatures of energetic electrons are furthemore observed in the absence of flares, where they may persist over several days. How these particles are accelerated is not definitely known. Candidate processes were reviewed by Melrose (1990) and Vlahos (1993).

Two kinds of diagnostics of particle acceleration and propagation have emerged: the analysis of particle observations in interplanetary space has to account for propagation and re-acceleration between the site of primary acceleration and the satellite. Reviews were given by Reames (1993) and Wenzel (this conference). The electromagnetic signatures probe the particles closer to the site of primary acceleration, but involve complex radiation processes. Recent reviews were given by Zirin et al. (1991), Brown (1991), Simnett (1991), Mandzhavidze & Ramaty (1993) and Vilmer (1993). The present paper lays emphasis on multi-wavelength studies and imaging observations. Section 2.1 attempts to identify the regions where particles are accelerated, and discusses the spatial and temporal fine structure as far as it is accessible to the observations. Section 2.2 deals with peculiar aspects of prolonged particle signatures during and outside flares. The propagation through the corona and towards interplanetary space is addressed in Sect. 3.

2 Particle Acceleration during and outside Flares

2.1 Impulsive Phase Acceleration

2.1.1 Sites of Particle Acceleration

Electromagnetic signatures reveal the sites where accelerated particles interact with the ambient plasma. Hard X-ray (HXR) and microwave emission are due to incoherent processes, respectively bremsstrahlung and gyrosynchrotron radiation. These emissions give to a certain extent a measure of the total energy content of suprathermal particles and therefore are the most direct diagnostics of the acceleration process. A similar argument is valid for nuclear gamma-ray line emission. Since the radiation processes are particularly efficient in a medium with high ambient density and magnetic field strength, the emission is not necessarily brightest near the sites where the particles are accelerated. In the case of collective processes (plasma emission at radio wavelengths), an appropriate distribution function must first be established. This may be achieved only after the accelerated particles have propagated over a considerable distance. Under special circumstances the site of acceleration can nevertheless be inferred. For instance, in the case of electron beams directed motions are revealed by drifts of the bright, short-lived emission in the dynamic spectrogram ("type III" burst). The emission is ascribed to the conversion of electrostatic waves, excited by the electron beam, to transverse waves at the plasma frequency or its harmonic. The frequency of the radio burst can hence be directly used to infer the density of thermal electrons at the site of emission, and the drift gives the direction of motion with respect to the density gradient.

HXR and microwave observations localize the *bulk* of the energetic particles in closed magnetic structures in the low atmosphere, some 10^3 to 10^4 km above the photosphere. Such evidence comprises HXR (e.g. Matsushita et al. 1993) and microwave imaging (Alissandrakis et al. 1988; Shevgaonkar & Kundu 1985; Dulk et al. 1986), as well as stereoscopic observations of hard X-ray bursts from two satellites, where the lower portion of the flaring region is located behind the solar limb for one of them (Kane 1983). The observations are consistent with different types of particle populations in different regions of closed magnetic structures: particles precipitating into the dense footpoints of magnetic loops and others radiating around the summits of loop structures.

That the energetic particles are indeed accelerated in the low corona can be inferred from radio spectrographic observations of fast drift bursts. The observations show that at high frequencies, emitted in the low corona, drifts towards higher frequencies prevail, revealing downward propagating electron beams (Stähli & Benz 1987; Allaart et al. 1990; Benz et al. 1992). The opposite is true at lower frequencies ("ordinary" drift, i.e. Benz & Zlobec 1978; Elgarøy 1980; Benz et al. 1983; Aschwanden & Benz 1986). This demonstrates that during impulsive flares electron beams are predominantly produced in a plasma where the ambient electron density is in the range 10^9 to some 10^{11} cm^{-3}. This range is broadly consistent with the height of HXR and microwave sources.

2.1.2 Height Extent of the Acceleration Region

Both compact and height-extended regions of acceleration are revealed by different types of radiative signatures in impulsive flares. Aschwanden et al. (1993) identified a neat dividing frequency between ordinary and reverse drift type III bursts during two flares. The small frequency gap allows for a density variation of only 6% within the volume of acceleration. Very compact accelerating sources were also inferred from the analysis of a purely millimetric microwave burst (Kaufmann et al. 1985). The absence of centimetric emission implies that the electrons have access only to regions of high magnetic field strengths ($\gtrsim 1000$ G) or high ambient density ($\gg 10^{11}$ cm^{-3}). Furthermore, the emitting volume must be small, with a characteristic size below 1000 km (de Jager et al. 1987; Klein 1987). Since the electrons, if accelerated at a remote site in the higher atmosphere, would emit centimetric microwaves in travelling to the site of millimetric emission, it was concluded that the particles were accelerated in the compact volume in the low atmosphere.

On the other hand, a height-extended range of particle acceleration is suggested by the comparison of the electron spectra measured in interplanetary space and inferred for the electrons interacting in the solar atmosphere through a bremsstrahlung model (Pan et al. 1984). Electrons of a few keV accelerated in the low corona are thermalized before reaching interplanetary space. Therefore escaping low-energy electrons must be accelerated at heights $\gtrsim 0.5$ R$_\odot$ above the photosphere. The extent of the nonthermal electron population during an impulsive flare down to a few keV was shown by X-ray and radio observations of a flare far behind the solar limb (Kane et al. 1992; Krüger et al. 1992), where only emission from above $2 \cdot 10^5$ km was visible and the normally dominating thermal flare plasma was masked. Besides the model dependent argument of Pan et al. (1984), direct evidence for electron acceleration at ~ 1 R$_\odot$ was inferred from spectrographic and heliographic radio observations (Klein & Aurass 1993).

The indications for acceleration in compact volumes or, alternatively, over an extended height range, are not necessarily contradictory, but reveal physical differences between individual events. Evidence for extended structures is more abundant than for compact acceleration volumes. The magnetic field topology of the flaring active region probably plays an important role.

2.1.3 Spatial Fine Structure of Flaring Regions

Imaging observations of energetic processes during flares show a complex spatial pattern: several distinct sources are successively or simultaneously activated as the flare evolves, as shown at dm/m- wavelengths (Raoult et al. 1985; Trottet et al. 1993), in X-rays (Machado et al. 1988; Sakao et al. 1993), centimetre (Kundu et al. 1982; Kattenberg & Allaart 1983; Willson et al. 1990) and millimetre waves (Herrmann et al. 1993). Different structures may correspond to different particle signatures: Chupp et al. (1993) showed for an impulsive flare that the most energetic photons (4-6, 10-25 MeV) rose some tens of seconds after the HXR above 100 keV. At that time a new feature, distinct from the pre-existing flare ribbons, brightened in the chromosphere (Hα), and a new radio source appeared in the overlying corona. Similar observations were reported by Wülser et al. (1990).

The whole-Sun HXR and gamma-ray observations of Chupp et al. (1993) imply that the acceleration of high-energy particles is delayed with respect to low energies. The images show that the delay is not due to the gradually increasing efficiency of a single acceleration process nor to the second step acceleration of a previously energized seed population. The observed delay refers to the spread of the flaring activity to different sub-structures of the flaring region, rather than to the acceleration process. Such observations are not in conflict with a global organisation e.g. due to large-scale current systems (Mandrini et al. 1991). They suggest that within such a large-scale configuration the structure on smaller, but still observable, spatial scales determines the efficiency of particle acceleration. The observable complexity of this kind has not yet been properly considered in attempts to model the emission of accelerated particles. HXR imaging at high energies and the new millimetric observations with imaging capability (Lim et al. 1992; Herrmann et al. 1993) are a promising tool to provide new constraints.

2.1.4 Short Time Scales and the Acceleration Process

The most direct measure of the energy content of suprathermal particle populations is their HXR emission. Since impulsive HXR emission comes from a dense medium, the accelerated particles lose their energy rapidly (an electron of 100 keV in a plasma with density $\simeq 10^{11}$ cm^{-3} has a collisional lifetime $\simeq 1$ s). The HXR time profile is therefore expected to reveal the intrinsic time scales of impulsive particle acceleration. Subsecond fluctuations were found in a minority of cases ($\simeq 10$ %) at energies above 30 keV (Kiplinger et al. 1983) and were also reported at energies above 100 keV in a few single event studies (Desai et al. 1987). More sensitive new observations (GRANAT, COMPTON GAMMA RAY OBSERVATORY) frequently reveal such fluctuations up to energies above 100 keV (Machado et al. 1993; Talon et al. 1993). The GRANAT data studied so far show that they are a ubiquitous feature of impulsive hard X-ray bursts (Talon et al. 1993; Trottet, private comm.). The relevant time scales may even reach down to tens of milliseconds (Murphy et al. 1993). This emphasizes that a rapid acceleration process acts during the impulsive phase. Since simultaneous short-lived peaks from some tens of keV to several MeV have not been temporally resolved with 2 s (Chupp 1984; Kane et al. 1986) or 1 s resolution (Talon et al. 1993), particles must be accelerated from subrelativistic to relativistic energies on subsecond time scales.

Imaging observations with 0.1 s time resolution (Westerbork Synthesis Radio Telescope, Kattenberg & Allaart 1983) have also shown that microwave sources varying on sub-second scale are generally present during impulsive flares. The rapidly varying emission occurs in compact ($< 10''$), bright sources which shows that they are due to nonthermal electrons. Several distinct sites of emission were often identified. If the rapid fluctuations are signatures of the acceleration, these observations again suggest that several sites of acceleration can be distinguished within a flaring volume, using presently available spatial resolution.

2.1.5 Observational Support for a "Statistical" Flare Scenario ?

It is tempting to identify the observed rapid fluctuations of the HXR or microwave emission with individual injections of accelerated particles into the solar atmosphere, and to consider an impulsive flare as a multitude of such individual energy releases.

In a few cases this idea could be substantiated because individual pulses were associated with well-identified type III bursts, i.e. electron beams (Dennis et al. 1984; Raoult et al. 1989; Aschwanden et al. 1993). The inferred durations of the elementary injections were ≃0.4 s. The hypothesis was further investigated through correlation studies between the rate of fluctuations and the intensity of wide-band continuum emission. Microwave observations were used by Correia & Kaufmann (1987), HXR and decimetric observations by Benz (1985), Aschwanden et al. (1990) and Aschwanden & Güdel (1992). A review was given by Benz & Aschwanden (1992).

Aschwanden et al. (1990) and Aschwanden & Güdel (1992) found a correlation between the count rate of moderately hard X-ray bursts ($h\nu \sim 30$ keV) and the rate of type III bursts or decimetric spikes. They interpreted this as evidence for elementary injections of accelerated particles with a given amount of energy per injection. The correlation of type III burst rate with the HXR counts is more convincing during the brightest phase of the impulsive emission than during either the early rise or the decay phase. In the light of Sect. 2.1.3 the local physical conditions at different sites of the flaring region may play a role. For example, the local density structure affects the HXR response to a beam, while the magnetic field topology determines whether a given site of acceleration is connected with open field lines which allow electron beams to escape (cf. Sect. 3).

It is not clear what exactly the observed correlations of HXR count rates with two different species of coherent emission tells us: type III bursts are thought to trace individual electron beams in these studies, but spike bursts also trace some kind of elementary energy injection, on ten times shorter time scales than type III bursts. The mechanism of spikes is less well understood than type III emission. Vlahos & Raoult (this conference) demonstrated that elementary beam injections into a fibrous corona generate both typical type III bursts and spike-like radio features. Hence part of the short time scales might be generated by the small-scale structure of the ambient corona, rather than the acceleration process.

2.2 Energetic Particle Signatures outside Impulsive Flares
2.2.1 Properties of Gradual Hard X-Ray/Radio Bursts

The original definition of gradual (or "extended") flares stems from soft X-ray observations (Pallavicini et al. 1977): gradual flares are long lasting, slowly evolving emissions from more extended sources, located at greater altitudes, than impulsive flares. They are often associated with coronal mass ejections. Electromagnetic signatures of energetic particles in gradual events were characterized by the smoothly evolving HXR emission (Frost & Dennis 1971), which generally displays a hardening photon spectrum in the course of the event. Such HXR/radio events are accompanied by extended soft X-ray bursts (Bai 1986). Reviews of the hard X-ray characteristics of such events were given by Tanaka (1987) and Dennis (1988). X-ray imaging at moderately hard energies ($h\nu \gtrsim 20$ keV) shows extended sources in the corona at heights of several 10^4 km, similar to the soft X-ray sources. The ambient electron density is of the order of 10^9 cm^{-3} (Tanaka 1987). When the photospheric magnetic field is extrapolated to those altitudes, the X-ray sources are found near the tops of extended magnetic loops (Sakurai 1985). Metric continuum bursts ("type IV", Pick 1986) indicate the time-extended injection of electrons into closed magnetic structures in the middle corona (height $\simeq 0.5$ $R_\odot$). Gradual phase emission of energetic particles hence occurs in a more dilute plasma with weaker magnetic field than the impulsive phase. The weak magnetic field also implies a lower turnover frequency of the microwave spectrum than in impulsive flares, and this is indeed observed.

Gradual soft or HXR/radio bursts probably do not form a separate class of flares. Harrison (1991) showed that the duration of soft X-ray brightenings does not display a bimodal distribution. Nakajima (1983) observed two co-existing, spatially distinct microwave sources with different behaviour: a slowly evolving one with time profile similar to $\gtrsim$ 700 keV photons, and a typically impulsive source which followed the evolution of the HXR ($\gtrsim$ 60 keV) emission. The gradual component of the event had a lower spectral turnover than the impulsive one, which points again to different magnetic field strengths. The observations suggest that following the degree to which regions with different density and magnetic field strength contribute to a given flare, its HXR/radio aspect will be either more impulsive or more gradual.

2.2.2 Evidence for a Distinct "Gradual" Acceleration Process ?

Two pieces of evidence were often quoted for a distinct acceleration process: 1. the hardening photon spectrum and the microwave richness of gradual HXR/radio events, i.e. the relatively high microwave flux density with respect to the HXR flux (Kai et al. 1985a; Bai 1986; Cliver et al. 1986); 2. the association with metric type II bursts, i.e. large-scale coronal shock waves (Frost & Dennis 1971; Ramaty et al. 1980). Furthermore, a smaller electron-to-proton ratio was found in the interplanetary space during gradual than during impulsive flares (e.g. Reames 1993).

The relevance of these features to the primary acceleration process in the corona is controversial. Hardening X-ray spectra and time delays between electromagnetic signatures of particles at different energies can be explained by the time-extended injection of protons and electrons into coronal traps where they undergo Coulomb collisions (Vilmer et al. 1982; Hulot et al. 1992; Bruggmann et al. 1993). The microwave richness compared with the impulsive phase is expected: at a given frequency gyrosynchrotron emission comes from electrons with higher energies when the magnetic field is weak than when it is strong. Therefore MeV electrons would be invisible at centimetre wavelengths during the impulsive phase, while contributing largely to the gradual phase emission. The "gradual" HXR and microwave signatures therefore do not imply a more efficient acceleration of electrons to high energies than during the impulsive phase. A causal relationship between the large-scale coronal shock wave and the gradual HXR/radio emission is also not confirmed by a closer inspection of the data (Klein et al. 1983; Kahler 1984; Trottet 1986; Cliver et al. 1986; Klein et al. 1988; Klein 1992).

Ramaty et al. (1993) derived spectra and total numbers of electrons and protons from the gamma-ray continuum in the range 0.3-1 MeV, assumed to be bremsstrahlung of relativistic electrons, and from the nuclear line emission. They did not find any systematic difference between impulsive and gradual flares as to the ratio of electron to proton fluxes, and concluded that the gamma-ray producing particles in both impulsive and gradual flares are accelerated by the same mechanism.

The short time scales characteristic of the impulsive phase are not found in long-lasting HXR events (Talon et al. 1993). Given the dilute ambient medium, considerably longer collisional lifetimes of the energetic particles are expected in the gradual source, such that even if the acceleration operated on sub-second scales, individual particle injections would be indistinguishable. Rapidly-varying radio emission does exist during gradual flares. While dm/m-λ type III bursts are in general not detected, at altitudes $\gtrsim$10 $R_\odot$ hectometric type III groups were found (Klein & Trottet 1993). Electron beams are thus also produced during gradual events. Rapidly-varying collective radio emissions are also observed at dm/m-λ (e.g. Chernov et al. 1975; Kuijpers 1980; Aurass et al. 1987; Mann et al. 1987; Aurass & Kliem 1992). It is not clear presently which of the spectral features refer to the acceleration process and which to the leaking of particles from coronal traps.

In summary, rather than referring to two separate classes of flares the terms "impulsive" and "gradual" seem to refer to extreme examples of a continuous distribution, which at least partly reflects the properties of the environment where the emissions occur. It seems plausible that a given acceleration mechanism evolves differently in different environments, but the observations or their interpretation have not yet attained a sufficient degree of sophistication. It is recalled that during the impulsive phase electron beam acceleration also occurs in the dilute plasma of the middle corona (cf. Sect. 2.1.2), without losing its "impulsive" character.

2.2.3 Particle Acceleration in the Late Phase or in the Absence of Flares

Suprathermal particles are also observed in active regions up to hours after or in the absence of flares. The oldest evidence stems from metric radio observations. For example, metric type III bursts were shown to be generated in relation to mass motions in the low solar atmosphere (Chiuderi-Drago et al. 1987). Storm continua and noise storms (cf. reviews by Kai et al. 1985b; Pick 1986) reveal the acceleration of electrons up to several tens of keV, over durations ranging from tens of minutes to several days. Noise storms are the most frequent manifestation of solar activity at metric wavelengths and demonstrate that particle acceleration with a low rate of energy release is a common process in the active corona.

In the interplanetary medium the most frequently detected suprathermal electron populations are beams with typical particle energies below 10 keV (Lin 1985, 1993), which are most often associated with decametric type III storms (e.g. Melrose & Dulk 1987). The extrapolation of the footpoints of the interplanetary field line which connects the electron streams with the corona suggests that metric noise storms in the middle solar corona are related to such events (Kayser et al. 1987).

A more dramatic manifestation of post-gradual energetic particle populations has been recently identified by gamma-ray observations: protons of some tens of MeV to $\sim$ 100 MeV can be stored or accelerated in the solar corona during several hours after flares (Leikov et al. 1993; Kanbach et al. 1993). The temporal evolution of the gamma-ray emission can be fitted by a trapped population of high-energy protons (Mandzhavidze & Ramaty 1992). Since long-lasting microwave emission accompanies some of these events (Akimov et al. 1993), acceleration over durations comparable to those of the metric storm continua is a plausible alternative.

Multifrequency observations of the coronal and active region plasma show that like flares or coronal mass ejections, the noise storms are related to the re-structuring of the plasma-magnetic field configuration (Švestka et al. 1982; Kerdraon et al. 1983; Stewart et al. 1986; Raulin et al. 1991; Raulin & Klein 1993). It is premature to force the long-lasting gamma-ray emissions and the metric/decametric storms into a single scenario. Nevertheless, the ensemble of these observations shows that the solar atmosphere does provide long-lived sources (~1 h) of suprathermal and even mildly relativistic particles for the heliosphere. Judging from the noise storm observations, the acceleration is part of the coronal response to the changing topology of the magnetic field in the course of the evolution of an active region. The main difference with an impulsive flare could be the rapidity by which the configuration changes and releases energy. Metre-wave storms share some phenomenologic aspects with flares, such as the formation of electron beams (which occur in the high corona during noise storms) and of short-lived, narrow-band bursts (≲1 s, 1 MHz) which accompany the continuum emission. These features deserve a deeper investigation of the similarities and differences of particle acceleration in different situations of the coronal evolution. This meets the motivation from theoretical work (Vlahos 1993; Einaudi, this conference) to consider phenomena on different energetic scales within a common framework.

3 Propagation of Energetic Electrons through the Corona and towards Interplanetary Space

The propagation of electrons over distances much greater than the dimension of an active region is demonstrated by radio observations. Electron beams propagating along open magnetic field lines can be tracked from the low corona to the Earth's orbit (cf. Pick & van den Oord 1990; Dulk 1990). Beams are also guided along closed magnetic structures from one active region to another (Labrum & Stewart 1970; Sheridan et al. 1973; Simnett 1982; Nakajima et al. 1985). This demonstrates that electrons can rapidly propagate within extended volumes of the corona, corresponding to cones of injection into the interplanetary space with opening angles of several tens of degrees (e.g. Pick & Ji 1986).

Whether or not particles gain access to magnetic field lines leaving the immediate vicinity of the acceleration site depends on the topology of the magnetic field near the site(s) of acceleration. Axisa (1974) found that a flare at the border of an active region has a significantly greater probability of being accompanied by type III bursts than when it occurs well in the interior of the region. He concluded that the electrons have easier access to open magnetic field lines when the acceleration occurs at the border of the active region. The process is a dynamic phenomenon which can change during a flare. Lantos et al. (1984), using imaging observations at centimetre and metre wavelengths, showed that minor changes (≲ 10″) of the structure of the flaring active region imply access to widely diverging open magnetic field lines in the solar corona. Such changes occur within some seconds to tens of seconds. Although these observations refer only to electrons and to the impulsive flare phase, they underline the general importance of the magnetic field topology on scales smaller than some arc seconds in giving energetic particles access to different sites in the solar atmosphere and to interplanetary space.

Coronal electron transport perpendicular to magnetic field lines is not apparent in the radio observations. Sources of metric noise storm emission stay confined over their lifetime for often several days, with no sign that the energetic electrons diffuse into the ambient corona over distances comparable with the opening angles cited above. A significant influence of the large-scale type II associated coronal shock waves on the transport of electrons is also not demonstrated: during the few gradual HXR/radio bursts where a joint study of HXR to hectometric radio emission was undertaken, evidence of electron beams propagating along open field lines to the high corona ($\gtrsim$10 $R_\odot$; Klein & Trottet 1993) was found, but none for the acceleration by the type II shock. It is important to keep in mind that no direct diagnostics of ion propagation in the corona are available. Conclusions from interplanetary ion observations which contradict those exposed here (e.g. Zirin et al. 1991; Reames 1993) may therefore refer to processes which are not detectable by the radio observations of the corona. However, where such conclusions are based on associations with coronal shock waves traced by type II emission, they deserve a critical re-consideration (cf. also the discussion by Simnett 1986).

4 Summary

Observations of electromagnetic signatures of particle acceleration and transport in the solar atmosphere have been reviewed, focussing on multi- wavelength studies with high spectral, temporal and spatial resolution. The results are summarized as follows:

1. Impulsive phase acceleration occurs in the corona, at heights ranging from a few 10^3 kms to about 1 $R_\odot$ above the photosphere. The bulk of the energetic particles seen during an impulsive flare are accelerated in the low corona, where the ambient electron density ranges between 10^9 and 10^{11} cm^{-3}. Significant acceleration especially of low-energy particles escaping to interplanetary space (e.g. electrons of a few keV) must also occur at greater altitudes during the same events.

2. Within individual flares, compact volumes of particle acceleration ($\leq 10^3$ km) have been inferred. Various observationally distinguishable elements of the flaring volume produce different particle populations in terms of energy and probably distribution functions.

3. Small-scale structures ($\leq$ some ") at and around the site(s) of acceleration govern the access of energetic electrons to more extended structures and eventually determine their ability to propagate over long distances in the corona and towards interplanetary space. The small-scale structures likely reveal the local magnetic field organization.

4. The activation of different sub-structures of a flaring volume and the access to different structures in the overlying corona can vary on time scales of some seconds. Particles can be accelerated on time scales $\leq$ 1 s up to relativistic energies. The observations are consistent with the idea that a variety of sites of acceleration contribute to build an impulsive flare.

5. Suprathermal particle populations may exist in the corona over hours after flares or in the absence of flares. This demonstrates that particle acceleration is a counterpart of coronal magnetic field evolution that is not restricted to the energy release during flares.

6. The coronal signatures of energetic particles do not reveal a clear difference between "impulsive" and "gradual" particle acceleration. The basic difference

appears related to the environment where the emission, and perhaps the acceleration, occur: gradual phase HXR and microwave sources are more dilute and have weaker magnetic fields than impulsive sources. Different gradual acceleration processes, e.g. by extended coronal shock waves, were not confirmed by more recent analyses.

Point 6 is in apparent conflict with interplanetary particle observations. The gradual signatures in interplanetary space are observed in the ion populations, while the electromagnetic signatures in the corona are more relevant to electrons. The different findings are also consistent with the idea that the interplanetary signatures are generated by acceleration processes at several $R_\odot$ above the photosphere and in the inner heliosphere, e.g. by shock waves in front of coronal mass ejections (e.g. Reames 1993). The understanding why these disturbances do not generate observable particle acceleration in the lower corona deserves some investigation.

Acknowledgements : The author is grateful to the many colleagues who made preprints of their work available. He acknowledges helpful discussions with G. Trottet, N. Vilmer and L. Vlahos, as well as G. Simnett's comments on the manuscript. He thanks the organizers of the Catania conference for their work.

References

Akimov, V.V., Leikov, N.G., Belov, A.V. et al. (1993): in Proc. Conf. High Energy Physics of Solar Flares, ed. by J. Ryan, in press

Alissandrakis, C.E., Schadee, A., Kundu, M.R. (1988): A&A **195**, 290

Allaart, M., van Nieuwkoop, J., Slottje, C., Sondaar, L.H. (1990): Solar Phys. **130**, 183

Aschwanden, M., Benz, A.O. (1986): A&A **158**, 102

Aschwanden, M.J., Benz, A.O., Schwartz, R. (1993): ApJ, in press

Aschwanden, M.J., Benz, A.O., Schwartz, R.A. et al. (1990): Solar Phys. **130**, 39

Aschwanden, M.J., Güdel, M.: (1992), ApJ **401**, 736

Aurass, H., Chernov, G.P., Karlicky, M., Kurths, J., Mann, G. (1987): Solar Phys. **112**, 347

Aurass, H., Kliem, B. (1992): Solar Phys. **141**, 371

Axisa, F. (1974): Solar Phys **35**, 207

Bai, T. (1986): ApJ **308**, 912

Benz, A.O. (1985): Solar Phys. **96**, 357

Benz, A.O., Aschwanden, M. (1992): in Eruptive Solar Flares, IAU coll. 133, ed. by B.V. Jackson, M.E. Machado, Z.F. Švestka (Springer, Berlin), p. 106

Benz, A.O., Bernold, T.E.X., Dennis, B.R. (1983): ApJ **271**, 355

Benz, A.O., Magun, A., Stehling, W., Su, H. (1992): Solar Phys. **141**, 335

Benz, A.O., Zlobec, P. (1978): A&A **63**, 137

Brown, J.C. (1991): Phil. Trans. R. Soc. Lond. A **336**, 413

Bruggmann, G., Vilmer, N., Klein, K.-L., Kane, S.R. (1993): Solar Phys., in press

Chernov, G.P., Korolev, O.S., Markeev, A.K. (1975): Solar Phys. **44**, 435

Chiuderi-Drago, F., Mein, N., Pick, M. (1987): Solar Phys. **103**, 235

Chupp, E.L. (1984): Ann. Rev. Astron. Astrophys. **22**, 359

Chupp, E.L., Trottet, G., Marschhäuser et al. (1993): A&A, in press

Cliver, E.W., Dennis, B.R., Kiplinger, A.L. et al. (1986): ApJ **305**, 920

Correia, E., Kaufmann, P. (1987): Solar Phys. **111**, 143

de Jager, C., Kuijpers, J., Correia, E., Kaufmann, P. (1987): Solar Phys. **110**, 317

Dennis, B.R. (1988): Solar Phys **118**, 49

Dennis, B.R., Benz, A.O., Ranieri, M., Simnett, G.M. (1984): Solar Phys. **90**, 383

Desai, U.D., Kouveliotou, C., Barat, C. et al. (1987): ApJ **319**, 567

Dulk, G.A. (1990) Solar Phys **130**, 139

Dulk, G.A., Bastian, T.S., Kane, S.R. (1986): ApJ **300**, 438

Elgarøy (1980): A&A **82**, 308

Frost, K.J., Dennis, B.R. (1971): ApJ **165**, 655

Harrison, R.A. (1991): Adv. Space Res. **11** no. 1, 25

Herrmann, R., Rolli, E., Correia, E., Costa, J.E.R. (1993): Solar Phys., in press

Hulot, E., Vilmer, N., Chupp, E.L., Dennis, B.R., Kane, S.R. (1992): A&A **256**, 273

Kahler, S. (1984): Solar Phys. **90**, 133

Kai, K., Kosugi, T., Nitta, N. (1985a): Publ. Astron. Soc. Japan **37**, 155

Kai, K., Melrose, D.B., Suzuki, S. (1985b): in Solar Radiophysics, ed. by D.J. McLean, N.R. Labrum (Cambridge Univeristy Press, Cambridge), p. 4125

Kanbach, G., Bertsch, D.L., Fichtel, C.E. et al. (1993): A&A Suppl. **977**, 349

Kane, S.R. (1983): Solar Phys. **86**, 355

Kane, S.R., Chupp, E.L., Forrest, D.J., Share, G.H., Rieger, E. (1986): ApJ **300**, L95

Kane, S.R., McTiernan, J., Loran, J. et al. (1992): ApJ **390**, 687

Kattenberg, A., Allaart, M. (1983): ApJ **265**, 535

Kaufmann, P., Correia, E., Costa, J.E.R., Zodi Vaz, A.M., Dennis, B.R. (1985): Nature **313**, 380

Kayser, S.E., Bougeret, J.-L., Fainberg, J., Stone, R.G. (1987): Solar Phys. **109**, 107

Kerdraon, A., Pick, M., Trottet, G. et al. (1983): ApJ **265**, L19

Kiplinger, A.L., Dennis, B.R., Emslie, A.G., Frost, K.G., Orwig, L.E. (1983): ApJ **265**, L99

Klein, K.-L. (1987): A&A **183**, 341

Klein, K.-L. (1992): in Proc. Conf. Solar Wind Seven, ed. by E. Marsch, R. Schwenn (Pergamon Press, Oxford), p. 635

Klein, K.-L., Aurass, H. (1993): Adv. Space Res. **13** no. 9, 295

Klein, K.-L., Trottet, G. (1993): in Proc. Conf. High Energy Physics of Solar Flares, ed. by J. Ryan, in press

Klein, K.-L., Trottet, G., Benz, A.O., Kane, S.R. (1988): in Plasma Astrophysics, ESA-SP 285, Vol.1, 157

Klein, L., Anderson, K., Pick, M., Trottet, G., Vilmer, N., Kane, S.R. (1983): Solar Phys. **84**, 295

Krüger, A., Hildebrandt, J., Kliem, B. et al. (1992): Solar Phys. **134**, 171

Kuijpers, J. (1980): in Radio Physics of the Sun, IAU Symp. no. 86, ed. by M.R. Kundu, T.E. Gergely (D. Reidel, Dordrecht), p. 341

Kundu, M.R., Schmahl, E., Velusamy, T., Vlahos, L. (1982): A&A **108**, 188

Labrum, N.R., Stewart, R.T. (1970): Proc. Astron. Soc. Australia **1**, 7

Lantos, P., Pick, M., Kundu, M.R. (1984): ApJ **283**, L71

Leikov, N.G., Akimov, V.V., Volzhenskaya, V.A. et al. (1993): A&A Suppl. **97**, 345

Lim, J., White, S.M., Kundu, M.R., Gary, D.E. (1992): Solar Phys. **140**, 343

Lin, R.P. (1985): Solar Phys. **100**, 537

Lin, R.P. (1993): Adv. Space Res. **13** no. 9, 265

Machado, M.E., Moore, R.L., Hernandez, A.M. et al. (1988): ApJ **326**, 425

Machado, M.E., Ong, K.K., Emslie, A.G. et al. (1993): Adv. Space Res. **13** no. 9, 175

Mandrini, C., Démoulin, P., Hénoux, J.C., Machado, M. (1991), A&A **250**, 541

Mandzhavidze, N., Ramaty, R. (1992): ApJ **396**, L111

Mandzhavidze, N., Ramaty, R. (1993): Nuclear Phys., in press

Mann, G., Karlicky, M., Motschmann, U. (1987): Solar Phys. **110**, 381

Matsushita, K., Masuda, S., Kosugi, T., Inda, M., Yaji, K. (1993): Publ. Astron. Soc. Japan **44**, L89

Melrose, D.B. (1990): Aust. J. Phys. **43**, 703

Melrose, D.B., Dulk, G.A. (1987): Physica Scripta **T18**, 29

Murphy, R.J., Share, G.H., Grove, J.E. et al. (1993): in Proc. Compton Symposium, in press

Nakajima, H. (1983): Solar Phys. **86**, 427

Nakajima, H., Dennis, B.R., Hoyng, P. et al. (1985): ApJ **288**, 806

Pallavicini, R., Serio, S., Vaiana, G.S. (1977): ApJ **216**, 108

Pan, L.-D., Lin, R.P., Kane, S.R. (1984): Solar Phys. **91**, 345

Pick, M. (1986): Solar Phys. **104**, 19

Pick, M., Ji, S.C. (1986): Solar Phys **107**, 159

Pick, M., van den Oord, G.H.J. (1990): Solar Phys **130**, 83

Ramaty, R., Colgate, S.A., Dulk, G.A. et al. (1980): in Solar Flares, ed. by P.A. Sturrock (Colorado Assoc. Univ. Press, Boulder), p. 117

Ramaty, R., Mandzhavidze, N., Kozlovsky, B., Skibo, J. (1993): Adv. Space Res. **13** no. 9, 275

Raoult, A., Correia, A., Lantos, P. et al. (1989): Solar Phys. **120**, 125

Raoult, A., Pick, M., Dennis, B.R., Kane, S.R. (1985): ApJ **299**, 1027

Raulin, J.P., Klein, K.-L. (1993): A&A, in press

Raulin, J.P., Willson, R.F., Kerdraon, A. et al. (1991): A&A **251**, 298

Reames, D.V. (1993): Adv. Space Res. **13** no. 9, 331

Sakao, T., Kosugi, T., Masuda, S. et al. (1993): Publ. Astron. Soc. Japan **44**, L83

Sakurai, T. (1985): Solar Phys. **95**, 311

Sheridan, K.V., McLean, D.J., Smerd, S.F. (1973): Ap Letters **15**, 139

Shevgaonkar, R.K., Kundu, M.R. (1985): ApJ **292**, 733

Simnett, G.M. (1982): ApJ **255**, 721

Simnett, G.M. (1986): Solar Phys. **104**, 67

Simnett, G.M. (1991): Phil. Trans. R. Soc. Lond. A **336**, 439

Stähli, M., Benz, A.O. (1987): A&A **175**, 271

Stewart, R.T., Brueckner, G.E., Dere, K.P. (1986): Solar Phys. **106**, 107

Švestka, Z., Dennis, B.R., Pick, M. et al. (1982): Solar Phys. **80**, 143

Talon, R., Barat, C., Dézalay, J.-P. et al. (1993): Adv. Space Res. **13** no. 9, 171

Tanaka (1987): Publ. Astron. Soc. Japan **39**, 1

Trottet, G. (1986): Solar Phys. **104**, 145

Trottet, G., Vilmer, N., Barat, C. et al. (1993): Adv. Space Res. **13** no. 9, 298

Vilmer, N. (1993): Adv. Space Res. **13** no. 9, 221

Vilmer, N., Kane, S.R., Trottet, G. (1982): A&A **108**, 306

Vlahos, L. (1993): Adv. Space Res. **13** no. 9, 161

Willson, R.F., Klein, K.-L., Kerdraon, A., Lang, K.R., Trottet, G. (1990): ApJ **357**, 662

Wülser, J.-P., Canfield, R.C., Rieger, E. (1990): in Proc. 3rd Max '91 Workshop, ed. by R.M. Winglee, A.L. Kiplinger (Boulder, Univ. Colorado), p. 149

Zirin, H., MacKinnon, A.L., McKenna-Lawlor, S.M.P. (1991): in Solar Interior and Atmosphere, ed. by A.N. Cox, W.C. Livingston, M.S. Matthews (Univ. Arizona Press, Tucson), p. 964

Proton Acceleration in Long Duration Flares

J. M. Ryan,   E. Bennett, and M. A. Lee

Space Science Center, University of New Hampshire, Durham, NH 03824,
USA

Abstract: The acceleration and trapping of protons necessary to explain long duration, high energy, gamma-ray solar flares is modeled as a single process within a coronal loop, where an injected proton spectrum is transported and accelerated by MHD turbulence. The loop is modeled as a leaky box where second-order Fermi-acceleration is a result of the turbulence. Protons accelerated in the coronal loop escape from the ends of the loop (precipitation) into the chromosphere or lower corona where they produce gamma-ray emission. The model provides a method for computing the proton spectrum as it evolves from an initial power law under coupled spatial and momentum diffusion. With reasonable coronal parameters the model yields time-extended relativistic proton precipitation, which peaks up to several hundred seconds after injection. Momentum diffusion is inversely proportional to spatial diffusion for an assumed Alfvén wave field within the coronal loop. Assuming a magnetic field on the order of 100 G, $\delta B/B$ need only be a few percent if the loop is 10^5 km in length in order to describe the temporal behavior of the high-energy gamma-ray flare of 3 June 1982.

1 Introduction

Long duration, high energy, solar gamma-ray flares can be characterized by extended (> 5 minutes) gamma-ray emission > 50 MeV, due, at least in part, to the production and subsequent decay of neutral and charged π mesons. The first clear example of such an event was the solar flare of 3 June 1982 (Chupp et al. 1987). After an impulsive phase lasting ~ 60 s, the nuclear line and bremsstrahlung gamma-ray spectral components decayed significantly. A relatively quiet period followed for ~ 100 s, after which high-energy gamma rays > 25 MeV were detected followed by high-energy neutrons. The gamma-ray emission has been interpreted by Forrest et al.(1986) as due to π-meson decays, both charged and neutral. Chupp et al.(1987) concluded from the neutrons measured by the Gamma Ray Spectrometer (GRS) and those measured at ground level (Debrunner et al. 1983) that proton acceleration and precipitation persisted for > 10 minutes after the impulsive phase. The *Compton Gamma Ray Observatory* has observed similar high-energy flares occurring on 4, 9, 11 and 15 June 1991 from Active Region 6659 (Murphy et al. 1993, Kanbach et al. 1993, and McConnell 1993). The longest measured duration (8 hours) is that of the 11 June 1991 flare (Kanbach et al. 1993). Gamma-ray and neutron emission in each case has been attributed to proton precipitation delayed relative to the impulsive phase.

The fundamental question is whether protons produced during the impulsive phase continue to be accelerated for extended periods, yielding the high-energy protons which characterize the gradual phase. If no new protons are injected into the system, a fraction of the proton population accelerated during the impulsive phase must be trapped in order to produce delayed gamma rays, placing greater requirements on the impulsive phase acceleration. Either this population bounces back and forth in a large coronal structure, much like protons in the earth's radiation belts, gradually escaping the structure through weak pitch angle diffusion before they drift out of the region altogether; or sufficient turbulence exists in the coronal structure after the occurrence of a large flare so that the protons escape by spatial diffusion while being accelerated by the second-order Fermi process. In the former case the scattering mean-free path λ is much longer than the loop length L, while in the latter case $\lambda \ll L$. The precise nature of the gamma ray and neutron emission is dependent upon the details of the precipitating proton population (Lingenfelter 1993), but the time profile of the prolonged high-energy gamma radiation should follow that of the precipitating high-energy protons. Using the former approach Mandzhavidze and Ramaty (1992a) have modeled the solar flare of 11 June 1991 with a decay timescale of $\sim 10^4$ s (Kanbach $et\ al.$ 1993). Their analysis places an upper limit on the turbulence within a large loop of $\delta B/B < 10^{-4}$ with $\lambda > 10^{16}$ cm.

Alternatively we have modeled the high-energy signature of these flares by the latter approach. An earlier model (Ryan and Lee 1991) assumed a high level of turbulence in the form of Alfvén waves in a coronal loop with $L = 10^5$ km, giving rise to energy-independent spatial diffusion of protons (Lee 1993). The associated momentum diffusion is then proportional to p^2, where p is the proton momentum. Ryan and Lee (1991) calculated the proton precipitation rate out of the loop assuming monoenergetic injection impulsive in space and time. These first results described both a prompt precipitation and the acceleration of a fraction of the protons to high energies, resulting in a delayed peaking of the high energy proton precipitation.

In this paper we explore a simpler description of the same process by assuming that the protons are trapped in a classical leaky box. This description dispenses with the details of the spatial transport and therefore does not resolve the time structure for times less than the initial characteristic escape time. However, it does adequately describe the high-energy precipitation above the π-meson production threshold and accommodates a more realistic injection spectrum.

2 The Model

Our model consists of a linear coronal loop filled with MHD turbulence into which we inject an impulsive spectrum of protons from an earlier acceleration. The quantity most closely related to the observable gamma-ray flux from the decay of π mesons is the proton loss rate from the loop above a momentum $\bar{p}$, where $\bar{p}$ is the lab-frame π-meson production threshold. Therefore, our goal is to produce an expression for the proton loss rate above momentum $\bar{p}$.

Within a leaky box model the time-dependent evolution of the particle omnidirectional distribution function $f(p,t)$ is described by the equation

$$\frac{\partial f}{\partial t} = p^{-2}\frac{\partial}{\partial p}\left[p^2 D(p)\frac{\partial f}{\partial p}\right] - \frac{f}{T} + Q(p)\delta(t),\tag{1}$$

where D is the diffusion coefficient in momentum space, $T\,(=L^2/\pi^2\kappa)$ is the characteristic escape time from the loop, κ is the characteristic spatial diffusion coefficient, and Q is the particle injection rate. Energy loss processes (collisions) are neglected, implying that the effects of matter within the coronal loop are negligible with respect to escape or acceleration. We choose the impulsive injection rate

$$Q(p) = \frac{N_0}{4\pi p_t^3}(\gamma - 3)\left(\frac{p}{p_t}\right)^{-\gamma}S(p - p_t),\tag{2}$$

where $S(x)$ is the standard Heaviside step function.

The leaky-box approximation is only valid when $t > T$, i.e. spatial gradients are minimized within a single diffusion time. Within quasilinear theory the product $D\kappa$ satisfies $D\kappa \propto p^2$. We further assume for simplicity that κ is a constant, so that $D = D_0 p^2$. The solution to Eq. (1) is

$$f(p,\tau) = \frac{N_0}{4\sqrt{\pi^3}p_t^3}(\gamma - 3)\exp\left\{-[\alpha - \gamma(\gamma - 3)]\tau\right\}$$

$$\times \left(\frac{p}{p_t}\right)^{-\gamma}\mathrm{erfc}\left[\frac{(2\gamma - 3)\tau - \ln\left(\frac{p}{p_t}\right)}{\sqrt{4\tau}}\right],\tag{3}$$

where $\tau \equiv D_0 t$, $\alpha \equiv (TD_0)^{-1}$, and $\mathrm{erfc}(x)$ is the complementary error function. The normalization N_0 is given by

$$4\pi \int_{p_t}^{\infty} Q(p)p^2\,dp = N_0.$$

Since this paper focuses on the gamma radiation emitted from the decay of π mesons, we may only consider protons above the π-meson production threshold, $\bar{p}$ ($\bar{p}c = 808$ MeV or E $= 300$ MeV). Integrating Eq. (3) over momenta $p > \bar{p}$ we obtain

$$\frac{n(\tau)}{T} \equiv \frac{4\pi}{T}\int_{\bar{p}}^{\infty} f(p,\tau)p^2\,dp = \frac{N_0}{2\pi T}\exp\left(-\alpha\tau\right)$$

$$\times\left\{\mathrm{erfc}\left[\frac{\ln\left(\frac{\bar{p}}{p_t}\right) - 3\tau}{\sqrt{4\tau}}\right] + \exp\left[(\gamma - 3)(\gamma\tau - \ln(\frac{\bar{p}}{p_t}))\right]\mathrm{erfc}\left[\frac{(2\gamma - 3)\tau - \ln\left(\frac{\bar{p}}{p_t}\right)}{\sqrt{4\tau}}\right]\right\},\tag{4}$$

where $n(\tau)/T$ is the high-energy proton loss rate per unit volume from the coronal loop, which is approximately proportional to the emission rate of π-meson-related gamma rays. The details of gamma-ray production are not contained in Eq. (4). Although the total inelastic cross section is relatively independent of energy compared to the cross sections for nuclear gamma-ray lines, the multiplicity of charged with respect to neutral pions is a strong function of energy just above the reaction threshold. Relativistic beaming and the transport of secondary electrons and positrons must also, in principle, be considered (Murphy *et al.* 1987, Mandzhavidze and Ramaty 1992b, Alexander and MacKinnon 1993). Nevertheless, if the geometry does not change, the time development of the total gamma-ray intensity should obey Eq. (4), although the shape of the gamma-ray spectrum may vary.

3 Discussion

The lower limit of the power-law spectrum p_t is taken to correspond to 1 MeV. The quantity α reflects the efficiency of acceleration relative to escape from the leaky box. The power-law index γ must be greater than 4 to avoid having an infinite proton energy density. Figure 1 gives the loss rate vs. τ from Eq. (4) for the case $\alpha = 0.4$ with $\gamma = 4$ and 6. We see that proton acceleration early in the event exceeds proton escape for momenta $p > \bar{p}$ yielding a loss rate which, even for an initially soft spectrum (p^{-6}), reaches a maximum at $\tau \approx 1$ or $0.4\,T$ ($\alpha = 0.4$). The decay of the flux asymptotically approaches $e^{-\alpha \tau}$ independent of γ.

It is also possible to model the time development of the nuclear gamma-ray line emission assuming that most of the gamma-ray lines from C, N, O and other heavy nuclei derive from a narrow part of the proton spectrum above the reaction threshold. The proton loss rate for a narrow band in momentum space (with $\alpha = 0.4$) corresponding to a proton energy of ~ 30 MeV ($pc = 44$ MeV) is given by Eq. (3) and also shown in Fig. 1. The precipitation of these particles peaks earlier ($\tau \approx 0.5$) followed by an asymptotic decay $e^{-(\alpha+9/4)\tau}/\sqrt{\tau}$. This faster decay occurs because in a narrow momentum band protons are lost not only through escape but also through diffusion to different parts of the spectrum. Thus, the decay for nuclear gamma-ray lines should be faster.

The spectrum continually evolves throughout this process. For $p \gg p_t$ and $\tau \gg 1$ the proton spectrum retains the same shape of the injection spectrum, i.e. $p^{-\gamma}$. As time progresses, though, the power-law behavior exists only at larger momenta. At lower momenta the spectrum has flattened out with the elimination of the discontinuity at p_t.

The simple model presented here is limited by the approximations that κ is independent of momentum and that spatial diffusion is not treated explicitly. Nevertheless the application of this model to specific solar flares is still instructive. The 3 June 1982 flare exhibited a distinctive delay in the onset of the high-energy radiation (Forrest *et al.* 1986). This behavior can be accounted for in this model by assuming that α is on the order of unity as illustrated in Fig. 1. An acceleration time of ~ 200 s with $\gamma \geq 4$ will yield a peaking of the high-energy flux a few minutes after the flare onset. Setting $T = 500$ s ($\alpha = 0.4$), the long decay will be replicated. In a loop 10^5 km long with an ambient magnetic field of 100 G. The scattering mean-free path of a 1 GeV proton will be ~ 60 km with $\delta B/B$ on the order of 6%, in the resonant part of the spectrum (well within reasonable solar parameters).

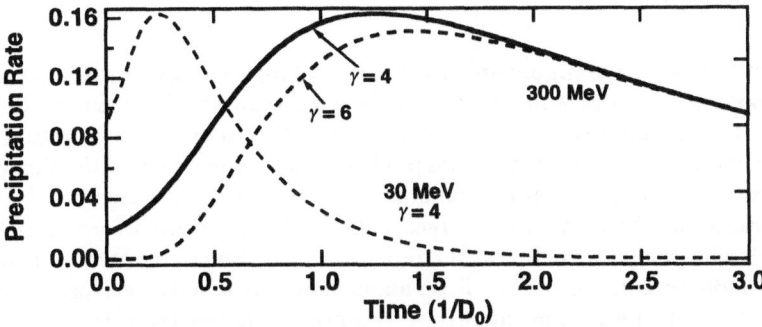

Fig. 1. The intensity-time profile of precipitating protons with momenta $p > \bar{p}$ for $\gamma = 4$ and 6 and with p corresponding to 30 MeV.

The flares of 11 and 15 June 1991 can also be understood. For both these events the escape time is measured, fixing T. The flare of 15 June 1991 had a decay time of ~ 1000 s (McConnell 1993) as observed in nuclear gamma-ray lines. This greater escape time requires a shorter λ (30 km at 1 GeV) and greater turbulence ($\delta B/B$ = 9%) in the resonant part of the spectrum. However, the flare of 11 June 1991 severely tests all models of prolonged gamma-ray emission. In this case the decay time of $\sim 10^4$ s requires an extremely large loop to avoid having $\delta B/B$ be too large (≈ 1). In order to keep $\delta B/B$ reasonable, we must choose the loop length to be on the order of 10^6 km (3 solar radii) with $B = 50$ G. This combination results in λ = 300 km and $\delta B/B = 4\%$. With a somewhat low Alfvén velocity of 1000 km-s^{-1} we have $\alpha = 2.5 \times 10^{-3}$, so that acceleration is more efficient than escape. The resulting intensity-time profile peaks at ~ 25 s, on the order of the duration of the impulsive phase. However, time structure for $t < T$ is not reliable in a leaky box approximation. Indeed $t = T$ corresponds to $\tau = 2.5$ in Figure 1, so that we are indeed pressing the validity of the leaky box.

Although the wave intensity in the resonant part of the spectrum (4%) is not large, the extrapolation to small wave numbers in a Kolmogorov spectrum can lead to large magnetic-field-fluctuation energy densities. Wave numbers as small as $1/L$ should be present in the wave field, but it is uncertain whether the spectrum has a Kolmogorov shape. It may be enhanced in the resonant part of the spectrum by energetic proton streaming instabilities.

Our interpretation of the long-duration emission from the 11 June 1991 flare contrasts with that of Ramaty and Mandzhavidze (1993). Their model requires low levels of turbulence in order to prevent protons from diffusing rapidly into the loss cone. Their required level of turbulence $(\delta B)^2 < 2 \times 10^{-8}$ ergs-cm^{-3} translates into a 1 GeV proton mean-free path on the order of 1000 AU. It is ironic that the turbulence must be at these extremes with the mean-free paths differing by 10^9. Solutions cannot be found in between these two limits. Either the level of turbulence must be extremely large or extremely small in order to trap 1 GeV protons for 10^4 s.

4 Conclusions

We have described the trapping and acceleration of high-energy protons in turbulent coronal loops with the protons impacting the solar chromosphere and producing π-meson-related gamma-ray emission. Using a leaky-box model with an initial proton power-law spectrum in momentum, we predict an initial increase in the flux of high-energy protons while acceleration dominates, followed by an exponential decay. The required amounts of turbulence are reasonable in loops on the order of 10^5 km in length and with magnetic fields on the order of 100 G. Since this level of turbulence is largely unobservable we can say little about how it is maintained for long periods of time. This may be a more attractive scenario than one requiring low levels of turbulence following large solar flares.

References

Alexander, D., MacKinnon, A.L.: 1993, submitted to *Solar Phys.*
Chupp, E.L. *et al.*: 1987, *Astrophys. J.* **318**, 913
Debrunner, H., Flückiger, E., Chupp, E.L., Forrest, D.J.: 1983, "The Solar Cosmic Ray Neutron Event on June 3, 1982", in *Proc. 18th Internat. Cosmic Ray Conf.*, **4**, p. 75
Forrest, D.J., Vestrand, W.T., Chupp, E.L., Rieger, E., Cooper, J., Share, G.H.: 1986, *Adv. Space Res.* **6**, 115
Kanbach, G. *et al.*: 1993, *Astron. Astrophys. Suppl. Ser.* **97**, 349
Lee, M.A.: 1993, "Stochastic Fermi Acceleration and Solar Cosmic Rays", in *High Energy Solar Phenomena*, ed. by J.M. Ryan and W.T. Vestrand, AIP **294**, in press
Lingenfelter, R.E.: 1993, "Solar Flare Neutrons and Gamma Rays.", in *High Energy Solar Phenomena*, ed. by J.M. Ryan and W.T. Vestrand, AIP **294**, in press
Mandzhavidze, N., Ramaty, R.: 1992a, *Astrophys. J. Lett.* **396**, L111
Mandzhavidze, N., Ramaty, R.: 1992b, *Astrophys. J.* **389**, 739
McConnell, M.: 1993, "An Overview of Solar Flare Results from COMPTEL", in *High Energy Solar Phenomena*, ed. by J.M. Ryan and W.T. Vestrand, AIP **294**, in press
Murphy, R.J., Dermer C.D., Ramaty, R.: 1987, *Astrophys. J.* **316**, L41
Murphy, R.J. *et al.*: 1993, "OSSE Observations of Solar Flares", in *Compton Symposium*, ed. by M. Friedlander, N. Gehrels, and D. Macomb, AIP **280**, 619
Ramaty, R., Mandzhavidze, N.: 1993, "Theoretical Models for High Energy Solar Flare Emissions", in *High Energy Solar Phenomena*, ed. by J.M. Ryan and W.T. Vestrand, AIP **294**, in press
Ryan, J.M.: 1986, *Solar Phys.* **105**, 365
Ryan, J.M., Lee M.A.: 1991, *Astrophys. J.* **368**, 316

Signatures of Proton Beams in the Lyα Profile: Sensitivity of the Diagnostics

M. Messerotti [1], M.T. Gomez [2], G. Severino [2]

[1] Astronomical Observatory, Via G.B. Tiepolo 11, I-34131 Trieste, Italy
[2] Astronomical Observatory, Via Moiariello 16, I-80131 Napoli, Italy

Abstract: We modelled the effect on the Lyα emission caused by an accelerated proton beam impacting onto the chromosphere to estimate its detection threshold as an effective diagnostics for protons. Chosen a model chromosphere and a H I model atom with 3 bound levels plus a continuum, we computed the background Lyα flux profile and the superthermal contribution in the red wing due to the beam. The profiles computed for different model parameters show that a source with a radius exceeding $1000\,\mathrm{km}$ and producing a total proton flux greater than $5\,10^{10}\,\mathrm{erg\,cm^{-2}\,s^{-1}}$ should be detectable in irradiance observations.

1 Introduction

Despite the important role played by protons in different physical processes on the Sun (see, e.g., Simnett 1986, 1991; Simnett and Haines 1990), actually no clear signatures for proton beam generation and propagation have been identified yet except γ-ray lines, for downward moving beams with energy of the order of $10\,\mathrm{MeV}$, and direct detection in the interplanetary medium, for outward moving ones.

Regarding the radio band some models invoke precipitating proton beams to explain plasma radiation in the GHz range (Smith and Benz 1991), escaping protons as exciters of slowly-drifting radio emissions (Benz and Simnett 1986) or neutral (proton+electron) beams generating Langmuir waves (Messerotti and Karlicky 1992), but the diagnostics are quite ambiguous. This means that a complementary diagnostic is needed in a different spectral band to have a better chance of identification during coordinated observations.

Orrall and Zirker (1976) suggested the possibility of observing non-thermal Lyα emission due to energetic protons in the range $(10\,\mathrm{keV}\text{-}1\,\mathrm{MeV})$, impacting onto the chromosphere; Canfield and Chang (1985) developed the model under different beam characteristics. Up to now no observational evidence of such emission has been found. Starting from a similar model, Messerotti et al. (1992) performed further computations of the Lyα profile for a given source size and different total proton fluxes, assuming a thermal energy spectrum for the beam particles. They found a rather high detectability threshold for the non-thermal emission in the red wing of the line, mainly dependent on the quiet Sun background and on the total proton flux.

In the present study we consider an extension of previous theoretical estimations by evaluating the sensitivity of the diagnostics to other model parameters, like the radius of the flare, which is assumed to coincide with the source of the superthermal emission, and the parameters fixing the shape of the spectral energy distribution of the incident protons. In Section 2 we describe the computational method and the model used in calculating the Lyα flux profile, in Section 3 we comment on the obtained results and in Section 4 we stress the related relevant points.

2 Computational Method and Atomic Data

To analyze the effect of a proton beam, downward moving after acceleration high in the atmosphere, on the Lyα emission, we have computed the background Lyα flux profile together with the superthermal contribution in the red wing caused by the accelerated proton beam.

2.1 Lyα Background Profile

To compute the background Lyα flux profile we used a H I model atom having 3 bound levels plus one continuum. The collisional bound-bound cross sections were taken from Sampson and Golden (1970) and Golden and Sampson (1971), and the bound-free cross-sections from Mihalas (1967).

The components of the background flux result from the quiet Sun, the active regions and the thermal part of the flare associated with the beam. The quiet Sun gives the main contribution by order of magnitudes (see Fig. 3 in Messerotti et al. 1992). Therefore we have taken into account only the quiet Lyα component. As a reference atmosphere for the quiet Sun, we have used the model C by Vernazza et al. (1981)(Hereafter called VAL-C).

The statistical equilibrium equations coupled to the radiation transfer equation for H I have been solved with the nonLTE code MULTI, implemented by Carlsson (1986) and based on Scharmer's approximate operator perturbation technique (Scharmer 1981, 1984; Scharmer and Carlsson 1985). Since the version of MULTI that we have used treats the line scattering processes with the assumption of complete frequency redistribution (CRD), while elastic collisions are so few in the solar chromosphere that scattering in the wings of Lyα is nearly coherent (PRD), we have reduced the resulting line emission in the wings ($\delta\lambda \geq 0.5$ Å), in order to match the PRD intensity profiles calculated by Vernazza et al. (1981, see their Fig. 27).

2.2 Lyα Emission Caused by the Beam

To compute the Lyα emission caused by the beam, we have followed essentially the approach proposed by Canfield and Chang (1985).

At the magnetic reconnection site a vertical downward proton beam is accelerated and then bombards the lower chromosphere. Interacting with the ambient medium, fast protons may become excited nonthermal H atoms by charge exchange following a collision, and then radiate in the Lyα red wing. This process can be treated in the approximation that a steady state is reached (Brown 1973; Orrall and

Zirker 1976). With the further assumption of optically thin emission (Orrall and Zirker 1976), the intensity contribution caused by the proton beam is

$$I_{\mathrm{p}} = \frac{(2\,m_{\mathrm{p}}\,h^2\,c^2)^{1/2}}{4\,\pi\,\lambda_{\mathrm{o}}^2}\,E^{1/2}\int_0^\infty F(E,z)\,R_{21}(E,z)\,n_{\mathrm{H}}\,dz \tag{1}$$

where
- λ_{o} is the Lyα line central wavelength,
- z is the distance downward from the top of the chromosphere,
- $F(E,z)$ the number flux of superthermal particles (protons and atoms) with energy E per unit energy interval, at level z,
- n_{H} the number density of ambient particles, at level z,
- $R_{21}(E,z)$ the Lyα emission rate per superthermal particle of energy E and per ambient particle, at level z.

We convert this intensity in flux assuming no limb-darkening and that the flare occurred at the disk center and covered a circular area of radius R_f.

The emission rate R_{21} is fixed by the statistical equilibrium of the protons in the beam. A steady state excitation and ionization balance that takes into account the radiative and collisional interactions of the beam protons with the ambient chromosphere has been considered. Charge exchange between beam protons and ambient H atoms, and spontaneous radiative recombination of beam protons create beam H atoms; at the same time, excitation and ionization of beam H atoms by collisions with ambient electrons, H atoms and protons, and spontaneous and stimulated radiative de-excitation of beam H atoms redistribute the beam H atoms between their energy levels. The dominant creation processes are the charge exchanges directly to the Lyα excited level or to the ground level followed by a collisional excitation. The dominant destruction process is spontaneous bound-bound emission (Canfield and Chang 1985).

The proton energy spectrum $F(E,z)$ is determined by the stopping of the beam in the chromosphere. This is described with a Coulomb collision model, and depends on the input proton energy spectrum at the top of the chromosphere $F(E_{\mathrm{o}},0)$. By analogy with fast flare electrons, we assumed that $F(E_{\mathrm{o}},0)$ is described by a power law, for energies greater than a cutoff E_{c}, that is

$$F(E_{\mathrm{o}},0) = \frac{\delta - 1}{E_{\mathrm{c}}}\,F_{\mathrm{T}}\left(\frac{E_{\mathrm{o}}}{E_{\mathrm{c}}}\right)^{-\delta} \tag{2}$$

δ being the spectral index and F_{T} the total input flux of the beam.

To perform analytically the integration involved in (1), Canfield and Chang adopted a simple pure-hydrogen model of the target chromosphere: uniform temperature T, total hydrogen number density n_{H}, and ionized fraction $x = n_{\mathrm{p}}/n_{\mathrm{H}}$. Orrall and Zirker stated that the atmospheric levels giving the maximum contribution to the emission due to protons having energy $E_{\mathrm{o}} \gg 10\,\mathrm{keV}$ are close to their stopping heights. Therefore, we adopted an uniform chromosphere for each value of the mean energy of the incident protons $< E > = E_{\mathrm{c}}\,(\delta - 1)/(\delta - 2)$, with the values of T, n_{H} and x interpolated on the VAL-C tables at the stopping height. We treat as free parameters the radius R_f of the flare, which is assumed to coincide with the source of the superthermal emission, and the parameters of the proton spectral energy distributions, that is the cutoff energy of the incident protons E_{c}, the spectral index δ, and the total beam flux F_{T}.

3 Results

Figure 1 shows the computed radiation flux in the red wing of the Lyα caused by the beam together with the quiet Sun background profile for comparison, for a number of values of the model free parameters. We assumed as standard the set $R_\mathrm{f} = 2000\,\mathrm{km}$, $E_\mathrm{c} = 100\,\mathrm{keV}$, $F_\mathrm{T} = 10^{11}\,\mathrm{erg\,cm^{-2}\,s^{-1}}$, $\delta = 4$, and show the effect of changing one parameter at a time in each panel of the figure. The emission is an increasing function of R_f, E_c (since at fixed total proton flux increasing the cutoff pushes the spectrum toward higher energies where the emission rate R_{21} reaches its maximum), and F_T, while it has a more complex behaviour with δ.

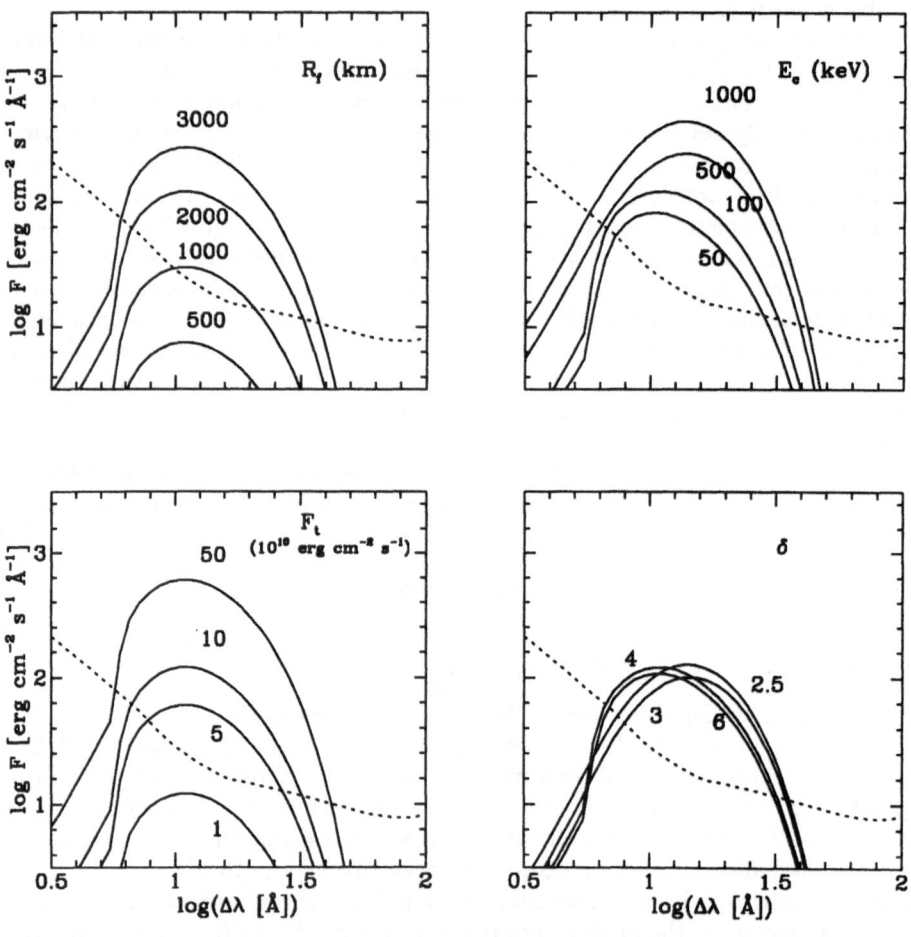

Fig. 1. Radiation flux in the red wing of the Lyα for different model parameters

4 Conclusions

Within the limit of the chosen values for the free parameters, it appears that a source with a size greater than 1000 km in radius which produces a total proton flux greater than $5\,10^{10}\,\mathrm{erg\,cm^{-2}\,s^{-1}}$, should be detectable in irradiance observations.

However, we must remark that there is an extra parameter which affects the sensitivity of the diagnostics, namely the kind of collision theory used to describe the stopping of the beam in the chromosphere. In fact, we have found that the collision model used by Canfield and Chang and by us produces an emission profile which can be a factor 6 greater than the emission profile computed with the collision theory used by Orrall and Zirker. Moreover, we assumed a specific geometry for the emitting source, i.e., circular shape, location at disk center and no limb darkening. Therefore our conclusions are strongly model-dependent.

Nevertheless, despite the caution imposed by all the limitations of our model, we argue that promising developments can be expected from future space observations in a deeper understanding of the role played by protons in the solar atmosphere. In particular, we cite the experiments onboard the SOHO satellite that will allow multiband observations, coordinated with ground-based instruments.

Acknowledgments

This work was partially supported by the Italian Space Agency (ASI), by the Consortium for the Magnetohydrodynamics of the Trieste University and by the Italian Ministry of University, Scientific Research and Technology (MURST).

We are in debt with Mats Carlsson who kindly supplied us with his nonLTE code.

References

Benz, A.O., Simnett, G.M. (1986): Nature 320, 508

Brown, J.C. (1973): Solar Phys. 31, 143

Canfield, R.C., Chang, C.-R. (1985): Ap. J. 295, 275

Carlsson, M. (1986): "A Computer Program for Solving Multi-Level Non-LTE Radiative Transfer Problems in Moving or Static Atmospheres", Report No. 33, Uppsala Astronomical Observatory

Golden, L.B., Sampson, D.H. (1971): Ap. J. 163, 405

Messerotti, M., Gomez, M.T., Severino, G. (1992): in Proc. 2nd Symp. on Plasma Dynamics - Theory and Applications, ed. by M. Tessarotto (MHD Consortium, Trieste University, Trieste, Italy), 87

Messerotti, M., Karlicky, M. (1992): in Proc 1st Symp. on Plasma Dynamics - Theory and Applications, ed. by M. Tessarotto (MHD Consortium, Trieste University, Trieste, Italy), 115

Mihalas, D. (1967): Ap. J. 149, 169

Orral, F.Q., Zirker, J.B. (1976): Ap. J. 208, 618

Sampson, D.H., Golden, L.B. (1970): Ap. J. 161, 321

Scharmer, G.B. (1981): Ap. J. 249, 720

Scharmer, G.B. (1984): in *Methods in radiative transfer*, ed. by W. Kalkofen, Cambridge University Press, Cambridge, 173

Scharmer, G.B., Carlsson, M. (1985): J. Comput. Phys. 59, 56

Simnett, G.M. (1986): Solar Phys. 106, 165

Simnett, G.M. (1991): Mem. S.A.It. 62, n. 2, 359

Simnett, G.M., Haines, M.G. (1990): Solar Phys. 130, 253

Smith, D.F., Benz, A.O. (1991): Solar Phys. 131, 351

Vernazza, J.E., Avrett, E.H., Loeser, R. (1981): Ap. J. Suppl. 45, 635

VII

New Instrumentation

Ground - Based Instrumentation

Franz - Ludwig Deubner

Institut für Astronomie und Astrophysik der Universität Würzburg
Am Hubland, D-97074 Würzburg, Germany

Abstract: In parallel with the development and deployment of solar space instrumentation a substantial effort is being made at many ground based observatories world-wide, not only to make better use of the available hours of sunshine, but also to reach out to new frontiers of solar research. Examples of novel instrumentation will be described, and some of the observational results will be discussed in the light of current problems in solar physics.

1 Introduction

In contrast to a widespread scepticism regarding the future of ground- based observational solar physics we are witnessing at present an amazing multitude of activities which in addition to modernization and upgrading of existing observational equipment are aimed at the development of novel instruments and techniques serving to fully exploit the potential of ground-based stations and their telescopes. We shall try, in this brief review, to describe and discuss some of this new equipment along with the problems, ideas and goals that justify such forceful rejuvenation of interest in ground-based solar research.

Recent observational papers in this field frequently emphasize even in the title the fact that they are based on high-resolution data. Foremost this epithet is applied with respect to spatial resolution. Careful site selection in combination with innovative telescope design as well as cunning data reduction has opened the sub-arcsec domain even for ground-based observations. We shall not enter here into the never-ending discussion of how spatial resolution should be defined quantitatively – the half width of the smallest detectable structure, the largest wavenumber which can be seen at some level above noise, or something else. We should, however, be aware of the relevant physical processes and parameters on the Sun, that either restrict or challenge our ability to evaluate quantitatively the spatial properties that we are interested in.

Such processes are summarized in Table 1 in the first column. It is clear that radiation exchange on the Sun will limit the distance down to which separate physical structures can be distinguished effectively. On the other hand, it is not quite clear whether the elementary physical processes responsible for solar activity can be satisfactorily described and analysed at this spatial scale, which happens to be quite close to the typical horizontal dimension of a magnetic flux tube, at the photospheric level. What role does the particle gyroradius play in a 3-D scenario with strong vertical energy exchange and (oscillatory) mass motion? It may well be that the value of $\approx 0.''1$ put down in Table 1 anticipates the state of the art at the end of this century rather than defines a meaningful physical limit.

Table 1

Spatial Resolution

photon mean free path flux tube dimension (gyroradius?)	$0.''1 - 0.''15$

Temporal Resolution

$c \approx 7\,km\,sec^{-1}$	$10\,sec$
seeing	$10\,msec$

Spectral Resolution

$\Delta \lambda \; \{K_3, K_2\}$	$15\,pm$
I, Q, U, V	$2.0 - 2.5\,pm$
$\Delta \lambda \; \{1\,km\,sec^{-1}$ at $450\,nm \}$	$1.5\,pm$

Of course, the requisite temporal resolution is not independent of the spatial resolution available. But, more importantly, it also depends strongly on the proper time scale of the physical process to be 'frozen'. A sound wave traverses $0.''1$ in about 10 sec, but high energy particles travel much faster, and 'freezing' of the ever moving terrestrial atmosphere requires exposure times as short as 10 msec. The main problem here appears to be logistic in nature rather than physical: one has to find ways of disposing of enormous amounts of unwanted low quality or redundant data in real time, in order to save storage space for those data one really wants to look at. High frequency resolution can only be obtained by means of extended observing runs. This exacerbites the storage problem just mentioned.

More difficult to estimate is the appropriate minimum spectral resolution, since thermal line broadening affects all spectral features. It may depend entirely on the signal/noise ratio whether the inversion of an observed spectral feature can be successfully attempted. Grating as well as Fourier transform spectrographs generally guarantee an intrinsic resolution that does not compromise the spectral information received from the Sun. With the advent of panoramic IR detectors the near infrared spectral region has become particularly attractive for studies of the solar magnetic fine structure. Quantitative polarimetric measurements of minute asymmetries in the Stokes line profile are indispensable in the world of flux tubes and penumbral filaments. Table 1, again, summarizes some typical figures.

In this context it appears worthwhile contemplating for a minute the intrinsic limitations set by sequential observing and recording. Simultaneity of exposure in a strict sense can be achieved either in two spatial dimensions (filtergram mode) or in one spatial and the spectral dimension (spectrographic mode). Completing one data cube with two spatial and one spectral dimension requires either sequential stepping of the filter transmission, or scanning of the solar surface. Either operation takes much more time than there is available with modern equipment, even under best seeing conditions, to freeze the 3-D image. The choice among the two different observing modes therefore depends on the observers' preferences for either pictorial veracity or preservation of the integrity of the local instantaneous line profile. Since in many applications the observed line profiles undergo inversion processes relying on the fidelity of subtle details of the spectral information, the use of high resolution spectrographs appears indispensable at least in these cases.

2 High Resolution Imaging (Approaching the Diffraction Limit)

2.1 Correlation Tracking

This technique of eliminating the 'image motion' component of terrestrial atmospheric seeing is a consequent further development of earlier devices like the classical limb guider which is being routinely used to compensate for guiding errors and (low frequency) shaking of the telescope by means of simple differential detector systems and an agile mirror. The principle works just as well with spots, and even with pores, allowing us to shrink the areal range from which the error signal is derived to the size of the isoplanatic patch. Of course, a residual distortion of the image which slowly increases with distance from the reference object is unavoidable.

A tracking system that does not depend on the availability of particular high contrast objects in the neighborhood of the area investigated on the Sun has recently been developed and put into operation (Rimmele et al. 1993) on Sacramento Peak. The 'correlation tracker' puts a high quality reference image of the observed scene into its memory, and derives, in real time, the error signal from the 2-D cross-correlation function of the actual solar scene and the reference image. The minimum contrast accepted for evaluation of the cross-correlation function may be set arbitrarily, say at 3 % r.m.s. and allows the observer to guide on the granulation proper if he wants to study the quiet Sun. The contrast discriminator can also be used to activate a frame-grabbing system, storing only the very best images within a given time interval.

The present correlation tracking system at the VTT in Izana (Tenerife) operates with a 60 mm diameter agile mirror and a Reticon array at a nominal time resolution of 2 msec. The residual r.m.s. image motion sampled within 40 msec exposures dropped to 0."05 – 0."3 while tracking, compared to 0."5 – 1."8 without tracking. At the Spanish IAC on Tenerife (Bonet et al. 1992) a new design has been finished for a correlation tracker that promises to operate even faster. This technique does not attempt to restore the original image with the spatial resolution of the telescope, but could of course be used in tandem with adaptive optics (see below).

2.2 Destretching (= Post Factum Correction)

If a high resolution (white light) reference image can be provided simultaneously and stored together with the cospatial spectral solar information, it is possible to eliminate numerically the distortion of the image geometry after the run by means of the socalled local correlation techniques which have been developed among others by November et al. (1987) in collaboration with the Lockheed/Palo Alto solar research group. Related algorithms have proven extremely useful in numerous observational studies of horizontal convective motion patterns. The technique does not require any special post focus hardware. Restoration of the telescopic resolution is not attempted either.

2.3 Speckles

Recently, speckle techniques have been utilized extensively, among others, by the Göttingen and the Zürich solar groups, with outstanding success (see also von der Lühe 1993) . They are based on correlation algorithms developed by Knox & Thompsen (1974) and Weigelt & Wirnitzer (1983), and are also applied after the exposure. In the solar case, with typical time scales of down to less then 10 sec this approach to image restoration needs high speed CCD's with video read out rates, and large digital memories.

De Boer (1992) and his coworkers have not only demonstrated that speckle techniques really work, but also have shown that they can accommodate complex 2-D scenes (field of view: 100″). They have also contributed exciting material to the discussion of the fine scale structuring by convective processes in active and quiet regions on the Sun. In particular, de Boer was able to show that many granules have a bright rim which was predicted by Nordlund & Stein (1989) in their 3-D simulation of convection. With a 10 nm band pass de Boer detects 'crinkles' in the white light photosphere, features that have mainly been seen in the light of Fraunhofer lines before. The speckle reconstruction of a pore has tiny notches at its rim, like a mini penumbra, and a light bridge of a large sunspot is decomposed into a chain of many subarcsec segments. It is not the aim of this report to discuss the many theoretical implications of these new findings.

Very interesting observations using speckle deconvolution have also been made by Keller & von der Lühe (1993), demonstrating that they have resolved spatial structure on the Sun with about 180 km diameter.

2.4 Adaptive Optics

The term in the heading of this subsection used to be the traditional term coined for systems of optical devices which are correcting in real time the wavefront aberrations (with the exception of a mean tilt, see Sec. 2.1) that are introduced by the terrestrial atmosphere, and by imperfections and changes of the telescopes' imaging properties. More recently the term 'live optics' has been suggested (cf. LEST design study, Engvold & Andersen 1990) to include specifically all elements of image correction from alignment control to seeing compensation.

The basic principles and components of such a system have been described in some detail in the above mentioned LEST document. Central parts of this system are the wavefront sensor and the adaptive mirror which can be either segmented, or monolithic but flexible. In either case a set of discrete actuators deforms the adaptive mirror in such a way that the error signal received from the wave front sensor is minimized. Such a scheme can work with a simple reference object like a sunspot or pore just as well as with correlation techniques.

At the present time, there has been comparatively little experience collected yet with working systems of 'live optics' in solar applications. One such test was carried out and described by the LPARL group (Acton & Smithson 1992) at the 76 cm Vacuum Tower Telescope on Sacramento Peak. They have measured, under various seeing conditions, the gain in image contrast as a function of wavenumber. The gain is 1 by definition under ideal seeing conditions, and has reached values of 3 to 4 at 0.7 to 1.0 cycles per arcsec with poor seeing.

Since the number of actuators increases as the area of the instrument's aperture over the square of r_0 (the Fried parameter) the number of actuators necessary for telescopes of the size of a LEST becomes larger than 100 (e.g. 169 in a hexagonal arrangement), requiring in turn a real time computing power that can only be provided with very efficient parallel computing. A continuous faceplate adaptive mirror with 61 actuators is presently being worked on; a new type of wavefront sensor based on a liquid crystal device is in development (Rabin 1993).

The efficiency of spatial deconvolution by wavefront sensing can, of course, also be tested by simulation. Such a test has been carried out by Keller & von der Lühe (1993) using the speckle reconstruction of a solar image as a test object. The conclusion was, that the isoplanatic patch, where wavefront deconvolution really works, is rather small (i.e. a few arc seconds in diameter) even under favourable seeing conditions. The development of techniques utilizing the interferometric information available from a 'focal volume' rather than just from the focal plane therefore merits the greatest interest.

3 Fast Narrowband Didimensional Imaging

As we have discussed briefly in the Introduction, the use of narrowband filters for spectral selection affords a rather direct approach to the restoration of 2-D images at the expense of the integrity of the line profiles. To optimize the information content available from this mode of observation one has attempted to meet the following requirements in the best possible way:

– Spectroscopic resolution close to $3 \cdot 10^{-6}$

– Simultaneity of exposures at several different spectral passbands
(to study effects of temperature, pressure, height etc. as in spectra)

– High temporal resolution
(to study the propagation of waves and other phenomena)

The second and third requirements can in principle be fulfilled by triggering exposures of large batteries of birefrigent filters in unison. This approach has been followed at the Huairu Solar Station of the Beijing Astronomical Observatory where a nine channel filtersystem is in operation for synoptic work and a 64-channel system is under construction.

At some other institutes (Arcetri, cf. Bonaccini & Stauffer 1990; Göttingen, cf. Bendlin et al. 1992, Bendlin 1993) the first requirement rather has had first priority in the design, and Fabry-Perot etalons have been applied both to achieve higher spectral resolution and higher photon throughput of the interferometers in order to reduce the exposure time. The spectral resolution is close to 2.0 nm, the exposure and readout time is short enough to fit 20 exposures of a $23'' \times 16''$ area into a 2 sec interval. Simultaneous white light exposures serve to provide reference images which can be used for alignment and destretching. Because of their high throughput and spectral resolution these filterheliographs are well suited for polarimetric studies.

4 Simultaneity (High Spatial Resolution with Spectral Purity)

4.1 Subtractive-Double-Pass Spectrometry

Two schemes have been invented and successfully probed in solar applications, that permit (almost) simultaneous acquisition of data in two spatial coordinates over a range of wavelengths, both using a grating spectrograph for spectral dispersion.

The first instrument, developed by P. Mein (1991) and his coworkers, the Multi-Slit-Subtractive-Double-Pass spectrometer (MSDP) of the Meudon Observatory obtains in a single shot, i.e. strictly simultaneously, two-dimensional information at a number of (in most cases) equidistant wavelength positions. The spatial information is continuous, line profiles derived from MSDP observations are not, but they are in principle intact as specified in the Introduction.

Upon taking a closer look at this scheme it seems fair to note that the near ideal performance of this instrument with regard to spatial resolution and simultaneity has been paid with a somewhat limited field-of-view in the spatial dimension parallel to the dispersion, and, connected to it, with reduced spectral resolution (3.0 to 6.0 pm in typical cases). This technique appears particularly suited for investigations in the chromospheric lines, in search for the chromospheric counterparts of dynamic events seen in the photosphere, for active regions and filaments, or penumbrae.

4.2 Rapid Spatial Scanning

Quite a different approach to 'simultaneity' was taken by Johannesson (1992) at the Swedish Telescope on La Palma, following the pioneering work of Scharmer (1987).

With the aim of preserving the spectral information complete and intact for reasons already discussed in the Introduction, Johannesson decided to adhere to the classical spectrogram as the basic piece of information, and to retrieve the missing information in the second spatial coordinate by rapid scanning; rapid enough to complete one scanning cycle (called a 'burst') within a few seconds of excellent seeing. Again, each spectral data set is complemented by a simultaneous set of white light slit jaw pictures, showing the exact position of the slit for each step within the burst.

The physical parameters to be studied are obtained directly from the spectra as usual, i.e. without regard to the spatial structure they belong to. However, parts of the best slitjaw pictures are merged into one single reference mosaic that covers the whole region of the scan, and serves to destretch the 2-D composite distribution, generated from the 1-D parameter slices, by local cross-correlation with the slit environment that belongs to each spectrum.

This technique guarantees high spectral resolution, and offers high spatial resolution in combination with sufficient spatial coverage. Real time frame selection is still possible. Comparison of the continuum images reconstructed from the spectral data with the white light reference image has convincingly proven that this technique works.

5 Stokes Polarimeters

High spatial and spectral resolution are prerequisites for Stokes polarimetry, and indeed most of the instruments described in the previous sections can easily be adapted to suit this purpose. There is, however, one further requirement to be met, namely strict simultaneity of the measurement of all Stokes parameters, i.e. of $I \pm V$, $I \pm Q$, $I \pm U$. Actually this follows already from the general requirement to preserve the full spectral information.

Of the many innovations accomplished in the last couple of years in this area let me first describe briefly the polarimeters developed at the ETH in Zürich by Keller et al. (1992) and Stenflo et al. (1992). The Zürich Imaging Stokes Polarimeter (ZIMPOL) I was intended as a prototype for a LEST polarimeter, and has been tested with the VTT in Izana (Tenerife) in 1992.

The analyser consists of two piezo-elastic modulators at 45° relative to each other, and a linear polarizer at 22.5°, followed by two beamsplitters. At the heart of the polarimeter we find special CCD cameras, one for each Stokes parameter, of which every second row has been masked. For demodulation the charges are shifted back and forth under the mask synchronously with the polarimetric modulation, such that the plus- and minus-signals are co-added separately. The modulation frequency of 50 kHz for V (100 kHz for Q and U) assures simultaneity of the measurements, since it is well above any seeing or adaptive optics frequency. Up to 10 images can be read out per second and stored. The sensitivity is 10^{-4}. The high frame rate of the system permits frame selection as well as speckle polarimetry.

The shortcomings of ZIMPOL I are easily recognized: There is a 50 % loss of light due to the masking on the chips, and a further reduction in efficiency by a factor 3 due to the beam splitters. Also, three separate cameras are needed with (potentially) different geometry etc.. Therefore ZIMPOL II which is presently under development includes two important modifications, intended to completely overcome these shortcomings. The new CCD operates in combination with lenslet arrays (f = 1 mm) that cover four pixel rows each, and focus light into one single row, leaving three storage rows and providing four image planes for the polarimetric analysis. The use of two phase locked modulators (50 kHz) together with a superachromatic quarter wave retarder in front of the polarizer then permits the extraction of all four Stokes parameters from these four image planes in 20 μsec.

Further recent development of Stokes polarimeters is reported from the US. The Advanced Stokes Polarimeter (NSO/HAO) operates with multiple CCDs, and aims chiefly at high resolution at several heights from the photosphere to the temperature minimum. It is integrated into the planned adaptive optics system at the Vacuum

Tower Telescope of Sacramento Peak Observatory, and is designed for community use.

The NASA/NSO Spectromagnetograph (Jones et al. 1992) in the Kitt Peak Vacuum Tower has replaced the old 512-channel diode array magnetograph which had two fixed exit slits, and does retain the full information on the line profile. This instrument also records routinely He I 10830 data together with the photospheric line profiles.

Finally, the Near Infrared Magnetograph (NSO) at the McMath - Pierce - Telescope (Rabin 1992) takes advantage of the larger splitting (in terms of Doppler broadening) of the Zeeman components at the larger wavelength ($\lambda = 1.565\,\mu$m). In addition to the magnetic flux derived from either the absolute area under the V - profile or from the amplitude of the fitted profile, the 'true' magnetic field strength can be obtained by measurements of the splitting proper. Spatial (2-D) mapping is done by stepping of the image.

I do not enter into the discussion of the effects of instrumental polarization, which has received particular attention in many recent observational papers. – A report on THEMIS, the French polarisation free telescope (Mein & Rayrole 1991), is given in a separate contribution to this meeting. Earlier this year, the concrete work for the THEMIS tower has begun on Tenerife.

6 Other Techniques and Projects (The Solar Interior and Corona)

Since it is impossible to do justice, in the given format, to every new instrumental development or observational technique in ground- based solar physics in the world, even at the modest level of this article, I conclude with a brief mention of some larger size projects which did not quite fit into the preceding sections, but which I find personally challenging and important. I owe my apology to all those who feel neglected or ill-represented by this review, and I take full responsibility for the selection I have made.

6.1 Helioseismology

Besides the enormous effort being put into the forthcoming SOHO mission by the solar community, there are still several important ground- based helioseismological projects and developments under way. Kitt Peak and Bartol scientists continue jointly the work on p - modes at the South Pole, investigating the velocity signal in the H_α line, and the brightness fluctuations both in H_α and K, formed at different heights in the atmosphere. Complementary to the GONG project, a new high degree helioseismograph has been installed at the Kitt Peak Vacuum Telescope as a pathfinder for the Solar Oscillations Investigation experiment onboard SOHO (Harvey 1993).

Of the ground-based networks, IRIS is now working with 3 stations, and has achieved a 60% duty cycle over 88 days reaching a sensitivity of $8\,\mathrm{mm\,sec^{-1}}$ at $l = 1$, $n = 11$ near 1.75 mHz (Fossat 1993). The GONG network is foreseen to become operational with 6 stations in late 1994 and to run for at least three years (Harvey et al. 1993).

6.2 Coronal Research

Of much interest with regard to ground based coronal observations is a program at the National Solar Observatory for the development of reflecting coronagraphs (MAC) utilizing the excellent experience with superpolished mirror surfaces, and aiming at a 2 - 3 m class daytime / nighttime coronographic telescope (Koutchmy & Smartt 1989).

Finally, a different aspect of the corona became routinely accessible by the new Radioheliograph (10″ resolution) of the Nobeyama Radioastronomical Observatory.

7 Conclusions

I feel there can be only one conclusion drawn in view of the vigorous development of instrumental techniques, of the multitude of new instruments already in operation or still on the drawing boards, and of the fascinating new results some of which we have briefly visited: solar research through ground-based observation is well and alive. We may confidently look forward to the coming years, when some of the machines we have presented will put out gigabytes of data of good quality every day the sun shines. Storing all these data in a sensible way might become one of the major challenges.

Acknowledgements

I am deeply indebted to a large number of colleagues who have responded to my inquiries not only by sending me recent papers and preprints, but also by providing a rich pictorial material that has formed the basis of my oral presentation. Without this support the review would not have been accomplished. Since I could be fairly sure of the cooperation of my European colleagues at this occasion, I feel urged to thank in particular J.Harvey, F.Hill, H.Jones, M.Penn, and D.Rabin for their prompt and generous help from overseas.

References

Acton, D.S., Smithson, R.C. (1992): Appl. Optics **31**, No.16, 3161
Bendlin, C. (1993): Ph.D. thesis, Göttingen
Bendlin, C., Volkmer, R., Kneer, F. (1992): Astron. Astrophys. **257**, 817
Bonet, J.A., Martin, C., Ballesteros, E., Fuentes, F.J., Lorenzo, F., Manescau, A.,
    Rodriguez, L.F., Zadrozny, A. (1992): LEST Technical Report (O. Engvold, O. Hauge,
    eds.) Univ. of Oslo; No. **51**
de Boer, C.R. (1992): Ph.D. thesis, Göttingen
de Boer, C.R., Kneer, F. (1992): Astron. Astrophys. **264**, L 24

Bonaccini, D., Stauffer., F. (1990): Astron.Astrophys. **229**, 272

Engvold, O., Andersen, T. (1990): in "Status of the Design of the Large Earth Based Telescope", LEST Foundation; p. 22

Fossat, E. (1993): in IRIS Newsletter (C. Régulo, ed.) IAC, La Laguna (Tenerife); No. **5**, p. 6

Harvey, J. (1993): priv.comm.

Harvey, J., Hill, F., Kennedy, J., Leibacher, J. (1993): Astron. Soc. Pacific (in print)

Johannesson, A., Bida, T., Lites, B., Scharmer, G.B. (1992): Astron. Astrophys. **258**, 572

Jones, H.P., Duvall, T.L.,Jr., Harvey, J.W., Mahaffy, C.T., Schwitters, J.D., Simmons, J.E. (1992): Solar Phys. **139**, 211

Keller, C.U., Aebersold, F., Egger, U., Powel, H.P., Steiner, P., Stenflo, J.O. (1992): LEST Technical Report (O. Engvold, O. Hauge, eds.) Univ. of Oslo; No. **53**

Keller, C.U., von der Lühe, O. (1993): in "Real Time and Post Factum Solar Image Correction" (R. Radick, ed.) Sunspot, New Mexico

Knox, K.T., Thompson, B.J. (1974): Astrophys.J. **193**, L 45

Koutchmy, S., Smartt, R.N. (1989): in "High Spatial Resolution Solar Observations" (O. von der Lühe, ed.) Sunspot, New Mexico; p. 560

Mein, P. (1991): Astron.Astrophys. **248**, 669

Mein, P., Rayrole, J.(1991): Adv. Space Res. **11** (No.5), 151

Nordlund, Å., Stein, R.F. (1989): in "Solar and Stellar Granulation" (R.J. Rutten, G. Severino, eds.) Kluwer, Dordrecht; p. 453

November, L.J., Simon, G.W., Tarbell, T.D., Title, A.M., Ferguson, S.H. (1987): in Proc. 2nd Workshop on "Theoretical Problems in High Resolution Solar Physics" (R.G. Athay, ed.) NASA Conf.Pub. **2483**, Boulder/Colorado; p. 121

Rabin, D. (1992): Astrophys J. **390**, L 103

Rabin, D. (1993): priv.comm.

Rimmele, T., Kentischer, T., Wiborg, Ph. (1993): in "Real Time and Post Factum Solar Image Correction" (R. Radick, ed.) Sunspot, New Mexico

Scharmer, G.B. (1987): in "The Role of the Fine Scale Magnetic Fields on the Structure of the Solar Atmosphere" (E.-H. Schröter, M. Vazquez, A.A. Wyller, eds.) La Laguna (Tenerife); p. 349

Stenflo, J.O., Keller, C.U., Powel, H.P. (1992): LEST Technical Report (O. Engvold, O. Hauge, eds.) Univ. of Oslo; No. **54**

von der Lühe, O. (1993): Astron. Astrophys. **268**, 374

Weigelt G.P., Wirnitzer, B. (1983): Opt. Letters **8**, 389

Space Instrumentation

P. Lemaire

Institut d'Astrophysique Spatiale Bâtiment 121, Université de Paris XI,
F-91405 Orsay Cedex (FRANCE)

Abstract: The capabilities of space instrumentation are in a continuous state of evolution with the improvement of new techniques such as the metallic multilayer coatings in the Extreme UltraViolet. After a brief review of a few of the new techniques that will provide improved detectors, a survey of current instrumentation is given (Yohkoh, HRTS, MSSA, NIXT). Several missions are in preparation (SOHO, CORONAS); a description of the payload capabilities of these is furnished. Finally, we speculate on the outlook for the new, improved instruments on future missions.

1 Introduction

Solar space research using rockets is 46 years old (the first rocket solar ultraviolet spectrograph was launched in October 1946 by the Naval Research Laboratory on a V-2 rocket, Baum et al. 1947) and dedicated solar satellites have been flown for the last 32 years (The NASA (National Aeronautics and Space Administration) Orbiting Solar Observatory OSO-1 was launched in March 1962). The continuing development of new techniques, e.g. multilayer coatings, has proved to be successful in improving the quality of space instrumentation in all wavelengths below 300 nm, resulting in a resolution which is comparable with the results obtained in visible ground-based observatories.

After a brief review of the advantages and disadvantages of space observations, the characteristics of missions in-orbit now (Yohkoh), ready to be launched (CORONAS-I) or in the integration stage (SOHO) are given, complemented by a description of some rocket instrumentation that has been flown during the last two years.

Then, some more general predictions are made for new missions which are in the study stage (MSV-0 and the proposed successor to Yohkoh) which hopefully will be started in the near future. Some elements of another possible future mission (HESP) are presented. In conclusion, some ideas for new missions, such as those that are being proposed within the ESA (European Space Agency) M3 call for mission concepts, are described.

2 Solar Space Observations

For more than 30 years we have been using space observation to break the limitation imposed by the earth's atmosphere on the wavelength range transmitted to ground-based observatories. In the light of new techniques developed to sound the solar interior, it appears that space observation also can open new domains for scientific investigation, such as helioseismology. In the following we analyze some advantages, disadvantages and limitations of solar space observations.

2.1 Wavelength Coverage

From space the whole electromagnetic wavelength domain is accessible on either side of the visible window. The real access to different parts of the spectrum begins at altitudes such as shown on Figure 1 where the altitude at which transmission from space electromagnetic radiation is reduced by a factor of 2 is plotted against wavelength. One can see that the X-rays and Ultraviolet solar radiations below 200 nm can only be observed at altitudes reached by rockets and satellites, while most of the γ- rays and infrared-submillimeter wavelengths can be detected at balloon altitudes (35-40 km).

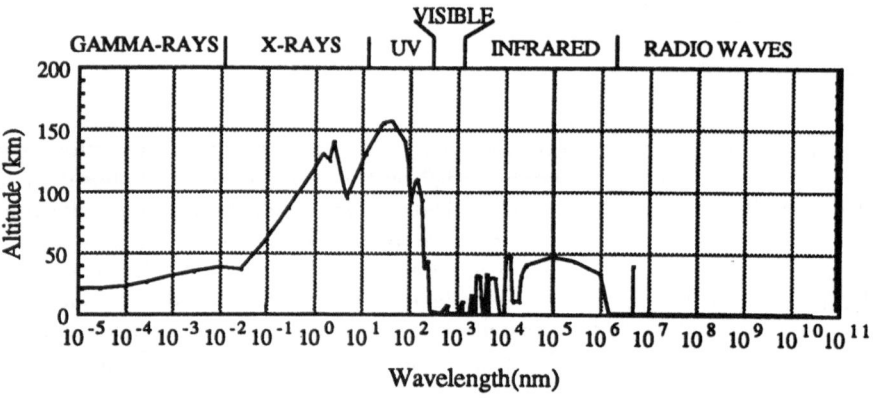

Fig. 1. Altitude where the space radiation is reduced by a factor of 2, as a function of wavelength. The spectral classification in different wavelength domains is indicated (from Goldberg 1969)

Up to now, apart from a few obscured windows, the infrared-submillimeter solar spectrum (15 μm - 1 mm) has been very coarsely explored from the ground with wide band filters and broad resolution on the disk. Instrumental difficulties have to be overcome: these include the size of the telescope and the resolution of the spectrometer. First, to obtain a high image quality the telescope size must be very large: a 51 m diameter telescope only provides one arcsec resolution at 200 μm. Then, in the far infrared, classical spectrometers with gratings or Fourier Transform Spectrometers with reasonable size do not provide the very high spectral resolution needed to obtain line profiles (a resolving power $\lambda/\Delta\lambda \simeq 30000$ is required). In this area the development of Heterodyne Spectroscopy in the vicinity of the frequency of a local oscillator, although limited to few frequencies, is a promising technique. The last and also the most sensitive problem is the detection system; much effort has been spent in this domain and we can expect substantial progresses in the near future.

As the exploration of wavelengths longer than the visible is still a prospective area for the future, this paper will concentrate on the shortest wavelengths that are used by existing instrumentation, or that to be flown in the very near future.

To represent the solar spectrum, it is convenient to use the differential number of photons as a function of wavelength, as now we have the ability to use photon counting devices from γ-rays up to a few microns. From Figure 2 it is clear that going from the visible to shorter wavelengths the number of photons per square centimeter per nanometer per second ($phot.cm^{-2}.nm^{-1}.s^{-1}$) decreases very sharply; this immediately demonstrates the difficulty of working at short wavelengths. The difficulty is worsened when the detection of line profiles is required at the same resolving power ($\lambda/\Delta\lambda$ or $\nu/\Delta\nu$) along the spectrum; at short wavelengths the $\Delta\lambda$ interval has to be very small, resulting in a reduced number of photons and the requirement for very high resolution instrumentation. Nevertheless, there is some advantage in using the shorter wavelengths, irrespective of the access to the outer solar atmosphere and the large temperature and density ranges encountered: this is the possibility of obtaining the same angular resolution as in the visible with a smaller telescope (to a first approximation the angular resolution of a circular telescope is given by 1.22 λ/D, where D is the diameter of the telescope).

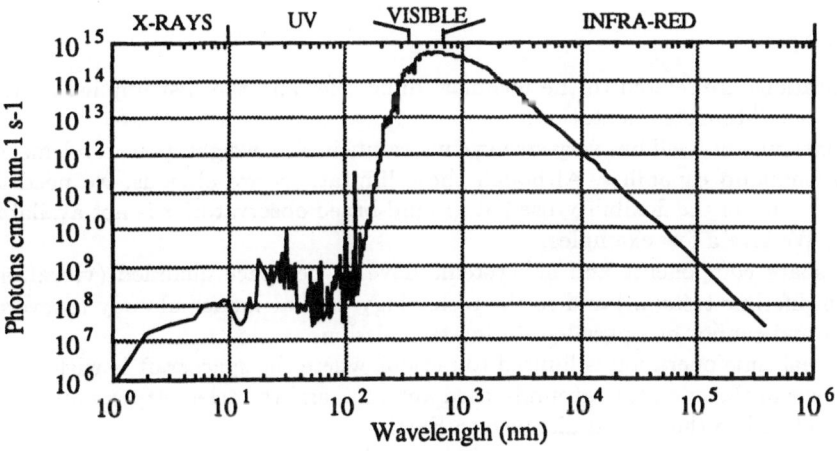

Fig. 2. Spectral irradiance distribution of the solar spectrum ($phot.cm^{-2}.nm^{-1}.s^{-1}$). The data used are taken from Donnelly et al. 1973 and Thekaekara 1974.

2.2 Space Vacuum

Besides the increase of the wavelength coverage the photons in space above 300 km travel through a medium with a constant index of refraction nearly equal to unity, with a density 2.10^{-10} that of the density at sea level. The small photometric fluctuation induced along the line of sight is at least 6 orders of magnitude smaller than the atmospheric photometric noise detected from ground level instrumentation. Going above 500 km improves this by a factor of 100. The consequence is that it is possible to detect very small brightness fluctuations of the solar emission (lines or continuum):

- to obtain a very homogeneous set of data only limited by instrument stability and photon noise,

- to collect an uninterrupted long time series for helioseismology provided we have made a proper selection of the spacecraft orbit.

2.3 Full Sun Coverage

At any given time From Earth we only have a partial view of the Sun, limited to the solar hemisphere in front of Earth. Using several spacecraft it is possible to have a full coverage of the Sun continuously. This can be achieved with two spacecraft combined with an Earth-orbiting spacecraft separated by 120° in the ecliptic plane and with two others looking over the solar poles. The original Out-of-Ecliptic mission was an attempt to survey both solar poles simultaneously; the Ulysses mission partially covers this objective but without an optical survey of the poles, and without the chance to obtain simultaneous observations. The Solar Probe is another project that was proposed to investigate the corona all around the Sun.

2.4 Limitations

The limitations are related to the technical difficulties and the cost implied by the launch of satellite instrumentation.

The instrument itself has very strong constraints in size, weight, power, telemetry and telecommand capability. Although these limitations are obvious, we need to recall that a lot of the flexibility used at ground-based observatories is not available in space. We give a few examples:

- all space components and subsystems have to be space qualified (vibration, radiation, lifetime tests, ...) and so the technology used at launch already is several years old and cannot be upgraded during the mission,

- the real time operation is limited to periods where the spacecraft is within the field-of-view of the operating ground-based antenna, and the telemetry capability is strongly related to the orbital distance to Earth.

Just as a ground-based telescope is not a perfect instrument with respect to its operations, the spacecraft platform pointing capability and stability are limited by mission constraints (mainly budget and orbital perturbations). One of the major perturbations to stability and operation in orbits around the Earth is the frequency of day/night cycling; typically a 500 km altitude spacecraft at low inclination has a period of about 96 min with around 60 minutes in sunlight.

3 A Current Mission - Yohkoh

The Yohkoh satellite has been developed and launched by ISAS (Institute of Space and Astronautical Science, Japan) on 30 August 1991. The general mission objectives and first results are summarized by Ogawara et al. 1991 and Acton et al. 1992. The detailed instrument specifications are given in Kosugi et al. 1991, Tsuneta et al. 1991, Yoshimuri et al. 1991 and Culhane et al. 1991. The main characteristics of the instruments on-board Yohkoh are presented in Table 1 .

Table 1. Main characteristics of the Yohkoh scientific payload.

Instrument	Wavelength coverage	Time res. (s)	Spectral res.	Field of view (")	Ang. res . (")	Detector	Ref.
SXT	.3-4.5nm	>.5	1-4.2 nm	2512^2	3.	CCD	Tsuneta
Visible	425-437nm	>.08	3 nm	2512^2	3.	CCD	et al
telescope	445-472nm		150nm				1991
HXT	15-100keV	>0.5	10-40keV	Full disk 64 Chan.	5	NaI(Tl)/PM	Kosugi et al. 1991
WBS	2-30 keV	>0.125	~10-20%	Full disk		BPC(128)	Yoshimuri
	20-600 keV					NaI(32)	et al. 1991
	.2-100 MeV					BGO(134)	
	5-300 keV					NaI+Si	
BCS	.176-.180nm	> 0.125	300	Full disk		GPC	Culhane
	.182-.189nm		to 8000			(Ar+Xe	et al. 1991
	.316-.319nm					+ CO_2	
	.501-.511nm						

The good stability of the pointing system (0.5 arcsec/second) permits a good angular resolution for the SXT where the exposure times are in the one second range. Combined with the high quality of the grazing incidence telescope, the SXT provides very high quality X-ray images.

The high spectral resolution achieved by the BCS illustrates the compromise between spectral resolution, spectral coverage and angular resolution.

4 The Next Mission to Fly - CORONAS I

The CORONAS I project is developed under the leadership of IZMIRAN (Institute for Terrestrial Magnetism, Ionosphere and Propagation of Radiowaves of the Academy of Sciences of Russia) and will be launched before the end of 1993. The general mission objectives are given in Oraevsky and Zhugzhda 1991. Detailed information on the instruments can be obtained in Kuznetsov et al. 1991, Rieger et al. 1991, and Sobelman et al. 1991. The principal parameters of the instruments are described in Table 2. As an example of the new technology used, it should be noticed that some of the TEREK UV Telescopes are normal incidence optics coated with metallic multilayers to select a very thin bandpass around EUV (Extreme UltraViolet) solar lines. The advantages are a better selection of temperatures in the corona compared to grazing incidence optics with filters, and a reduced level of scattered light given by better polishing.

Table 2. Characteristics of the CORONAS-I payload

Instrument	Wavelength coverage	Time res. (s)	Spectral res.	Field of view (")	Ang. res. (")	Detector	Ref.
DIFOS photometer	520nm 710nm 400-100nm	16	100nm 120nm 600nm	Full disk		diodes	Oraevsky et al. 1991
SUFR uv radiometer	0.3-2.5 nm 0.3-12 nm 121.6nm	18	2.2nm 11.7nm	Full disk		PM	Oraevsky et al. 1991
SORS radiospectro.	.05-30 MHz 25-300 MHz	0.1	10 kHz	Full disk		2	Oraevsky et al. 1991
VUSS uv spectro.	20-58nm		400	Full disk		Ne gas	Oraevsky et al. 1991
TEREK uv telescope	.5-2.5nm 17-18nm 30.4nm			Full disk corona	20 7 30	ICCD CCD	Sobelman et al. 1991
RES-C X ray spectro.	19-20.5nm .841-.843nm .185-.187nm		.02pm .07pm 3.pm	Full disk	120 6x6 6x90	ICCD	Sobelman et al. 1991
DIOGENESS BS BF	.2835-.3356nm 2-8 keV 10-160keV	0.1 1 0.1		Full disk			Oraevsky et al. 1991
HELIKON X γ-ray	10-50keV 50-250keV 250-800keV	.002		Full disk		scintil 64 chan.	Oraevsky et al. 1991
IRIS	2-300keV	0.01		Full disk		12 chan.	Oraevsky et al. 1991
SKL particles	0.1-500MeV	∼0.01					Kuznetsov et al. 1991

The polar orbit of the spacecraft and the instantaneous good stabilization will permit long series of observations for photometric studies and good angular resolution with the ultraviolet imaging telescopes.

2.2 The Following Mission - SOHO

The ESA/NASA SOHO (SOlar and Heliospheric Observatory) mission is in development, and it is currently scheduled to be launched in July 1995 and, after 4 months of navigation to be positioned in a halo orbit around the L_1 lagrangian point between the Earth and the Sun. The mission objectives, with the main parameters of each instruments are described in the SOHO Mission report 1989. In Table 3 we summarize the main characteristics of the instruments.

The uninterrupted view of the Sun at the L_1 point gives SOHO a unique opportunity to make helioseismology measurements; this important fact is utilised by GOLF, VIRGO and MDI. The high stability of the spacecraft (1 arcsec during 15 min) is fully taken in account by the UV spectrometers (CDS and SUMER) and the set of coronal imagers (EIT, UVCS, and LASCO).The other instruments are well-suited for studying the heliosphere and the extended corona up-to the vicinity of Earth (SWAN, CELIAS, COSTEP and ERNE).

As an example of the continuous improvement of the capability of telescopes in the UV we present in Figure 3 a comparison of the telescope figures of the HCO Skylab SO-54 grazing incidence telescope (X-ray, Vaiana et al. 1977), the SXT/Yohkoh

Table 3. Characteristics of the SOHO payload

Instrument	Wavelength coverage	Time res. (s)	Spectral res.	Field of view (")	Ang. res. (")	Detector	Ref.
GOLF	588.9-589.7	~40	~.1 pm	Full disk		PM	Gabriel et al. 1989
VIRGO CRM/PM0	All lambda			Full disk		cavities	Fröhlich et al. 1989
SPM	335-500-865	10	5	Full disk		diodes	
LOI	500	10	5	Full disk	12 pix	diode array	
MDI/SOI	<.1	0.2	.01	2000^2	1.4	CCD 1024^2	Scherrer et al. 1989
SUMER	50.-80. 80.-160.	>.02	2 pm 4 pm	.3×120 to 1×300	>1 ×1	MAMA 360×1024	Wilhelm et al. 1989
CDS	15.5-78.7 31.-38. 51.7-63.3	>.25	3-7 pm 8pm 0.14	>2×2 >2×240	2×2 >2×2	4CMA-2000 MCP/CCD 1024^2	Patchett et al. 1989
EIT	17.1,19.5 28.4,30.4	>2	1.	3000^2	2-3	CCD 1024^2	Delaboudinière et al. 1989
UVCS	49.9-52.1 61.0-62.5 100.-104.2 120.-125.0	>10	>0.1	1.5-10R$_\odot$	7	MAMA 360×1024	Kohl et al. 1989
LASCO C1,C2,C3	400-800 FPI 530.3	few sec.	>1	1.1-30R$_\odot$ .04	>2.8 1.1-3R$_\odot$	3 CCD 1024^2	Michels et al. 1989
SWAN	121.6	1	1 pm	2π	$1°×1°$	ICCD (5×5)	Bertaux et al. 1989
CELIAS	.1 keV	>4					Hovestadt et al 1989
COSTEP	-330 Mev						Kunow et al 1989
ERNE particles							Torsti et al. 1989

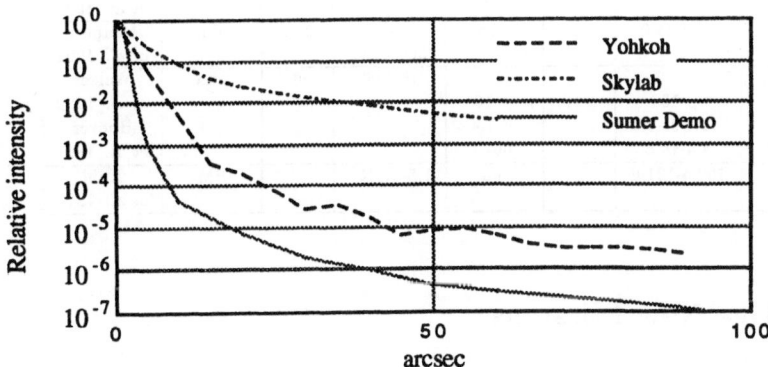

Fig. 3. Point spread profiles of Skylab SO-54 Vaiana et al. 1977, Yohkoh SXT Tsuneta et al. 1991, and Soho SUMER (1992)

grazing-incidence telescope (X-ray, Tsuneta et al 1991) and the SUMER/SOHO normal- incidence telescope.

6 Current Rocket Instruments

The rocket program is critical for the development, testing and improvement of new instruments; and at the same time it may provide exciting new results. All the instruments described in Table 4 have flown on Black Brant rockets with 0.2-0.3 arcsec stabilization. HRTS has also flown on SPACELAB 2 (1 arcsec stabilization) and UVCS on SPARTAN (arcsec stabilization). With the exception of the UVCS instrument all the others have used photographic films as the recording medium. The short duration of the observation and the need for collection of a large quantity of data during this short time has biased the choice towards the photographic detector, with all the inconvenience inherent in this choice; in the future, the increasing capability of large bidimensional photoelectric arrays and large telemetry capacity may lead to a more widespread use of electronic detectors.

Table 4. Characteristics of some rockets payloads

Instrument	Wavelength coverage	Time res. (s)	Spectral res.	Field of view (")	Ang. res. (")	Detector	Ref.
NRL/HRTS(1975, ...)	117.5-171. 158.-162.	>3. >3.	.005 4.	1×1000 600×1000	1×1 1×1	film	Brueckner et al. 1978
LPSP/TRC (1979, ...)	121.6-160.- 220.	.1-1.	8.5-13	1000²	1×1	film	Bonnet et al. 1980
HCO/NIXT(1989, ...)	0.14	1-60	.14	2500²	.7×.7	film	Spiller et al. 1991
SERTS (1987, ...)	22.5-44.0	>2.	.005	.5×48 300×480	2×2	film	Neupert et al. 1992
MSSTA (1991 ..) 7R-C telescopes 2 Cassegrain tel. 8 Herschelian tel.	15.-155. 17.-21.5 4.4-14.5	>10	.6-10 .6-1.4 .06-.7	≃4000	≃0.7	film	Walker et al. 1992 Hoover et al. 1991
UVCS WLC	102.6-121.6		.03	1.5-3.5R⊙		PM	Kohl et al. 1978

7 Prospective New Missions

This section includes the next possible Japanese Solar mission (which if selected would become Solar B) and the the long term NASA program to study the Mechanisms of Solar Variability (MSV) and the high energy solar physics mission (HESP).

7.1 The Proposal for the Next Japanese Mission

This very ambitious mission (Hirayama 1993) is currently under study in Japan with collaborations open to other countries. It is hoped that it may begin in 1994 and be launched at the turn of the century. The major objective is 'the resolution of the coronal and chromospheric heating by investigating the dynamics of photospheric (and chromospheric) magnetic flux tubes and the resultant heating of the chromosphere and the corona' (Hirayama 1993). The basic payload proposed consists of the following:

- a diffraction-limited optical telescope of 80cm size aperture of 0.1- 0.2 arcsec resolution, with a set of focal plane instruments,

- a soft X-ray telescope.

To complement this core payload other instruments, related to the scientific objective, are envisaged such as EUV spectrometer and UV/EUV imagers.

7.2 MSV

This is a long term program to study the Sun developed by NASA (MSV 1992, 1993). During the first 5-year phase (MSV-0) the goal is the construction of a 1 meter class solar balloon-borne telescope, which would then have multiple long-duration flights. This telescope will operate in the visible part of the spectrum and will have a 0.1 arcsec resolution capability in order to explore the small-scale structures of the solar photosphere. It will be complemented in the UV and X-ray by a series of rocket payloads with arcsecond or better resolution. The second phase (MSV-1) will be a satellite mission, planned to start development in the late 1990s.

7.3 HESP

The High Energy Solar Physics mission is waiting for a new start in 1995 at NASA (HESP 1993). The scientific objectives are the detailed study of solar flares through the Hard X-Ray and Gamma-Ray continuum produced by accelerated electrons. The goal is to have high spatial resolution, in the 1-2 arcsecond angular range; 1-2 keV resolution in the 10 keV-10 MeV spectral range; and tens of milliseconds in the temporal domain. Complementary instruments working in the UV-Soft X-ray range will provide complementary data on the erupting flares. This mission is optimized for the next solar maximum and it is proposed to fly it on three "Lightsats" (small satellites) that can be launched by cheap rockets:

- two satellites, spin-stabilized, into an equatorial orbit with a High Energy Imaging and an Imaging Bragg Spectrometer, and an Ultra- High Energy Gamma-Ray and Neutron Spectrometer

- a 3-axis stabilized satellite in a Sun-synchronous orbit, with EUV/Soft X-Ray imaging telescope and spectrometer.

8 Conclusions

The continuity of space solar physics missions seems to be assured for the near future. From Yohkoh, CORONAS-I to SOHO a well-balanced set of instruments has been and is been developed. Extrapolating to the longer-term future is more hazardous given the projections of the global situation of space sciences in the years to come. It is now evident that any mission in solar physics can be only realized in the context of international cooperation, with all the constraints implied in this type of very large cooperation. The few future missions presented here have a good probability of being accomplished. In the long term range, from the recent call for ideas proposed by ESA for the Medium size (M3) missions, at least two solar missions are proposed:

- an update of the Solar Probe mission to sound in-situ the corona at a few solar radii,
- a new version of the SIMURIS (Damé et al 1993) interferometer.

After the selection process (end of sepember 1993) anyone solar mission has been selected

Acknowlegements. This paper has benefited from my participation in the Oslo Workshop on Future Space Solar Physics Missions in January 1993, where I was invited by P. Maltby. I also want to thank B. Dennis and W. Wagner for unpublished information on the MSV and HESP programs. A part of my travel to the 7th European Meeting on Solar Physics was paid by the Direction des Relations Internationales du Centre National de la Recherche Scientifique.

References

Acton, L., Tsuneta, S., Ogawara, Y., Bentley, R., Bruner, M., Canfield, R., Culhane, G., Doschek, G., Hiei, E., Hirayama, T., Hudson, H., Kosugi, T., Lang, J., Lemen, J., Nishimura, J., Makisima, K., Uchida, Y., Watanabe, T. (1992): "The Yohkoh Mission for High-Energy Solar Physics", Nature **58** pp 618-625.

Baum, W.A., Johnson, F.S., Oberly, J.J., Rockwood, C.C., Strain, C.V., Tousey, R. (1946): "Solar Ultra-violet Spectrum to 88 kilometers", Phys. Review **70**, 781.

Bertaux, J.L., Pellinen, R., Chassefière, E., Dimarellis, E., Goutail, F., Holzer, T.E., Kelhä, V., Korpela, S., Kyrolä, E., Lallement, R., Leppälä, K., Leppelmeier, G., Liede, I., Rautonen, K., Torsti, J. (1989): "'SWAN'- A Study of Solar Wind ANisotropies", in The SOHO Mission, Scientific and Technical Aspects of the Instruments, ed. by V. Domingo, **ESA SP-1104**, pp 63-68.

Bonnet, R.M., Bruner, E.C. Jr., Acton, L.W., Brown, W.A., Decaudin, M. (1980): "High-resolution Lyman-alpha filtergrams of the Sun", Astrophys. J., **237**, L45-L50.

Culhane, J.L., Hiei, E., Doschek, G.A., Cruise, A.M., Ogawara, Y., Uchida, Y., Bentley, R.D., Brown, C.M., Lang, J., Watanabe, T., Bowles, J.A., Deslattes, R.D., Feldmen, U., Fludra., A., Guttridge, P., Henins, A., Lapington, J., Magraw, J., Mariska, J.T., Payne, J., Phillips, K.J.H., Sheather, P., Slater, K., Tanaka, K., Towndrow, E., Trow, M.W., Yamaguchi, A. (1991): "The Bragg Crystal Spectrometer for Solar-A", Solar Phys. **136** pp 89-104.

Damé, L., Martic, M., Rutten, R.J. (1993): "Prospects for Very- High-Resolution Solar Physics with the Simuris Interferometric Mission", in Scientific Requirements for Future Solar-Physics Space Missions, ed. by P. Maltby and B. Battrick, **ESA SP-1157** pp. 119-144

Delaboudinière, J.P., Gabriel, A.H., Artzner, G., Michels, D., Dere, K., Howard, R., Catura, R., Stern, R., Lemen, J., Neupert, W., Cugnon, P., Koecklenbergh, A., Van Dessel, E.L., Jamar, C., Maucherat, A., Chauvineau, J.P., Marioge, J.P. (1989): "'EIT'- Solar corona synoptic observations from SOHO with an Extreme- ultraviolet Imaging Telescope", in The SOHO Mission - Scientific and Technical Aspects of the Instruments, ed. by V. Domingo, **ESA SP-1104** pp. 43-48

Donnelly, R.F., Pope, J.H. (1973): "The 1-3000 Å Solar Flux for a Moderate Level of Solar Activity for Use in Modeling the Ionosphere and Upper Atmosphere", in NOAA Technical Report ERL 276-SEL 25 (U.S. Department of Commerce, Boulder) pp. 1-40

Fröhlich, C., Andersen, B.N., Berthomieu, G., Crommelynck, D., Delache, Ph., Domingo, V., Jimenes, J., Jones, A.R., Roca Cortès, T., Wehrli, Ch. (1989): "'VIRGO'- The solar monitor experiment on SOHO, in in The SOHO Mission - Scientific and Technical Aspects of the Instruments, ed. by V. Domingo, **ESA SP-1104** pp. 19-23

Gabriel, A.H., Bocchia, R., Bonnet, R.M., Cesarsky, C., Christensen-Dalsgaard, J., Damé, L., Delache, PH., Deubner, F.L., Foing, B., Fossat, E., Fröhlich, C., Gorisse, M., Dough, D., Grec, G., Hoyng, P., Pallé, P., Paul, J., Robillot, J.P., Roca Cortès, T., Stenflo, J.L., Ulrich, R.K., van der Raay, H.B. (1989): "'GOLF'- Global Oscillations at Low Frequencies for the SOHO mission, in The SOHO Mission - Scientific and Technical Aspects of the Instruments, ed. by V. Domingo, **ESA SP-1104** pp. 13-17

Goldberg, L. (1969):"Ultraviolet Astronomy", Scientific American June 1969, pp. 92-102

HESP- High Energy Solar Physics Mission 1993,"Report of the HESP Science Study Group" by Lin, R., Dennis, B., Canfield, R., Crannell, C., Davis, J., Doschek, G., Emslie, G., Fisher, R., Forrest, D., Haerendel, G., Reiger, E., Hudson, H., Hurford, G., Vedrenne, G., Lavigne, J.M., Ling, J., Pick, M., Vilmer, N., Ramaty, R., van Beek,H. F., Wagner, W.

Hirayama, T. (1993):" The Next Japanese Solar satellite", in Scientific Requirements for Future Solar-Physics Space Missions, ed. by P. Maltby and B. Battrick, **ESA SP-1157** pp. 105-113

Hoover, R.B., Walker, A.B.C. Jr., Lindblom, J., Allen, M., ONeal, R., DeForest, C. (1991): "Solar observations with the Multi-Spectral Solar Telescope Array", in Multilayer and Grazing Incidence X- Ray/EUV Optics, SPIE vol. **1546** pp. 175-187

Hovestadt, D., Geiss, J., Gloecker, G., Möbius, E., Boschsler, P., Gliem, F., Ipavich, F.M., Wilken, B., Axford, W.I., Balsiger, H., Bürgi, A., Coplan, M., Dinse, H., Galvin, A.B., Gringauz, K.I., Grünwaldt, H., Hsieh, K.C., Klecker, B., Lee, M.A., Managadze, G.G., Marsch, E., Neugebauer, M., Rieck, W., Scholer, M., Stüdemann,W. (1989): "'CELIAS'- Charge, Element and Isotope Analysis System for SOHO" in The SOHO Mission, Scientific and Technical Aspects of the Instruments, ed. by V. Domingo, **ESA SP-1104**, pp. 69-74

Kohl, J.L., Noci, G., Hartmann, L.W., van Ballegooijen, A.A., Raymond, J.C., Weiser, H., Withbroe, G.L., Antonucci, E., Huber, M.C.E., Geiss, J., Livi, S., Gloeckler, G., Rosner, R., Hollweg, J.V., Tondello, G. (1989): "'UVCS' - An UltraViolet Coronagraph Spectrometer for SOHO", in The SOHO Mission-Scientific and Technical Aspects of the Instruments, ed. by V. Domingo, **ESA SP-1104** pp. 49-54

Kohl, J.l., Reeves, E.M., Kirkham, B. (1978): " The Lyman Alpha Coronograph", in New instrumentation for Space Astronomy, ed. by K. van De Hucht qnd G.S. Vaiana, Pergamon Press, pp. 91-94

Kosugi, T., Makishima, K., Murakami, T., Sakao, T., Dotani, T., Inda, M., Kai, K., Masuda, S., Nakajima, H., Ogawara, Y., Sawa, M., Shibasaki, K. (1991): "The Hard X-ray telescope (HXT) for the Solar-A Mission", Solar Phys. **136** pp. 17-36

Kunow, H., Fischer, H.,Green, G., Müller-Mellin, R., Wibberenz, G., Holweger, H., Evenson, P., Meyer, J.-P., Hasebe, N., von Rosenvinge, T., Reames, D., Medina, J., Witte, M., Matsoka, M., Mardsen, R.G., Sanderson, T.R., Wenzel, K.P., McKenna- Lawlor, S., Sequeiros, J., Doke, T., Kikuchi, J. (1989): "'COSTEP'- COmprehensive SupraThermal and Energetic Particle analyser for SOHO", in The SOHO Mission, Scientific and Technical Aspects of the Instruments, ed. by V. Domingo, **ESA SP-1104**, pp. 75-80

Kuznetsov, S.N., Gotselyuk, Y.V., Podorolsky, A.N., Tyumin, S.P., Fisher, S., Kudela, K. (1991): "Scientific goals and Specific Devices of the 'Solar Cosmic Rays'(SCR) Scientific Set of Instruments in the 'CORONAS' Project", CORONAS information No 3, ed. by K. Pflug (Institute for Astrophysics in Postdam) pp. 1-6

MSV - Mechanisms of Solar Variability Program (1992): "The Mechanisms of Solar Variability (MSV) program", Report of a Worshop convened by the OSSA Space Physics Subcommittee jan. 7- 9, 1992; 1993: "MSV program would study solar change", in Space Physics Newsletter, ed. P. Glosser, vol. **5**, issue 1, pp. 3-4

Michels, D.J., Schwenn, R., Howard, R.A., Bartoe, J.-D.F., Antiochos, S.K., Brueckner, G.E., Cheng, C.-C., Dere, K.P., Doscheck, G.A., Mariska, J.T., Sheeley, N.R.Jr., Socker, D.G., Daly, P.W., Inhester, B., Keller, H.U., Kramm, J.R., Rosenbauer, H., Lamy, P., Llebaria, A., Maucherat, A., Parker, E.N., Kahler, S.W., Koutchmy, S.L., Smartt, R.N., Wagner, W.J., Bougeret, A.L., Pick, M., Noëns, J.C., Giese, R.H., Koomen, M.J., Giovane, F., Misconi, N.Y., Simnett, G.M., Eyles, C., Bedford, D., Priest, E., Lallement, R. (1989): " LASCO - A white Light And Spectrometric COronagraph for SOHO", in The SOHO Mission- Scientific and Technical Aspects of the Instruments, ed. by V. Domingo, **ESA SP-1104** pp. 55-61

Neupert, W.M., Epstein, G.L., Thomas, R.J., Thompson, W.T. (1992):"An EUV imaging spectrograph for High-resolution observation of the solar corona ", Solar Phys. **137** 87

Ogawara, Y., Takano, T., Kato,T., Kosugi, T., Tsuneta,S. , Watanabe, T., Kondo, I., Uchida, Y. (1991): "The Solar-A Mission: an overview", Solar Phys. **136** pp. 1-16

Oraevsky, V.N., Zhugzhda, Y.D. (1991): "Project CORONAS-I: Orbital Observations of the Solar Activity and Oscillations", CORONAS information No 1, ed. by K. Pflug (Institute for Astrophysics in Postdam) pp. 1-11

Patchett, B.E., Harrison, R.A., Sawyer, E.C., Aschenbach, B., Culhane, J.L., Doschek, G, Gabriel, A.H., Huber, M.C.E., Kjeldseth-Moe, O., Mason, H.E., McWhirter, R.W.., Parkinson, J.H., Poland, A.I., Priest, E.R., Schmitt, J.H.M.M., Thomas, R.J., Timothy, J.G., Tondello, G., Trumper, J. (1989): "'CDS' - The Coronal Diagnostic Spectrometer", in The SOHO Mission - Scientific and Technical Aspects of the Instruments, ed. by V. Domingo, **ESA SP-1104** pp. 39-42

Rieger, E., Kanbach, G., Marschhäuser, H. (1991): "Co- Investigator proposal for the SONG-Experiment of the SKL-Complex on the CORONAS-I Satellite: Temporal and Spectral Investigation of the Solar Flares in the Energy Rage 0.1 to ≈ 100MeV", CORONAS information No 4, ed. by K. Pflug (Institute for Astrophysics in Postdam) pp. 1-6

Scherrer, P.H., Hoeksema, J.T., Bogart, R.S., Walker, A.B.C. Jr., Title, A.M., Tarbell, T.D., Wolfson, C.J., Brown, T.M., Christensen- Dalsgaard, J., Gough, D.O., Kuhn, J.R., Leibacher, J.W., Libbrecht, K.G., Noyes, R.W., Rhodes, E.J. Jr., Toomre, J., Zweibel, E.G., Ulrich, R.K. Jr. (1989): "'SOI'- The Solar Oscillation Imager for SOHO", in in The SOHO Mission - Scientific and Technical Aspects of the Instruments, ed. by V. Domingo, **ESA SP-1104** pp. 25-30

Sobelman, I., Zhitnik, I., Ignatiev, A., Korneev, V., Krutov, V., Kuzin, S., Lomkova, V., Mitrofanov, A., Oparin, S., Pertzov, A., Slemzin, V., Tindo, I., Urnov, A., Salashenko,

S., Valnicek, B., Gudec, R., Rybansky, M., Koutchmy, S., Zharkova, V. (1991): "XUV and Optical Observations of the Sun by Means of TEREK-C Telescope/Coronograph and RES-C Spectroheliometer aboard the CORONAS-I Satellite", CORONAS information No 5, ed. by K. Pflug (Institute for Astrophysics in Postdam) pp. 1-11

Spiller, E., Wilczynski, Golub, L., Nystrom, G. (1991): "The Normal Incidence Soft X-Ray, λ=63.5Å, Telescope of 1991", in Multilayer and Grazing Incidence X-Ray/EUV Optics, SPIE vol. **1546** pp. 168-174

Thekaekara, M.P. (1974): "Extraterrestrial Solar Spectrum, 3000-6100 Å at 1-Å Intervals", Appl. Opt. **13** 518

Torsti, J., Aurela, A., Kehlä, V., Leppelmeier, G., Leppälä, K., Liede, I., Lumme, M., Osborne, J.L., Pekllinen, R., Saarikko, H., Tenhunen, H., Urpo, S., Valtonen, E., Wolfendale, A.W., Äystö, J. (1989): "'ERNE'- Energetic and Relativistic Nuclei and Electron experiment", in The SOHO Mission, Scientific and Technical Aspects of the Instruments, ed.by V. Domingo, **ESA SP-1104**, pp. 81-84

Tsuneta, S., Acton, L., Bruner, M., Lemen, J., Brown, W., Caravalho, R., Catura, R., Freeland, S., Jurcevich, B., Morrison, M., Ogawara, Y., Hirayama, T., Owens, J. (1991):" The Soft X-ray Telescope for the Solar-A Mission": Solar Phys. **136** pp. 37-67

Vaiana, G.S., Van Speybroek, L., Zombeck, M.V., Krieger, A.S., Silk, J.K., Timothy, A. (1977): Space Sci. Instr. **3** 19

Walker, A.B.C. Jr., Hoover, R.B., Barbee, T.W. Jr. (1992): "The Multi-Spectral Solar Telescope Array: Initial Results and Future Plans", in Multilayer and Grazing Incidence X-Ray/EUV Optics for Astronomy and Projection Lithography, ed. by R.B. Hoover and A.B.C. Jr. Walker, SPIE vol. **1546** pp. 500-514

Wilhelm, K., Axford, W.I., Curdt, W., Gabriel, A.H., Grewing, M., Huber, M.C.E., Jordan, S.D., Lemaire, P., Marsch, E., Poland, A.I., Richter, A.K., Thomas, R.J., Timothy, J.G., Vial, J.-C., (1989): "'SUMER' - Solar Ultraviolet Measurements of Emitted Radiation", in The SOHO Mission - Scientific and Technical Aspects of the Instruments, ed. by V. Domingo, **ESA SP-1104** pp. 31-37

Yoshimori, M., Okudaira, K., Hirashima, Y., Igarashi, T., Akasaka, M., Takai, Y., Morimoto, K., Watanabe, T., Ohki, K., Nishimura, J., Yamagami, T., Ogawara, Y., Kondo, I. (1991):"The Wide band Spectrometer on the Solar-A", Solar Phys. **136** pp. 69-88

The Yohkoh Mission:
Instruments and Recent Results

R.D.Bentley [1] and the Yohkoh Team [2]

[1] Mullard Space Science Laboratory, Holmbury St. Mary, Dorking,
Surrey RH5 6NT, United Kingdom
[2] Institute for Space and Astronautical Science, Yoshinodai 3-1-1,
Sagamihara-Shi, Kanagawa 229, Japan

Abstract: The Yohkoh spacecraft was launched on 30 August, 1991 by the Japanese Institute of Space and Astronautical Science. The instruments, which include hard and soft X-ray imaging telescopes, a Bragg spectrometer and wide band full sun spectrometers, (which cover the energy range 2 kev to 10 MeV), were described. Results obtained from these instruments were presented with special emphasis on data from the soft X-ray telescope which views the Sun in the 0.2 - 3 keV range with an angular resolution of 2.5 arc sec. Videos made from SXT datas were also presented at the meeting.

1 The Yohkoh Mission

The Yohkoh spacecraft was launched into low-Earth orbit on August 30, 1991 from the Kagoshima Space Centre. The principle aim of the Yohkoh mission is the observation of energetic phenomena related to solar flares in X- and gamma-rays using a coordinated set of instruments. Yohkoh, which is the only solar mission launched near the maximum of the current solar cycle, has made many new and interesting observations and the Yohkoh Team is making concerted efforts to establish collaborations for both observations and data analysis.

The Yohkoh spacecraft weighs 400 kg and is approximately 1m × 1m × 2m in size. The payload is 3-axis stabilized, pointing at Sun centre with an accuracy of $\lesssim 1$ arcsec. Its slightly elliptical orbit is expected to give the mission a lifetime of more than 6 years. Between station contacts, the observations are stored on board in a 10 Mbyte Bubble Data Recorder (BDR); data is dumped at both contact with the Kagoshima Space Centre and ground stations of the NASA Deep Space Network. A detailed overview of the mission is given by Ogawara et al. (1991, 1992). A description of the reformatted Yohkoh database and the observing log is given by Morrison et al. (1991).

2 The Scientific Payload

Yokoh carries the following payload of scientific instruments: the Soft X-ray Telescope (SXT), Hard X-ray Telescope (HXT), the Bragg Crystal Spectrometer (BCS), and the Wide Band Spectrometer (WBS). The technical details of these instruments are described in papers in a special edition of Solar Physics (1991, Vol. 136) and only a brief outline is given here.

2.1 Soft X-ray Telescope (SXT)

The SXT (Tsuneta et al., 1991) is a grazing-incidence telescope operating in the X-ray energy band 3 to 60 Å. Detection is by a 1024 × 1024 pixel CCD which is located behind two filter wheels and a shutter device. Filter and exposure selection are controlled by a dedicated microprocessor that is commanded from tables loaded from the ground; the different filters are used to select different energy bands. The SXT instrument views the whole Sun, but during flare mode only selected parts of the image are relayed to the ground in order increase the cadence at which the instrument can operate; in this mode a time resolution of 0.5 s can be achieved.

The resolution of the SXT is around 2.5 arcsec, similar to that of Skylab, but the much lower scattering from the SXT mirror results in much sharper images than those from Skylab; additionally, the use of a CCD rather than film allows a much better time resolution than was possible with Skylab.

2.2 Hard X-ray Telescope (HXT)

The HXT (Kosugi et al., 1991) is a Fourier synthesis telescope. It consists of 64 bi-grid modulation collimators, each backed with a NaI(Tl) crystal attached to a phototube. The individual sub-collimators measure spatially-modulated photon counts providing 32 complex "Fourier" components that are combined using a maximum entropy technique on the ground to form images. The HXT instrument views the whole Sun, while the synthesis aperture, determined by the grid pitch, is about 2 × 2 arcmin; the angular resolution of the reconstructed images is about 5 arcsec. When the spacecraft is in flare mode, images are made in four energy bands (15–24–37–57–100 keV) with a temporal resolution of 0.5 s.

2.3 Bragg Crystal Spectrometer (BCS)

The BCS (Culhane et al., 1991) consists of four bent-crystal spectrometers with position-sensitive proportional counters. The wavelength bands covering S XV, Ca XIX, Fe XXV and Fe XXVI lines were chosen to get information about the temperature and motion of high temperature plasmas produced in solar flares. The BCS is equipped with its own queue memory to store initial-phase data with a time resolution of up to 0.125 s; data accumulated in the queue is read out to the main data processor at a fixed rate. The data accumulation of the BCS is controlled by a dedicated microprocessor.

The sensitivity of the BCS is up to 10 time that of a similar instrument on the Solar Maximum Mission.

2.4 Wide Band Spectrometer (WBS)

The WBS (Yoshimori et al., 1991) consists of four separate detectors: the Soft X-ray Spectrometer (SXS), the Hard X-ray Spectrometer (HXS), the Gamma-Ray Spectrometer (GRS), and the Radiation Belt Monitor (RBM). The SXS, HXS and GRS view the whole Sun in the energy bands 2–30 KeV, 20–400 KeV and 0.2–100 MeV using a variety of detection methods; the temporal resolution of the different detectors varies but lies in the range 0.125 and 4 s. The RBM views perpendicular to the solar direction and is used to automatically switch off those instrument detectors that use high voltage supplies.

3 Recent Results

Highlights of analysis work made by different members of the Yohkoh team were presented with particular emphasis on SXT datas, on the coordinated use of the Yohkoh instruments and on observations made in collaboration with ground-based observatories.

Many of these highlights appeared in a special issue of the Publication of the Astronomical Society of Japan (1992, Vol. 44, 5); others will appear in the proceedings of a meeting held at ISAS in February 1993.

A video of images from the SXT was also shown.

4 Concluding Remarks

The instruments on the Yohkoh mission were selected to produce a coordinated set of observations. During the 21 months that the Yohkoh spacecraft has been in orbit, many new and exciting observations have been made, and the Yohkoh Team hopes that these will lead to many new advances in the field of solar physics.

References

Culhane, J.L., Hiei, E., Doschek, G., Cruise, A.M., Ogawara, Y., Uchida, Y., Bentley, R.D., Brown, C.M., Jang, J., Watanabe, T., and 17 co-authors: 1991, Solar Physics, Vol. 136, 89

Kosugi, T., Makishima, T., Sakao, T., Dotani, T., Inda, M., Kai, K., Masuda, S., Nakajima, H., Ogawara, Y., Sawa, M. and Shibasaki, K.: 1991, Solar Physics, Vol. 136, 17

Morrison, M.D., Lemen, J.R., Acton, L.W., Bentley, R.D., Kosugi, T., Tsuneta, S., Ogawara, Y., and Watanabe, T.:, 1991, Solar Physics, Vol. 136, 105

Ogawara, Y., Takano, T., Kato, T., Kosugi, T., Tsuneta, S., Watanabe, T., Kondo, I., and Uchida, Y.: 1991, Solar Physics, Vol. 136, 1

Ogawara, Y., Acton, L.W., Bentley, R.D., Bruner, M.E., Culhane, J.L., Hiei, E., Hirayama, T., Hudson, H.S., Kosugi, T., Lemen, J.R., Strong, K.T., Tsuneta, S., Uchida, Y., Watanabe, T., and Yoshimori, M.: 1992, Pub. Aston. Soc. Japan, Vol. 44, 5, L41

Tsuneta, S., Acton, L., Bruner, M., Lemen, J., Brown, W., Caravalho, R., Catura, R., Freeland, S., Jucevich, B., Morrison, M., Ogawara, Y., Hirayama, T., and Owens, J.: 1991, Solar Physics, Vol. 136, 37

Yoshimori, M., Okudaira, K., Hirasima, Y., Igarashi, T., Akasaka, M., Takai, Y., Morimoto, K., Watanabe, T., Ohki, K., Nishimura, J., Yamagami, T., Ogawara, Y., and Kondo, I: 1991, Solar Physics, Vol. 136, 69

High Resolution Solar Observations: Spectropolarimetry with THEMIS

E. Landi Degl'Innocenti [1], J. Rayrole [2], P. Mein [2]

[1]Dipartimento di Astronomia e Fisica dello Spazio, Largo E. Fermi 5, I-50125 Firenze, Italia
[2]Observatoire de Paris, Section de Meudon, F-92195 Meudon Cedex, France

Abstract: Solar observations now require many capabilities: high resolution to detect fine flux tubes, polarization-free optics to measure the vector magnetic field, spectral range including many lines to disentangle thermodynamic from magnetic signatures. The site of Canary Islands, the active optics, the Cassegrain telescope, the long spectrographs and the universal filter of THEMIS fulfill many conditions for major advances in the near future of solar physics.

1 The THEMIS Project

THEMIS is an acronym which stands for *Télescope Héliographique pour l'Etude du Magnétisme et des Instabilités Solaires*. It started something like ten years ago as a uniquely french project, but, approximately two years ago, the italian solar physics community joined the project through the signature of a formal agreement (Convention) within the french *Conseil National de la Recherche Scientifique* (CNRS) and the italian *Consiglio Nazionale delle Ricerche* (CNR).

According to this Convention, CNR is participating to the project through: a) a financial contribution; b) the construction of a technologically sophisticated dome; and, c) the implementation of post-focus instrumentation with a filter of very high spectral resolution developed by the solar spectroscopy group of the *Dipartimento di Astronomia e Scienza dello Spazio* of the University of Firenze and of the *Osservatorio Astrofisico di Arcetri*.

Consequently, THEMIS is now a french-italian project, France contributing for 80% of the cost, and Italy for the remaining 20%. The observing time will be, obviously, divided between the french and italian communities according to the same percentages.

2 Scientific Goals

The main scientific goal of the THEMIS project is the study of solar magnetism at very high spatial resolution ($\leq$ 0.3 arcsec). Indeed, it is nowadays fairly well understood that solar magnetic fields are concentrated, outside sunspots, in tiny elements, whose lateral dimensions are well beyond the resolution limit that can be attained with the polarimetric instruments presently available.

With a very high polarimetric accuracy, an excellent spatial resolution, and the possibility of observing simultaneously the Stokes parameters profiles of many spectral lines, THEMIS will provide a unique breakthrough in the study of solar magnetism.

The accurate spectro-polarimetric observations provided by THEMIS will also be of extreme importance in the study of the following related topics, namely:

• Energy transport mechanisms and velocity fields in magnetic concentrations

• Magneto-hydrodynamic turbulence and magnetoconvection

• Onset of instabilities, flares, and electric currents

• Fundamental processes in atomic physics (radiative transfer for polarized radiation, Hanle effect, impact polarization, depolarizing collisions, etc...).

3 Main Technical Capabilities

The main technical capabilities of the THEMIS project have already been described in some detail by Mein and Rayrole (1989) and by Rayrole and Mein (1993). We just summarize here what has already been pointed out in these papers.

3.1 Telescope and Polarimeter

THEMIS is provided with an evacuated Cassegrain telescope with azimuthal mount and a Ritchey-Chrétien primary mirror of 90 cm. As the fundamental goal of the project is the measurement of vector magnetic fields through the signatures observed in polarized radiation, normal incidence optics is obviously necessary before the polarimetric analysis. Moreover, extreme accuracy in the construction of the entrance window is necessary. The birefringence of the window has to be kept as low as 100 Å at the edges of the window, with perfect cylindrical symmetry. Many modes of polarization analysis will be possible by exchanging automatically the analyzer and the related optics.

3.2 Image Quality

As already stated in Sec. 2, the study of solar magnetism implies high spatial resolution. This is made possible by:
a) the selection of an appropriate site (Izaña, Tenerife) whose good seeing conditions have been carefully tested by an extensive site-testing campain, promoted several years ago in the framework of JOSO, and b): the correction of image motion by means of a tilting mirror activated by a granulation tracker at frequencies exceeding 100 Hz.

3.3 Spectroscopy

The telescope is provided with high flexibility spectroscopy. This will make possible the simultaneous observation of two Stokes parameters profiles ($I \pm Q$, $I \pm U$, or $I \pm V$) in many (typically 10) spectral ranges, arbitrarily distributed along the spectrum. A long predisperser (with three exchangeable gratings) and an echelle-spectrograph will provide this unique possibility. Additive or subtractive modes will also be possible.

The spectra will be recorded by several bi-dimensional CCD cameras (288×384 pixels). Typically one will be able of observing 10 spectral ranges in 2 polarization directions, at 384 wavelengths in 288 points along the slit (which implies 2.2×10^6 data!) every few seconds.

4 Observing Modes

Several observing modes will be possible with the THEMIS telescope. These are the following:

• *Multi-line spectroscopy*
This observing mode is probably the most important. The possibility of observing the Stokes parameters profiles simultaneously in several spectral lines is expected to provide an important breakthrough in our understa nding of the morphology and the physics of magnetic concentrations at very small scales. It will not be surprising if new, unexpected phenomena will happen to be discovered on a serendipitous basis.

• *Fast two-dimensional spectroscopy*
The operating mode is the so-called MSDP (Multichannel Subtractive Double Pass). The principle of this spectroscopic method and its diagnostic content are described in Mein (1977). The possibilty of operating THEMIS in MSDP mode is due to the fact that the predisperser has a focal length close to the focal length of the main echelle-spectrograph. This type of spectroscopy is well adapted for the study of fast events and for coordinated observing campains among different observatoires.

• *Maps of longitudinal magnetic field (Magnetograph mode)*
The telescope can also be operated in the so-called magnetograph mode with the use of a slit in subtractive spectroscopy. By keeping the telescope fixed, a strip on the Sun is scanned taking advantage of the diurnal motion. Typically, 8 strips of 4 arcmin are scanned to have a full coverage of the Sun, which requires approximately 20 mn. The longitudinal component of the magnetic field and the Doppler velocity are deduced through the method of the center-of-gravity.

This observing mode is very fast and will be of great advantage in coordinated observations between the ground and space-probes (like for instance the SOHO mission).

• *Two-dimensional spectroscopy through the italian filter*
The italian filter that will be installed on the THEMIS telescope results from the combination of a tunable Universal Birefringent Filter (UBF) and a tunable Fabry-Perot Interferometer (FP). Through this combination it is possible to obtain an image of the solar surface with a field of view of the order of 1 arcmin or less in a narrow wavelength interval having a FWHM of the order of 20 mÅ at 5000 Å.

By tuning the two instruments across the profile of a spectral line one can obtain, sequentially, various images of the same region at different wavelengths (and/or in different polarization directions). By means of suitable de-stretching algorithms (November, 1986) it is the possible to recover a two-dimensional "map" of line profiles, one for each resolution element of the solar surface. An extensive description of the filter is given in Cavallini (1993).

5 Conclusion

THEMIS is scheduled to be operative in three years from now. Its main characteristics are: a polarization-free, evacuated Cassegrain telescope, a technologically sophisticated polarimeter, and a high quality spectroscopic system with long spectrographs. These characteristics, joined with the goodness of the site, the presence of active optics and the addition of a tunable, universal filter of high spectral resolution will make THEMIS one of the most powerful instruments for the study of solar magnetism in the following years.

References

Cavallini, F.: 1993, *Proceedings of the 37th Annual Meeting of the Italian Astronomical Society*, (in press)

Mein, P.: 1977, Solar Phys. **54**, 45

Mein, P., Rayrole, J.: 1989, in O. Von der Lühe (ed.), *High Spatial Resolution Solar Observations*, 10th Sacramento-Peak Summer Workshop, pag. 12

November, L.J.: 1986, Applied Optics, **25**, 392

Rayrole, J. Mein, P.: 1993, in H. Zirin, G. Ai, and H. Wang (eds.), *The magnetic and Velocity Fields of Solar Active Regions*, IAU Colloquium No. 141, pag. 170

Closing Session

Conference Summary

Carole Jordan

Department of Physics (Theoretical Physics), University of Oxford,
1 Keble Road, Oxford, OX1 3NP, UK

1 Introduction

This Seventh European Meeting on Solar Physics has been extensive in scope. The title *Advances in Solar Physics* has allowed contributions on all parts of the Sun, from the core to the interplanetary medium. It is encouraging to reflect on the progress that has been made in understanding the wide variety of physical problems that the Sun presents, since this series of meetings began. During this meeting we have heard 15 Invited Reviews, 29 Invited contributions and have viewed 183 posters, about half of which were also presented orally. To attempt a full summary is obviously an invidious task, and my selection of material is necessarily restricted. Professor Pecker made his own perceptive selection of important topics before the meeting began. I have arranged this Summary according to the seven main sessions of the meeting, but have not done justice to the one on New Instrumentation, since I had to miss it in order to prepare this presentation.

2 Solar Internal Structure

2.1 The Solar Core

For many years the only measurements of the flux of solar neutrinos were those made by Davis and his colleagues using the $^{37}Cl + \nu \rightarrow {}^{37}A$ reaction, in their experiment in the Homestake mine. These are high energy neutrinos arising from the ^8B branch of the proton-proton cycle. A *solar neutrino problem* became apparent since the then standard solar model predicted a flux which was about a factor of three larger than that observed. A number of possible modifications to the solar models were put forward. Bahcall (1989) has reviewed these observations and theories. However, developments over the past few years, in measuring neutrinos of different energies, in helioseismology and in the calculation of opacities in the solar interior now suggest that the problem does not lie in the current Standard Solar Model (hereafter, SSM) but in the neutrino physics.

Bellotti gave a comprehensive review of the status of measurements of solar neutrino fluxes. Additional results are now available from the Kamiokande water

Cherenkov detector (Hirata *et al.* 1990) and the gallium detectors, GALLEX (Anselmann *et al.* 1992) and SAGE (Abazov *et al.* 1991). In particular, he stressed the relation between the neutrinos observed in the different experiments and the nuclear reactions producing them. The measurements from GALLEX and SAGE, which can detect low energy neutrinos produced in the first step of the proton-proton cycle, have added significantly to our knowledge in this respect. He pointed out that the flux measured by the GALLEX experiment is slightly larger than the flux predicted by the proton-proton reaction alone, but that overall the results from the four existing experiments still show a deficit compared with the predictions of the SSM. However, he cautioned that some of the nuclear reactions do not have their cross-sections measured at the relevant solar energies. This topic was taken up in more detail by Paternó and Scalia, who reported results based on new cross-sections for several relevant reactions.

Dziembowski summarized the assumptions inherent in the SSM. He pointed out that the model by Bahcall & Pinsonneault (1992), which uses up-to-date opacities and nuclear reaction rates, and takes account of the gravitational settling of helium, comes closest to the ideals of an SSM. The recent models all produce about 8 Solar Neutrino Units (SNU), as detectable by the Homestake Cl experiment, so a discrepancy of about a factor of 4 still remains. He stressed the importance of recent calculations of opacities, even in the core. In particular, the inclusion of additional lines of Fe in the OPAL code (Iglesias, Rogers & Wilson 1992) has *raised* the opacity, leading to an *upwards* revision of the core temperature and of the neutrino fluxes predicted for the Cl and Ga experiments. The abundance of iron thus becomes the main source of uncertainty in the calculated neutrino flux. Bahcall & Pinsonneault (1992) adopted the meteoritic value from Anders and Grevesse (1989), which in fact now agrees with the most recent photospheric abundance, (7.51 ±0.01) recommended by Grevesse, Noels & Sauval (1992).

Another recent development concerns the effects of gravitational settling of elements in the core, particularly of helium (Cox *et al.* 1989; Proffitt & Michaud 1991). Including gravitational settling again *increases* the core temperature and the predicted neutrino flux. Of other effects, rotation would have to be so large that it can be excluded, but the upper limit to the magnetic field imposed by helioseismological measurements does not rule out the importance of magnetically induced circulation. Settling of heavier elements has not yet been included but is also expected to act in a direction to increase the predicted neutrino fluxes.

Dziembowski then discussed the information that can be learned about the core from helioseismology. Using the adiabatic sound speed as the point of comparison (see also section 2.2) there is remarkably good agreement between the values derived from the observations and the predictions of recent models (Bahcall & Pinsonneault 1992; Dziembowski *et al.* 1993, to be published). He regarded the present results within r = 0.1 $R_\odot$ to be uncertain and stressed that the accuracy of the low l-mode data needs to be improved. This point was also made by Thompson, in the context of the rotation rate of the core, which can be inferred from the *splittings* of low-degree modes, but at present the data inversion is subject to large uncertainties (see e.g. Gough 1990; Thompson 1993).

He concluded that it is unlikely that the cause of the differences between the observations and theory lies in the SSM, and that, within the present uncertainties, the MSW theory of neutrino type conversion (Mikheyev & Smirnov 1985; Wolfstein 1978) could explain these differences (Bahcall & Bethe 1990; Langacker 1993). In the long term there is the prospect that a more precise SSM could constrain the neutrino mass difference and the mixing angle.

2.2 The Radiative Zone

The effects of gravitational settling and the diffusion of elements were also common themes of the talks on the radiative zone.

Thompson described the techniques by which very accurate measurements of the frequencies of global oscillations can be used to provide detailed information on the solar interior through comparisons of the inferred and theoretical adiabatic sound speeds, which depend on the temperature. The frequencies also depend on the mean molecular weight, and hence on element abundances and the equation of state. Until recently the sound speed deduced from the observations was greater than that predicted by the theory (Christensen-Dalsgaard et al. 1985). Although using the new OPAL opacities reduces the discrepancy in the radiative zone, significant differences still remain. Also, the models, even with the new opacities, require an initial abundance of helium of $Y = 0.28$, while the helioseismology observations in the outer convective zone suggest Y = 0.23 to 0.24 (Dziembowski et al. 1991; Kosovichev et al. 1992). However, including the settling and diffusion of helium (and heavier elements) into the models removes both these difficulties (Christensen-Dalsgaard et al. 1993; Guzik & Cox 1993).

There has been increasing evidence that the p-mode frequencies vary with time (Elsworth et al. 1990; Libbrecht & Woodard 1990), and are correlated with variations in the photospheric magnetic field (Woodard & Libbrecht 1993; Bachmann & Brown 1993). The variations also appear to have a latitude dependence (Libbrecht & Woodard 1990). Thompson concluded that the changes as a function of frequency imply that the variations are caused *in the highest layers of the atmosphere*, in the photosphere, or even the chromosphere. Their likely cause is variations in the magnetic field, which can affect acoustic modes, and/or the structure of the chromosphere (Evans & Roberts 1992).

Kosovichev and Vauclair also presented evidence for mixing in the radiative interior. Vauclair discussed the rotation-induced mixing of lithium which could account for the lithium depletion observed in the Sun (Charbonnel, Vauclair & Zahn 1992), and this effect is confirmed by improvements to the theory of meridional circulation and angular momentum transport (Zahn 1992). Although the gravitational settling of lithium is less important, settling of helium does lead to about a 10% depletion in the outer convection zone.

3 Generation of Large-Scale Magnetic Fields

The mean-field theory of the solar dynamo was reviewed by Schmitt. In these models the dynamo action arises from terms which depend on differential rotation (the ω effect) and helical turbulence (the α effect). Although early mean-field theory (e.g. Steenbeck & Krause 1969) looked encouraging, in that it could account for many aspects of solar cycle surface phenomena (e.g. the butterfly diagram, Hale's polarity rules, phase relations between the field components), more recent work, including observations of solar oscillations, which can be used to determine $d\omega/dr$, has shown several of the basic assumptions to be unrealistic. Parker (1975) pointed out the problem of storing large amounts of magnetic flux in the convection zone for times of the order of the solar cycle. Schüssler (1993) has shown that it is difficult to account for observations of the polarity and tilt of bi-polar active regions if the magnetic field originates throughout a turbulent convection zone. Helioseismology has shown that most of the convection zone rotates like the surface, with a region of steep radial gradient only near the *base* of the convection zone. There have been various suggestions for overcoming the difficulties presented by the absence of a radial gradient, including an anisotropic α term and meridional circulation. However, locating the dynamo action in the region of overshoot between the radiative and convective zones overcomes the problem of buoyancy, and fields as strong as 10^4 to 10^5 G could be stored there (see e.g. Spiegel & Weiss 1980; Schüssler 1993). Models based on the helioseismology observations show that to match the solar cycle the turbulent diffusivity would have to be reduced in this region, and that the α effect would need to be concentrated near the equator. Parker (1993) has suggested a model combining the above aspects of the overshoot model with a traditional dynamo operating in the rest of the convective zone.

Schmitt also stressed the possible importance of a *dynamic*, rather than a turbulent, α effect. This could arise when magnetic instabilities lead to magnetostrophic waves (Cattaneo & Hughes 1988; Acheson & Gibbons 1978). He also discussed flux-tube dynamos (Schüssler 1993) and the excitation of a spectrum of dynamo modes by a randomly fluctuating α effect, which can be compared with the spherical harmonic decomposition of the surface magnetic fields (Hoyng *et al.* 1993).

Brandenburg presented numerical hydromagnetic simulations of the convection zone, which, although idealized, allow important aspects of the dynamo process and the behaviour of magnetic fields to be investigated (see also Nordlund *et al.* 1992). Some results were presented with the aid of a very impressive video. A "seed" field is amplified by the continual stretching, twisting and folding of the magnetic field lines by the convective flow. The gravitational stratification creates downdrafts, that are set into rotation by the Coriolis force. The fields tend to be sucked into and wound around downdrafts (see also Brandenburg *et al.* 1991). Saturation occurs when the magnetic flux is forced to slip through the plasma and the work against the curvature force is transformed to Joule heating. The magnetic field is formed into flux-tube like structures, which are mainly vertical in the bulk of the region, but are mainly horizontal near the lower boundary. Although the magnetic field is strongest at the base of the layer, where it accumulates, the dynamo action may take place throughout the convection zone. Many of the above features also appear in the 3D simulations of magneto-convection by Nordlund & Stein (1990) (see review by Démoulin). Brandenburg stressed that the simulations are not yet entirely realistic, and in particular do not produce strong buoyant magnetic fields. However, bi-polar regions do appear. Although the simulations produce fields in small, flux tube structures, other work on the magnetic energy spectrum shows the development with time of large scale magnetic fields (Nordlund *et al.* 1992; Brandenburg 1992). Investigations of the α effect yield the latitude dependence, which deviates from a simple $\cos \theta$ law (Rüdiger & Brandenburg 1993). Although computed mean-field dynamos (Rüdiger & Brandenburg 1993) lead to the poloidal and toroidal fields being in phase, in contrast to the observations at low latitude, Brandenburg suggested that this might be due to the observations of B_r and B_ϕ probing different depths in the convection zone. The present simulations need to be extended to larger volumes, including spherical geometry, but they are already producing many useful insights.

Démoulin introduced his talk on the control of the corona by magnetic fields (see Section 4) with an outline of the main features of magneto-convection. Like Brandenburg, he stressed that the convection zone contains gentle ascending motions but strong concentrated downdrafts. The downdrafts originate around photospheric granules, and become organized on larger scales at greater depths. Although Nordlund & Stein (1990) propose that there is an *inverse* cascade of scales, with the larger scale convection being driven by the smaller scale motions, Zahn (1988) suggests the opposite, with the energy from the supergranules driving the granules.

Belvedere reminded us that dynamos operating in other stars may be quite different from that of the Sun. Amongst the main-sequence stars, the Sun is a relatively slow rotator, and the dynamo action could differ in rapidly rotating dwarfs. The magnetic fields of close binary stars may be more complex, and for a given rotation rate, the different convection zone structure of evolved stars, compared with dwarfs, could also lead to significant differences in the dynamo action. However, rotation related activity occurs in main-sequence stars, later than about FV 0, and in late-F to early K giants. For the dwarfs, Noyes *et al.* (1984) found that the Ca II emission line flux can be correlated with the Rossby number ($Ro = P_{rot}/\tau_c$, where τ_c is the turnover time at the base of the convection zone), and similar studies of other emission line fluxes have since been carried out. Observations of photospheric magnetic fields, B_{ph}, and area filling factors, f_{ph} (see review by Saar 1991), show that $B_{ph}f_{ph}$ and f_{ph} increase with the rotation rate, and also correlate with Ro. However, not everyone regards Ro as an appropriate parameter (Rutten & Schrijver 1986). The filling factor increases from about 1 to 2% in the Sun, to about 10% in moderately active stars, and above 50% in RS CVn systems and BY Dra stars. The relation between stellar age and rotation is also important in the context of magnetic braking, a topic which has received much attention since the early work of Skumanich (1972).

Extensive observations of variations over stellar cycles have been carried out in recent years (e.g. Baliunas & Vaughan 1985; Saar & Baliunas 1992). At present there is no clear correlation between the length of a stellar activity cycle and parameters such as P_{rot}, Ro and age, but cycle periods do appear to increase with the fractional depth of the convection zone. However, dynamos can act in both a cyclic or chaotic manner, and some 25% of the stars in the above samples show chaotic variations.

Belvedere went on to review non-linear modelling of stellar dynamos and pointed out that in this approach stellar activity would be interpreted in terms of deterministic chaos (Belvedere *et al.* 1990). He also summarized the results of modelling a non-linear $\alpha - \omega$ dynamo acting in a thin layer at the base of the convection zone, and using the internal rotation derived from helioseismological observations (Brown *et al.* 1989, Libbrecht 1989). Belvedere *et al.* (1991) found that this could reproduce the main features of the solar activity cycle, such as the direction of the propagation of dynamo waves for latitudes above and below $35°$.

The stellar theme was continued by Montesinos, who presented comparisons between the predictions of recent calculations, using a simple mean-field dynamo and two different approximations for the filling factors and $d\omega/dr$, and the dependence of the observed fields and filling factors on Ro (Montesinos & Jordan 1993). He suggested that the apparent saturation of the mean magnetic field at small values of Ro might be due to a change in the internal rotation law at about Ro = 0.4.

4 Coupling between the Interior and Corona

Alissandrakis reviewed observations that show the role of the magnetic field in controlling the corona, and gave a detailed account of methods which use microwave or radio emission to determine the magnetic field in various coronal structures. Measurements of thermal gyroresonance emission in the microwave region have given direct measurements of the field over sunspots (e.g. Akhmedov *et al.* 1990; Lee *et al.* 1993; Alissandrakis *et al.* 1993), at heights between the base of the transition region and low corona, depending on the harmonics observed. Values of between 1800 G and 720 G have been measured. Radio maps of active regions can also be used, together with modelling of the temperature structure, to investigate the magnetic fields present (e.g. Alissandrakis *et al.* 1980; Bogod *et al.* 1992). Non-thermal gyrosynchrotron emission has been used to find the fields in active region and flare loops. The fields range from around 15 to 20 G, high (1.5×10^5 km) in active region loops to a few hundred Gauss in flare loops (e.g. Preka-Papadema & Alissandrakis 1988). He also summarized the information that can be found from the different types of radio bursts, at various frequencies. The polarization of Type III bursts leads to magnetic fields of between about 50 G at 435 MHz and 8 G at 164 MHz (Mercier 1990). The frequency drift of Type II bursts implies fields of a few Gauss between 0.2 and 1 $R_\odot$. Although it is not possible to measure directly the magnetic field in the general corona, some information can be found from the magnitude of tangential discontinuities in the electron density, apparent in white light eclipse images. These suggest that B is about 0.8 G at 1.5 $R_\odot$, and 0.12 G at 2.4 $R_\odot$ (Koutchmy 1972). The methods of extrapolating photospheric magnetic fields into the corona have also been developed to deduce coronal fields from K-coronameter data (Bagenal & Gibson 1991). Long wavelength radio maps reveal the structure of large scale features, such as large coronal loops, coronal holes and streamers (Lantos *et al.* 1987; Kundu *et al.* 1987). Finally, he reminded us that if the fields are force free, and the plasma β is low, then the flux tubes are essentially independent, and pressure changes in one tube should not affect the others. An important, long standing question is: out of all the flux tubes present, why, at any one time, are some heated, but others are not?

The concentration of the surface magnetic fields by convective motions was discussed by Démoulin. Simulations (e.g. Galloway & Proctor 1983; Nordlund & Stein 1990), have shown how the magnetic field is concentrated into regions of downdraft until the field and gas become mutually exclusive, so that the resulting field attains the equipartion value (observations of stellar magnetic fields (see Saar 1990), also give values close to the equipartion value). He gave a comprehensive account of the ways in which the time-dependent surface magnetic fields can lead to a complex coronal field topology and the formation of current sheets on the magnetic field separatrices. Although analytical solutions of the formation of current sheets driven by photospheric motions have been calculated for separatrices in 2D (e.g. Low 1992; Vekstein & Priest 1992, and earlier papers), he considered that at present a general proof of the formation, in this way, of current sheets in 3D does not exist.

Magnetic reconnection is increasingly viewed as the key to coronal heating, but it is difficult to determine which of the several possibilities is the most important. Reconnection in 2D has been recently reviewed by Priest (1992), and its relevance to solar phenomena has been discussed by Forbes (1992). The more recent models are characterized by faster reconnection speeds than in the early models (see e.g. Jardine *et al.* 1992; Priest & Forbes 1992a). Regarding recent work on 3D reconnection, Démoulin pointed out the complementary nature of the approaches taken by Hesse & Schindler (1988) (where reconnection stems from an increased resistivity), Priest & Forbes (1989), Lau & Finn (1990) (who stress the role of velocity fields) and Priest & Forbes (1992b) (who show that reconnection can take place without null points by a "flipping" of field lines).

Démoulin also discussed the production of solar flares in terms of the formation of current sheets on separatrices. A number of authors have found observational evidence for the presence of photospheric current concentrations near flare ribbons (see e.g. Démoulin *et al.* 1993). The energy for the flare could be stored either in field aligned currents, the flare being triggered by the current sheet becoming unstable, or in twisted flux tubes, with a flare occurring when the twisted tube reaches the location of a separatrix. In either case the magnetic energy would be released at the separatrices. He showed several examples of field configurations in relation to flaring active regions observed during the Solar Maximum Mission.

5 Large Scale Structure of the Corona

Einaudi introduced his discussion of coronal heating mechanisms with a brief review of the range of structures apparent in images of the corona, i.e. coronal holes, active regions, the "quiet" sun. He preferred to group these into regions with "closed" or "open" magnetic field lines, since the heating processes may depend on the field geometry. He reminded us that the total energy requirements estimated for these different regions are *time averaged* values. In any magnetic heating process the energy is ultimately provided by the Poynting flux, generated by the interaction of the photospheric velocity fields with the magnetic fields. The typical velocities of the granulation are 0.25 to 2 km s^{-1}, and of the supergranulation 0.3 km s^{-1}. With a magnetic field of, say, 100 G, there is ample energy flux available. The coronal response to these perturbations, on a time scale τ, depends on the ratio τ/τ_A where τ_A is the Alfvén time scale, given by the length of the structure and the Alfvén speed within it. If $\tau/\tau_A \gtrsim 1$, then the response is quasistatic, and current dissipation (reconnection) is of interest, but if $\tau/\tau_A \lesssim 1$, then a periodic response (waves) can result.

While the closed magnetic fields of active regions, and their complexity, allows in principle many different processes, the simple fields of coronal holes might favour the production of MHD waves (see recent reviews in Ulmschneider *et al.* 1991, and Einaudi *et al.* 1993). Of the slow, fast and Alfvén waves, only the latter are expected to carry sufficient energy to the corona, provided the wave periods are not too large.

There is indeed evidence for Alfvén waves in the solar wind (Belcher & Davis 1971; Mangeney *et al.* 1991). However, to obtain dissipation steep spatial gradients are required over small length scales, which could be generated by non-linear effects or interactions with an inhomogeneous background field. Phase mixing (Heyvaerts & Priest 1983) and resonant absorption (Mok & Einaudi 1985) are both of interest. Heating by reflected waves may also be important, particularly in open field regions and large (quiet coronal) loops (Velli 1993).

Heating from energy released in the slow evolution of magnetic fields has been investigated by a number of authors since it was first proposed by Gold (1964) (e.g. Sturrock & Uchida 1981; Parker 1983). The same type of physics is involved in models which invoke large-scale turbulence, although the magnetic field is described in a different way. The heating essentially comes from a turbulent relaxation of the field (Heyvaerts & Priest 1984; van Ballegooijen 1986; Gomez & Ferro-Fontan 1992).

Small active region loops and, in particular, flares present further problems, since the above theories relate to long-lived structures. The shorter timescales require the magnetic field causing the heating to be structured on a scale below one meter, placing strong constraints on the duration, size and number of current sheets present (see e.g. Berger 1991).

Einaudi envisaged the following unifying scenario: the photospheric motions lead to a "turbulent" corona, with dynamical small scale structures and local instabilities. Only a small fraction of the large number of current sheets present need to be active at a given time, and we observe only a time average of the heating events. It might, therefore, be more appropriate to take a "statistical mechanics" approach, rather than a thermodynamic approach. The corona may be close to instability, in a self-organized critical state, and always "ready to go" (e.g. Bak *et al.* 1987). The question of whether flares are a statistical fluctuation or require an external trigger, remains, and the physics of the elementary events is not yet known.

6 Small Scale Dynamics of the Transition Region and Corona

The observational signatures of "jets" and "explosive events" were described by Hénoux. Dramatic examples of these were observed in transition region lines during the first two flights of the Naval Research Laboratory's High Resolution Telescope and Spectrograph (HRTS) (Brueckner & Bartoe 1983). Jets are characterized by blue-shifted lines, and in the lines of C IV the projected velocities reach to about 400 km s^{-1}, far larger than the sound speed of 47 km s^{-1} at 10^5 K. Since the magnetic field strength is not known, one cannot make comparisons with the Alfvén speed. The coronal forbidden transitions in the HRTS spectral range are too weak at the location of the C IV jets to determine whether or not there are any corresponding shifts. Jets are not very common and Dere (1992, 1993) estimates that they do not play a major role in the coronal mass and energy balance. Jets appear to be associated with nearby, lower velocity red-shifts (see Brueckner & Bartoe 1983), and the changes in their structure with time could indicate motions within a small loop. There are also smaller flows observed only in chromospheric lines, which are located in supergranulation cell interiors. Explosive events have been discussed by

Dere *et al.* (1989). They are characterized by high velocity (100 km s^{-1}) shifts in both the blue and red wings, at, or near, the same location, and seem to occur at the *edge* of the supergranulation boundaries, not at the location of the highest magnetic field (Porter & Dere 1991). Their size is small (2 arcsec) since little extension is seen along the slit, even over time periods of 200s. Combining the electron density measured from the O IV] lines with the volume emission measure indicates that only about 2×10^{-4} of the volume is filled with material at 10^5 K (Dere *et al.* 1982). It is possible that the explosive events are related to the squeezing together of field lines or to reconnection. Brueckner *et al.* (1988) have found that some explosive events are related to emerging flux regions. However, the explosive events are too rare to be the prime cause of coronal heating.

Zaitsev discussed the plasma physics of explosive phenomena and reviewed the advantages and disadvantages of various flare models. Single loop models, in which heating might occur by a tearing mode instability following twisting by footpoint motions (Spicer 1977), have difficulty in matching the rapid rise times observed, because of the large Reynolds number, and require the presence of thin current sheets. It is easier to match the observed rise times in loop coalescence models (Sweet 1958; Tajima *et al.* 1982), but there are unanswered questions concerning the actual conditions of the reconnection, the behaviour in three dimensions and what happens when asymmetric loops coalesce. The emerging flux model (Heyvaerts, Priest & Rust 1977) can be regarded as a variant of the loop coalescence model. Sturrock's (1968) central helmet streamer model, shares the problems of any basic reconnection theory. He pointed out that all the above flare models depend on the formation of a single current sheet, and provided the thickness of the sheet is sufficiently small the rapid rise time of flares can be matched. However, they have problems with the particle acceleration, the heating of large volumes and the cross-field transport theory.

In statistical models, flares occur spontaneously, by a variety of processes, in the presence on many tangled field lines (Parker 1988; Vlahos 1989). It is easier to fill large volumes and accelerate particles in these models, but so far much of the analysis has been quantitative in nature and detailed calculations of times scales and the number of events needed, are required. The circuit model, which does not depend on a single current sheet, goes back to work by Alfvén and Carlquist (1967), and involves large currents generated by photospheric convective motions over the time scale of the flux emergence. He described recent work by Zaitsev and Stepanov (1992). The flare could be triggered by instabilities in a prominence, and after inter-action with the loop containing the current, energy would be dissipated by ion-atom collisions and Joule heating. Some fraction of the energy would go into accelerat-ing the particles. The non-classical resistivity could be important, and the most powerful energy release could occur in the chromosphere. Overall, Zaitsev favoured the statistical model, which has much in common with the heating model discussed by Einaudi, and cited observations which could support this model. However, he stressed that more quantitative work is needed.

7 Propagation of Energetic Particles in the Corona

The acceleration and storage of energetic particle in the corona was discussed by Wentzel. A number of advances have been recently made in this area due to improved instrumentation, further data analysis, both of individual events and statistical properties, correlations of observed properties and the refinement of theories. The events can be grouped into two main categories; gradual events (main proton events), which can last for days, and impulsive events (Cane et al. 1986). If the gradual events are associated with flares, the long time scale must be explained, hence the concepts of coronal "diffusion" and "storage". The gradual events are observed over a wide range of longitude, but the time to maximum depends on the viewing angle, and azimuthal transport of material is implied (Cane et al. 1988; Kallenrode et al. 1992). They have a lower mean charge state (corresponding to about 2×10^6 K), than the impulsive events. The relative element abundances reflect the conditions in the coronal source. The gradual events are thought to be caused by coronal mass ejections, giving rise to shocks in the overlying corona, which are the source of the solar energetic particles. The impulsive events are rich in ^3He and have high $^3He/^4He$ abundance ratios. Elements with low first ionization potentials have enhanced abundances compared with the photosphere, and resonant wave-particle interactions in the flare are the likely cause of these enhancements. The impulsive events are observed over only a narrow range of longitude and the mean charge state corresponds to about 2×10^7 K (see e.g. Luhn et al. 1987; Reames 1992). Thus they involve the actual flare material.

Wentzel also discussed particle acceleration models (see Melrose 1992) and the production of gamma-rays. In the impulsive events streaming electrons can excite electromagnetic ion cyclotron waves, which in turn accelerate ^{3} He and heavy ions to MeV energies (Temerin & Roth 1992). In the gradual events a fast coronal mass ejection drives a shock wave in the overlying corona, which can accelerate the particles (Lee & Ryan 1986).

Proton acceleration in long duration solar flares was addressed in more detail by Ryan. High energy γ rays and neutrons are detected for 10 to 100 min after the decay in X-rays, so an explanation must be found for the trapping of protons on these time-scales. He proposed that the protons are trapped in coronal structures by means of turbulence, which then accelerates them to high energies via the second-order Fermi mechanism in the Alfvén wave field.

Klein discussed the electromagnetic signatures of particle acceleration and propagation. He stressed how the different wavelengths used in observing reflect conditions in different parts of flaring, or coronal loop structures. For example, the hard X-ray and microwave emission show that most of the energetic particles are localized and accelerated in the low corona (Matsushita et al. 1993; Alissandrakis et al. 1988). Observations of frequency drifts in radio emission support this view and provide evidence of both upward and downward propagation of particles. They also show that in the impulsive phase the electron beams are produced where N_e has values between 10^9 to 10^{10} cm^{-3} (Benz et al. 1983; 1992). However, there is also direct evidence of electron acceleration at heights above 0.5 $R_\odot$, indicating a range of field topologies. Klein stressed the complex fine structure of flaring regions, and the production of emission at different wavelengths, locations and times as the flare proceeds. E.g. radiation at dm and m wavelengths occurs as different structures are activated (Trottet *et al.* 1993) and the delay in producing hard X-rays and gamma rays can also be attributed to the spread of activity through the different sub-structures of the flaring region (Chupp *et al.* 1993). The fluctuations observed at high energies suggest that there are elementary episodes of acceleration (e.g. Talon *et al.* 1993). Klein also discussed the properties of gradual hard X-ray and radio bursts, and concluded that there is probably a range of events, from impulsive to gradual, whose different properties arise from the particular environment of a given flare. He suggested that the apparent delay of particle acceleration for up to hours after a flare (e.g. as shown by metric noise storms), could be caused in part by the response of the corona to changes in the topology of the active region fields. Similarly the field geometry allows particles to reach regions distant from the flare, and interplanetary space (e.g. Lantos *et al.* 1984; Klein & Trottet 1993).

8 Future Work and Conclusions

Since I have not followed in detail work on the solar core and radiative zone, I found the progress in these areas particularly impressive. The new opacities are making a considerable impact. Element abundances are of course important in the interpretation of lines at the levels of the photosphere and above, but I had not appreciated the significance of the iron abundance in the solar interior. The existing helioseismological observations have provided strong constraints on models of the interior, and we now await new observations from the ground and from the SOHO satellite (see Domingo, this volume), and await with interest further tests of neutrino physics.

I was also impressed by the developments in the theory and simulation of the way that magnetic fields behave in the convection zone. The wide range of phenomena associated with magnetic activity, observable on the surface of the Sun, offers the hope of understanding at least the solar dynamo. The new instrumentation described by Deubner and Landi Degl'Innocenti should provide improved observations which will be of benefit in this respect, as well as in the understanding of the overlying chromosphere and corona.

The question of coronal heating remains unresolved, although the magnetic field is clearly implicated. The main issue is whether Alfvén waves or current sheets play the dominant role, although both may be important depending on the geometry. Theoretical work is progressing well, but there is a continued need for high resolution observations. The value of high resolution images has been demonstrated by Yohkoh. Again, observations from SOHO, particularly of non-thermal line widths in the high transition region and inner corona, and the other space missions described by Lemaire, should provide additional constraints.

There is every hope that when we next met in three years time, at least some of the outstanding issues will have been resolved.

References

Abazov, A.I. et al. (1991): Phys. Rev. Letts. **67** 3332

Acheson, D.J., Gibbons, M.P. (1978): J. Fluid Mech. **85** 743

Akhmedov, Sh.B., Bogod, V.M., Korzhavin, A.N., Aurass, H., Hildebrandt, J., Krüger, A. (1990): Sol. Phys. **129** 351

Alfvén, H., Carlquist, P. (1967): Sol. Phys. **1** 120

Alissandrakis, C.E., Kundu, M.R., Lantos, P. (1980): Astron. Astrophys. **82** 30

Alissandrakis, C.E., Schadee, A., Kundu, M.R. (1988): Astron. Astrophys. **195** 290

Alissandrakis, C.E., Gelfreikh, G.B., Borovik, V.N., Korzhavin, A.N., Bogod, V.M., Nindos, A., Kundu, M.R. (1993): Astron. Astrophys. **270** 509

Anders, E., Grevesse, N. (1989): Geochim. Cosmochim. Acta **53** 197

Anselmann, P. et al. (1992): Phys. Letts. **B285** 390

Bachmann, K.T., Brown, T.M. (1993): Astrophys. J. Letts. **411** L45

Bagenal, F., Gibson, S. (1991): J. Geophys. Res. **96** 17663

Bahcall, J.N. (1989): Neutrino Astrophysics, (Cambridge University Press, Cambridge)

Bahcall, J.N., Bethe, H. (1990): Phys. Rev. Lett. **65** 2233

Bahcall, J.N., Pinsonneault, M.H. (1992): Rev. Mod. Phys. **64** 885

Bak, P., Tang, C., Wiesenfeld, K. (1987): Phys. Rev. Letts. **59** 381

Baliunas, S.L., Vaughan, A.H. (1985): Ann. Rev. Astron. Astrophys. **23** 379

Belcher, J.W., Davis, L. (1971): J. Geophys. Res. **76** 3534

Belvedere, G., Proctor, M.R.E., Pidatella, R.M. (1990): Geophys. Astrophys. Fluid Dyn. **51** 263

Belvedere, G., Proctor, M.R.E., Lanzafame, G. (1991): Nature **350** 481

Benz, A.O., Aschwanden, M. (1992): in Eruptive Solar Flares, IAU Colloq. **133**, ed. by B.V. Jackson, M.E. Machado & Z.F. Svestka, (Springer-Verlag, Berlin), p. 106

Benz, A.O., Berold, T.E.X., Dennis, B.R. (1983): Astrophys. J. **271** 355

Berger, M.A. (1991): in Advances in Solar System MHD, ed. by E.R. Priest & A.W. Hood, (Cambridge), p. 241

Bogod, V.M., Gelfreikh, G.B., Willson, R.F., Lang, K.R., Opeikina, L.V., Shatilov, V., Tsvetkov, S.V. (1992): Sol. Phys. **141** 303

Brandenburg, A. (1992): Phys. Rev. Lett. **69** 605

Brandenburg, A., Jennings, R.L., Nordlund, A., Stein, R.F., Tuominen, I. (1991): in Proc. IAU Colloq. **130**, The Sun and Cool Stars: activity, magnetism, dynamos, ed. by I. Tuominen, D. Moss & G. Rüdiger, (Springer-Verlag, Berlin), p. 86

Brown, T.M., Christensen-Dalsgaard, J., Dziembowski, W., Goode, P.R., Gough, D.O., Morrow, C.A. (1989): Astrophys. J. **343** 526

Brueckner, G.E., Bartoe, J.-D. (1983): Astrophys. J. **272** 329

Brueckner, G.E., Bartoe, J.-D., Cook, J.W., Dere, K.P., Socker, H., Kurokawa, H., McCabe, M. (1988): Astrophys. J. **335** 986

Cane, H.V., McGuire, R.E., von Rosenvinge, T.T. (1986): Astrophys. J. **301** 448

Cane, H.V., Reames, D.V., von Rosenvinge, T.T. (1988): J. Geophys. Res. **93** 9555

Cattaneo, F., Hughes, D.W. (1988): J. Fluid Mech. **196** 323.

Charbonnel, C., Vauclair, S., Zahn, J.-P. (1992): Astron. Astrophys. **255** 191

Christensen-Dalsgaard, J., Proffitt, C.R., Thompson, M.J. (1993): Astrophys. J. Letts. **403** L75

Christensen-Dalsgaard, J., Duvall, T.L., Gough, D.O., Harvey, J.W., Rhodes, E.J. (1985): Nature **315** 378

Chupp, E.L., Trottet, G., Marschhäuser et al. (1993): Astron. Astrophys., in press

Cox, A.N., Guzik, J.A., Kidman, R.B. (1989): Astrophys. J. **342** 1187

Démoulin, P., van Driel-Gesztelyi, L., Schmieder, B., Hénoux, J.C., Csepura, G., Hagyard, M.J. (1993): Astron. Astrophys. **271** 292

Dere, K.P. (1992): in Electromagnetic Coupling of the Solar Atmosphere, ed. by D.S. Spicer & P. McNeice (AIP, New York), p. 63

Dere, K.P. (1993): Adv. Space Res., in press

Dere, K.P., Bartoe, J.-D., Brueckner, G.E. (1982): Astrophys. J. **259** 366

Dere, K.P., Bartoe, J.-D., Brueckner, G.E. (1989): Sol. Phys. **123** 41

Dziembowski, W.A., Pamyatnykh, A.A., Sienkiewicz, R. (1991): Mon. Not. R. astr. Soc. **249** 602

Einaudi, G., Chiuderi, C., Califano, F. (1993): Adv. Space Res. **13** 85

Elsworth, Y., Howe, R., Isaak, G.R., McLeod, C.P., New, R. (1990): Nature **345** 322

Evans, D.J., Roberts, B. (1992): Nature **355** 230

Forbes, T. (1992): in Proc. IAU Colloq. **133**, Eruptive Solar Flares, ed. by B.V. Jackson, M.E. Machado & Z.F. Svestka, (Springer-Verlag, Berlin) p. 79

Galloway, D.J., Proctor, M.R.E. (1983): Geophys. Astrophys. Fluid Dyn. **24** 109.

Gold, T. (1964): in The Physics of Solar Flares, ed. by W. Hess (NASA SP-50) p. 389

Gomez, D.O., Ferro-Fontan, C.F. (1992): Astrophys. J. **394** 662

Gough, D.O. (1990): in Proc. IAU Colloq. **121**, Inside the Sun, ed by. G. Berthomieu & M. Cribier, (Kluwer, Dordrecht), p. 451.

Grevesse, N., Noels, A., Sauval, A.J. (1992): in Coronal Streamers, Coronal Loops, and Coronal and Solar Wind Composition, (ESA SP-348) p. 305

Guzik, J.A., Cox, A.N. (1993): Astrophys. J. **411** 394

Hesse, M., Schindler, K. (1988): J. Geophys. Res. **93, A6** 5559

Heyvaerts, J., Priest, E.R. (1983): Astron. Astrophys. **117** 220

Heyvaerts, J., Priest, E.R. (1984): Astron. Astrophys. **137** 63

Heyvaerts, J., Priest, E.R , Rust, D.M. (1977): Astrophys. J. **216** 123

Hirata, K.S. et al. (1989): Phys. Rev. Letts. **63** 16

Hoyng, P., Schmitt, D., Teuben, L.J.W. (1993): Astron. Astrophys. submitted

Iglesias, C.A., Rogers, F.J., Wilson, B.G. (1992): Astrophys. J. **397** 717

Jardine, M., Allen, H.R., Grundy, R.E., Priest, E.R. (1992): J. Geophys. Res. **97, A4** 4199

Kallenrode, M.-B., Cliver, E.W., Wibberenz, G. (1992): Astrophys. J. **391** 370

Klein, K.-L., Trottet, G. (1993): in Proc. Conf. High Energy Physics of Solar Flares, ed. by J. Ryan, in press

Kosovichev, A.G., Christensen-Dalsgaard, J., Däppen, W. Dziembowski, W.A., Gough, D.O., Thompson, M.J. (1992): Mon. Not. R. astr. Soc. **259** 536

Koutchmy, S. (1972): Sol. Phys. **24** 374

Kundu, M.R., Gergely, T.E., Schmahl, E.J., Szabo, A., Loicono, R., Wang, Z., Howard, R.A. (1987): Sol. Phys. **108** 113

Langacker, P. (1993): in Unified Symmetry in the Small and in the Large (Proc. Conference at Coral Gables, Florida, January 1993), in press

Lantos, P., Pick, M., Kundu, M.R. (1984): Astrophys. J. Letts. **283** L71

Lantos, P., Alissandrakis, C.E., Gergely, T., Kundu, M.R. (1987): Sol. Phys. **112** 325

Lau, Y.T., Finn, J.M. (1990): Astrophys. J. **350** 672

Lee, J.W., Hurford, G.J., Gary, D.E. (1993): Sol. Phys. **144** 45

Lee, M.A., Ryan, J.M. (1986): Astrophys. J. **303** 829

Libbrecht, K.G. (1989): Astrophys. J. **336** 1092

Libbrecht, K.G., Woodward, M.F. (1990): Nature **345** 779

Low, B.C. (1992): Astron. Astrophys. **253** 311

Luhn, A., Klecker, B., Hovestadt, D., Möbius, E. (1987): Astrophys. J. **317** 951

Mangeney, A.M., Grappin, R., Velli, M. (1991): in Advances in Solar System MHD, ed. by E.R. Priest & A.W. Hood, (Cambridge), p. 327

Matsushita, K., Masuda, S., Kosugi, T., Inda, M., Yaji, K. (1993): Publ. Astron. Soc. Japan **44** L89

Melrose, D.B. (1992): in Particle Acceleration in Cosmic Plasmas, AIP Conf. Proc. **264** p. 3

Mercier, C. (1990): Sol. Phys. **130** 119

Mikheyev, S.P., Smirnov, A. Yu. (1985): Sov. J. Nucl. Phys. **42** 913

Montesinos, B., Jordan, C. (1993): Mon. Not. R. astr. Soc. **264** 900

Mok, Y., Einaudi, G. (1985): J. Plasma Phys. **33** 199

Nordlund, A., Stein, R.F. (1990): in Proc. IAU Symp. **138**, Solar Photosphere: Structure, Convection and Magnetic Fields, ed. by J.O. Stenflo, (Kluwer, Dordrecht), p. 191

Nordlund, A., Brandenburg, A., Jennings, R.L., Rieutord, M., Ruokolainen, J., Stein, R.F., Tuominen, I. (1992): Astrophys. J. **392** 647

Noyes, R.W., Hartmann, L.W., Baliunas, S.L., Duncan, D.K., Vaughan, A.H. (1984): Astrophys. J. **279** 763

Parker, E.N. (1975): Astrophys. J. **198** 205

Parker, E.N. (1983): Astrophys. J. **264** 642

Parker, E.N. (1988): Astrophys. J. **330** 474

Parker, E.N. (1993): Astrophys. J. **408** 707

Porter, J.G., Dere, K.P. (1991): Astrophys. J. **370** 775

Preka-Papadema, P., Alissandrakis, C.E. (1988): Astron. Astrophys. **191** 365

Priest, E.R. (1992): in Proc. IAU Colloq. **133**, Eruptive Solar Flares, ed. by B.V. Jackson, M.E. Machado & Z.F. Svestka, (Springer-Verlag, Berlin), p. 15

Priest, E.R., Forbes, T. (1989): Sol. Phys. **119** 211

Priest, E.R., Forbes, T. (1992a): J. Geophys. Res. **97**, **A11** 16757

Priest, E.R., Forbes, T. (1992b): J. Geophys. Res. **97**, **A2** 1521

Proffitt, C.R., Michaud, G. (1991): Astrophys. J. **380** 238

Reames, D.V. (1992): in Particle Acceleration in Cosmic Plasmas, AIP Conf. Proc. **264** p. 213

Rüdiger, G., Brandenburg, A. (1993): Astrophys. J. submitted

Rutten, R.G.M., Schrijver, C.J. (1986): in Cool Stars, Stellar Systems and the Sun, ed. by M. Zeilik & D. Gibson, (Springer-Verlag, Berlin), p. 19

Saar, S.H. (1990): in Proc. IAU Symp. **138**, Solar Photosphere: Structure, Convection and Magnetic Fields, ed. by J.O. Stenflo, (Kluwer, Dordrecht), p. 427

Saar, S. H. (1991): in Proc IAU Colloq. **130**, The Sun and Cool Stars: activity, magnetism, dynamos, ed. by I. Tuominen, D. Moss & G. Rüdiger, (Springer-Verlag, Berlin) p. 389

Saar, S.H., Baliunas, S.L. (1992): in Proc. 4th Solar Cycle Workshop, ed. by K.L. Harvey, ASP Conf. Ser. **27** p. 150

Schrijver, C.J. (1993): in Proc. IAU Colloq. **137**, Inside the Stars, ed. by W.W. Weiss, A. Baglin, ASP Conf. Ser. **40** p. 591

Schüssler, M. (1993): in Proc. IAU Symp. **157**, The Cosmic Dynamo, ed. by F. Krause, K.-H. Rädler, G. Rüdiger, in press

Skumanich, A. (1972): Astrophys. J. **171** 565

Spicer, D.F. (1977): Sol Phys. **53** 305

Spiegel, E.A., Weiss, N.O. (1980): Nature **287** 616

Steenbeck, M., Krause, F. (1969): Astron. Nachr. **291** 49

Sturrock, P.A. (1968): Astron. J. **73** 79

Sturrock, P.A., Uchida, Y. (1981): Astrophys. J. **246** 331

Sweet, P.A. (1958): in Electromagnetic Phenomena in Cosmical Plasmas, (CUP), p. 123

Tajima, T., Brunel, F., Sakai, J. (1982): Astrophys. J. Letts. **258** L45

Talon, R., Barat, C., Dézalay, J.-P., et al. (1993): Adv. Space Res. **13(9)** 171

Temerin, M., Roth, I. (1992): Astrophys. J. Letts. **391** L105

Thompson, M.J. (1993): in GONG 1992: Seismic Investigation of the Sun and Stars, ASP
   Conf. Ser. Vol. **42**, ed. by T.M. Brown, p. 141
Trottet, G., Vilmer, N., Barat, C. et al. (1993): Adv. Space Res. **13(9)** 298
Ulmschneider, P., Priest, E.R., Rosner, R. (1991): eds. Mechanisms of Chromospheric and
   Coronal Heating, (Springer-Verlag, Berlin).
van Ballegooijen, A.A. (1986): Astrophys. J. **311** 1001
Vekstein, G.E., Priest, E.R. (1992): Astrophys. J. **384** 333
Velli, M. (1993): Astron. Astrophys. **270** 304
Vlahos, L. (1989): Sol. Phys. **121** 431
Vorontsov, S.V., Baturin, V.A., Pamyatnykh, A.A. (1992): Mon. Not. R. astr. Soc. **257** 32
Wolfenstein, L. (1978): Phys. Rev. D **17** 2369
Woodard, M.F., Libbrecht, K.G. (1993): Astrophys. J. Letts. **402** L77
Zahn, J.-P. (1988): in Solar and Stellar Physics, ed. by E.H. Schröter & M. Schüssler,
   (Springer-Verlag, Berlin) p. 55
Zahn, J.-P. (1992): Astron. Astrophys. **265** 115
Zaitsev, V.V., Stepanov, A.V. (1992): Sol. Phys. **139** 343

Lecture Notes in Physics

For information about Vols. 1–394
please contact your bookseller or Springer-Verlag

New Series m: Monographs